Lectures in Applied Mathematics
Volumes in This Series

Proceedings of the Summer Seminar, Boulder, Colorado, 1960

1 **Lectures in Statistical Mechanics,** *G. E. Uhlenbeck and G. W. Ford with E. W. Montroll*
2 **Mathematical Problems of Relativistic Physics,** *I. E. Segal with G. W. Mackey*
3 **Perturbation of Spectra in Hilbert Space,** *K. O. Friedrichs*
4 **Quantum Mechanics,** *R. Jost*

Proceedings of the Summer Seminar, Ithaca, New York, 1963

5 **Space Mathematics. Part 1,** *J. Barkley Rosser, Editor*
6 **Space Mathematics. Part 2,** *J. Barkley Rosser, Editor*
7 **Space Mathematics. Part 3,** *J. Barkley Rosser, Editor*

Proceedings of the Summer Seminar, Ithaca, New York, 1965

8 **Relativity Theory and Astrophysics. 1. Relativity and Cosmology,** *Jürgen Ehlers, Editor*
9 **Relativity Theory and Astrophysics. 2. Galactic Structure,** *Jürgen Ehlers, Editor*
10 **Relativity Theory and Astrophysics. 3. Stellar Structure,** *Jürgen Ehlers, Editor*

Proceedings of the Summer Seminar, Stanford, California, 1967

11 **Mathematics of the Decision Sciences, Part 1,** *George B. Dantzig and Arthur F. Veinott, Jr., Editors*
12 **Mathematics of the Decision Sciences, Part 2,** *George B. Dantzig and Arthur F. Veinott, Jr., Editors*

Proceedings of the Summer Seminar, Troy, New York, 1970

13 **Mathematical Problems in the Geophysical Sciences. 1. Geophysical Fluid Dynamics,** *William H. Reid, Editor*
14 **Mathematical Problems in the Geophysical Sciences. 2. Inverse Problems, Dynamo Theory, and Tides,** *William H. Reid, Editor*

Proceedings of the Summer Seminar, Potsdam, New York, 1972

15 **Nonlinear Wave Motion,** *Alan C. Newell, Editor*

Proceedings of the Summer Seminar, Troy, New York, 1975

16 **Modern Modeling of Continuum Phenomena,** *Richard C. DiPrima, Editor*

Proceedings of the Summer Seminar, Salt Lake City, Utah, 1978

17 **Nonlinear Oscillations in Biology,** *Frank C. Hoppensteadt, Editor*

Proceedings of the Summer Seminar, Cambridge, Massachusetts, 1979

18 **Algebraic and Geometric Methods in Linear Systems Theory,** *Christopher I. Brynes and Clyde F. Martin, Editors*

Proceedings of the Summer Seminar, Salt Lake City, Utah, 1980

19 **Mathematical Aspects of Physiology,** *Frank C. Hoppensteadt, Editor*

Proceedings of the Summer Seminar, Chicago, Illinois, 1981

20 **Fluid Dynamics in Astrophysics and Geophysics,** *Norman R. Lebovitz, Editor*

Proceedings of the Summer Seminar, Chicago, Illinois, 1982

21 **Applications of Group Theory in Physics and Mathematical Physics,** *Moshe Flato, Paul Sally and Gregg Zuckerman, Editors*

22 **Large-Scale Computations in Fluid Mechanics, Parts 1 and 2,** *B. Engquist, S. Osher and R. C. J. Somerville, Editors*

23 **Nonlinear Systems of Partial Differential Equations in Applied Mathematics, Parts 1 and 2,** *Basil Nicolaenko, Darryl D. Holm and James M. Hyman, Editors*

Nonlinear Systems of Partial Differential Equations in Applied Mathematics

Part 2

Volume 23 - Part 2

Lectures in Applied Mathematics

Nonlinear Systems of Partial Differential Equations in Applied Mathematics

Edited by
Basil Nicolaenko
Darryl D. Holm
James M. Hyman

1986
American Mathematical Society, Providence, Rhode Island

The proceedings of the SIAM-AMS Summer Seminar on Systems of Nonlinear Partial Differential Equations in Applied Mathematics were prepared by the American Mathematical Society with partial support from the following sources: National Science Foundation Grant DMS8402204 U. S. Army Research Office Contract DAAG 29-84-M-0343. (The views, opinions, and/or findings contained in this report are those of the authors and should not be construed as an official Department of the Army position, policy, or decision, unless so designated by other documentation.) The Center for Nonlinear Studies at Los Alamos National Laboratory Contract 5-LT4-N6232-1, through a grant from the Department of Energy.

1980 *Mathematics Subject Classification* (1985 Revision). Primary 35L65, 35L67, 35Q20; Secondary 58-XX, 76-XX, 80-XX.

Library of Congress Cataloging-in-Publication Data

SIAM-AMS Summer Seminar on Systems of Nonlinear Partial Differential Equations and Applications (1984 : College of Santa Fe)

Nonlinear systems of partial differential equations in applied mathematics.

(Lectures in applied mathematics, ISSN 0075-8485; v. 23)

"Proceedings of the SIAM-AMS Summer Seminar on Systems of Nonlinear Partial Differential Equations and Applications"—Verso t.p.

Hosted by the Los Alamos Center for Nonlinear Studies and held at the College of Santa Fe, 1984.

Includes bibliographies.

1. Differential equations, Partial—Congresses. I. Nicolaenko, Basil, 1943– II. Holm, Darryl D. III. Hyman, James M. IV. Society for Industrial and Applied Mathematics. V. American Mathematical Society. VI. Center for Nonlinear Studies (Los Alamos National Laboratory) VII. Title. VIII. Series.

QA377.S4917 1984 515.3'53 85-15107

ISBN 0-8218-1123-1 (alk. paper)

ISBN 0 8218-1126-6—Part II

Printed in the United States of America.

The paper used in this book is acid-free and falls within the guidelines established to ensure permanence and durability.

Contents

Part 1

Dispersive Systems

Lectures in Applied Mathematics
Volume 23, 1986

ON THE INFINITELY MANY STANDING WAVES OF SOME NONLINEAR SCHROEDINGER EQUATIONS

C.K.R.T. Jones[1]

ABSTRACT. The proof that there are spherically symmetric standing waves of nonlinear Schroedinger equations with a prescribed number of zeroes is sketched. A situation is described in which these are not all unique; in fact in this example the number of solutions increases with the number of zeroes. The question of the stability of these solutions is discussed and a mechanism is described for forcing an instability.

1. INTRODUCTION. The nonlinear Schroedinger equations I shall consider have the form

$$iu_t = \Delta u + f(uu^*)u \qquad (1.1)$$

where u^* = the complex conjugate of the u, f is a smooth function, $f:R \to R$ and Δ is the Laplacian in R^n, some n. A standing wave is a solution of the form $\exp(iat)u(x)$. It therefore satisfies the equation:

$$\Delta u + f(uu^*)u + au = 0. \qquad (1.2)$$

I shall assume that $u(x)$ is real. (1.2) then becomes:

$$\Delta u + f(u^2)u + au = 0 \qquad (1.3)$$

which has been studied extensively, when $a < 0$. The

1980 Mathematics Subject Classification 35B05
[1]Supported by the National Science Foundation under grant #MCS8200392.

interesting nonlinearities $f(u)$ are ones that grow like a power of u at infinity, for instance one could think of $f(u) = |u|^{s/2}$. From the work of Pohozaev, it is known that there is a subtle relationship between the power and the dimension of the underlying space n. Pohozaev [12] proves that for the particular case of $f(u) = |u|^{s/2}$ and $a < 0$, there are no solutions if $s > 4/(n-2)$. The earliest results on the existence of solutions to (1.3), satisfying a condition at x = infinity, are due to Nehari [11] and others. Berger [3] applied a variational method to prove the existence under various assumptions on the nonlinear term $f(u)$, of infinitely many solutions in space dimension 3. Strauss [14] proved such a theorem in a more general setting with assumptions including one that the power growth of $f(u)$ is less than $4/(n-2)$, where n is the space dimension. Berestycki-Lions [2] extended Strauss's result to cover fairly general nonlinear terms, so long as n is not equal to 2. I believe the 2d. case has now been covered by Berestycki.

In each of the above mentioned results, the solutions are forced to be shperically symmetric. This is necessary for the variational proofs to ensure some compactness by fixing the center. It has long been suspected that the solutions are determined, in some sense, by the number of zeros they have in the radial variable. Ryder [13], following on the work of Nehari, proved such a result in space dimension 3, but the general case has remained open. Spherically symmetric solutions of (1.3) satisfy the equation:

$$u_{rr} + ((n-1)/r)\, u_r + f(u^2)u + au = 0. \qquad (1.4)$$

I shall develop a geometric (ODE) way of viewing this equation and sketch the proof of the theorem that there are solutions with any prescribed number of zeroes. For the sake of comprehensibility, I shall restrict the discussion to the special case of f being an exact power. For details of the more general conditions and theorem, see Jones and Kuepper [7]. Consider the equation:

$$u_{rr} + ((n-1)/r)u_r + |u|^s u + au = 0 \qquad (1.5)$$

together with the boundary conditions:

$$u_r(0) = 0 \qquad (1.6)$$
$$\lim_{r\to+\infty} u(r) = 0.$$

THEOREM. If $a < 0$ and $s < 4/(n-2)$, $n \neq 1$, then (1.5) posesses a solution with m zeroes, for any specified m, satisfying (1.6). If $n = 2$ these solutions exist for any s.

A difficult question is whether these solutions are unique. There is a result due to Coffman [5] for the positive solution, in dimension 3 and extended by McLeod-Serrin [10]. As far as this author knows there are no results for these solutions with more zeroes. In section 3, I shall describe a mechanism for non-uniqueness that leads to a rather curious result. The nonlinearity involved is not an exact power and these solutions are probably not physically important but the result is rather curious.

An even deeper question is whether these standing waves are stable relative to the associated nonlinear Schroedinger equation. Since the problem is conservative, the linearised stability approach may not suffice for proving stability for the full equation. However, it is an important aspect of the stability question to see whether all the eigenvalues of the linearised problem must lie on the imaginary axis or if there is something that forces the existence of a real eigenvalue. By the symmetries in the equation, it is easy to see that if λ in R is an eigenvalue so is $-\lambda$.

In the last section I shall describe a method for proving the existence of real eigenvalues. This uses some geometry of the equations linearised at the solution of interest. These equations induce a flow on the space of Lagrangian planes and the Maslov index can be used to see when real eigenvalues are forced to exist. I shall then apply this to prove that some of the extra solutions found above are unstable.

2. THE PHASE SPACE. If one thinks of $r \to +\infty$ in eq. (1.5), the result is the same equation one gets by setting $n = 1$. This is not surprising as it says that spherically symmetric solutions are asymptotically one-dimensional. The one-dimensional equation is, x in R:

$$u_{xx} + |u|^s u + au = 0. \tag{1.7}$$

The phase portrait for this equation can be seen in the $q = 1$ plane of fig. 1.

In order to bring this behavior out for (1.5), firstly rewrite it as a system and introduce r as an dependent variable. Then replace r in its role as a dependent variable by $q = r/(r+1)$, this leads to the system:

$$\begin{aligned} u' &= v \qquad\qquad ' = d/dr \\ v' &= -((n-1)(1-q)/q)v - |u|^s u - au \\ q' &= (1-q)^2. \end{aligned} \tag{1.8}$$

If we set $q = 1$ in (1.8), we get back the one-dimensional system (1.7). However the system (1.8) is still singular at $q = 0$, where we have the other boundary condition. To correct this, introduce another independent variable which has the effect of multiplying (1.8) by q:

$$\begin{aligned} u' &= qv \qquad\qquad ' = d/dt \\ v' &= -(n-1)(1-q)\, v - q(|u|^s\, u + au) \\ q' &= q(1-q)^2. \end{aligned} \tag{1.9}$$

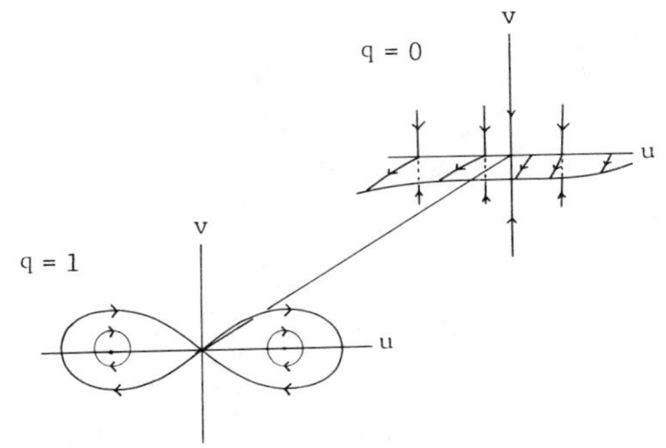

Fig. 1

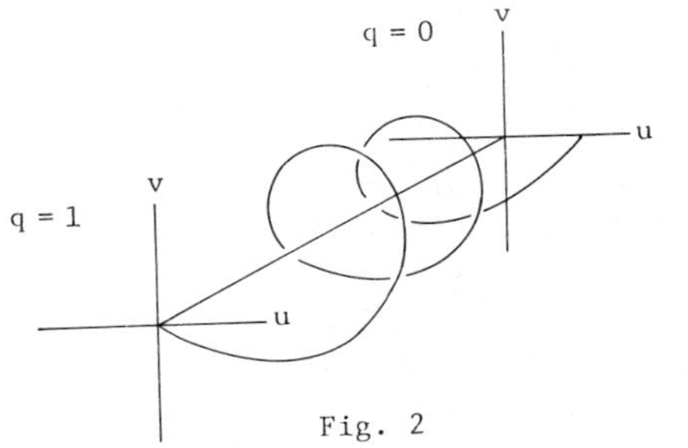

Fig. 2

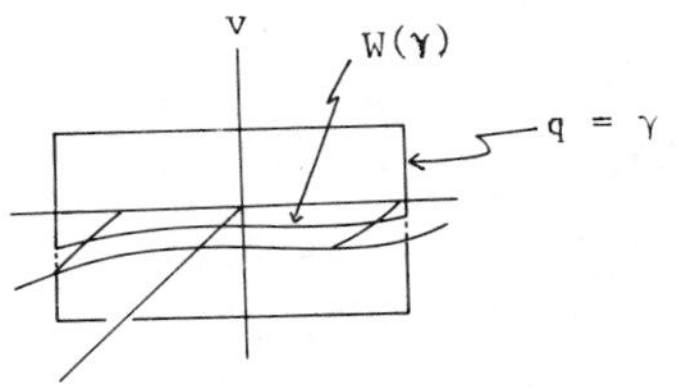

Fig. 3

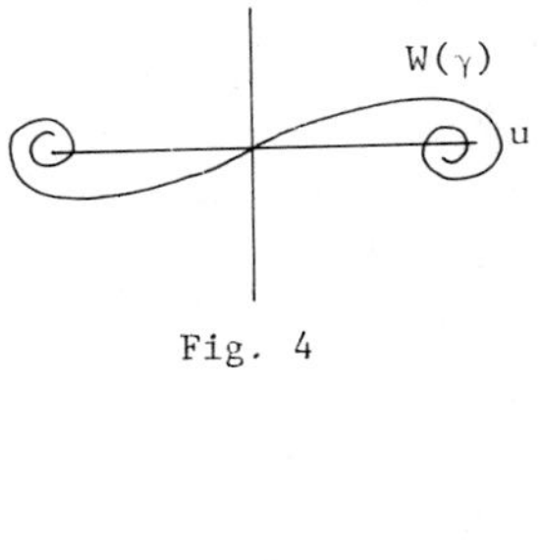

Fig. 4

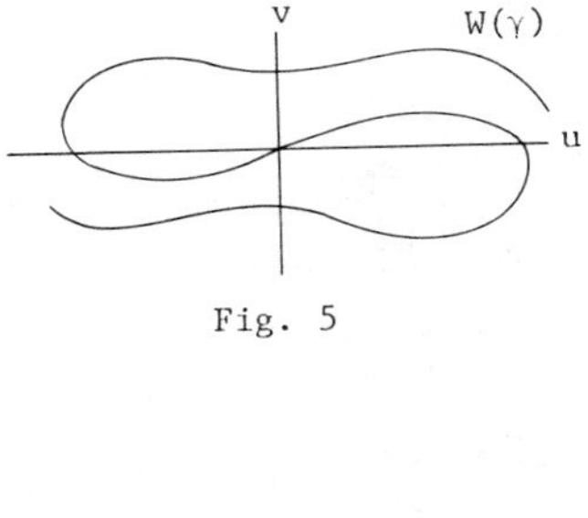

Fig. 5

In (1.9) the planes $q = 1$ and $q = 0$ are both invariant. The former carries the flow of the 1d. problem while the latter is very simple. The line $v = 0$ is a line of critical points and each vertical line is invariant consisting of trajectories tending towards the critical point on that line, see fig. 1.

A solution satisfying the boundary conditions is easily seen to be one that tends (as $t \to -\infty$) to the line $v = 0$ and to the critical point $(0,0,1)$ as $t \to +\infty$. For the associated solution to have a certain number of zeroes, it is necessary and sufficient that this trajectory in phase space oscillate around the invariant curve $u = v = 0$ that many times. I therefore want to find solutions of the form depicted in fig. 2.

The set of critical points at $q = 0$, namely $v = 0$ has a two dimensional unstable manifold which points out in the direction of increasing q. I shall call this set W, see fig. 1. Any initial condition lying on this set leads to a solution that satisfies the boundary condition at $r = 0$. I shall study this manifold near $q = 0$ and apply the flow until it is near $q = 1$ and see which solutions go to the critical point $(0,0,1)$.

The difficulty in this approach is that the critical set $v = 0$ is not compact. Consider the set $W(\gamma) = W \cap \{q = \gamma\}$, see fig. 3. Where q is very close to 0, one would guess that this set was approximately a straight line but this is not the case. To understand the behavior of the above set is the subtlety of the problem and this is where the relationship between the space dimension and the nonlinear growth comes in.

To elucidate this behavior, it turns out that a scaling is necessary. Consider (1.5) again and perform the transformation:

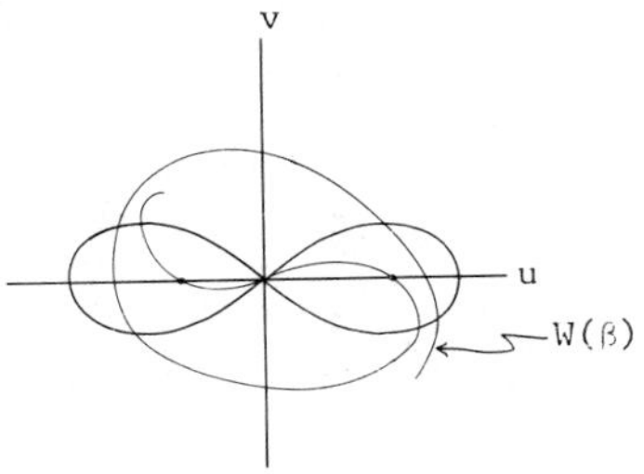

Fig. 6

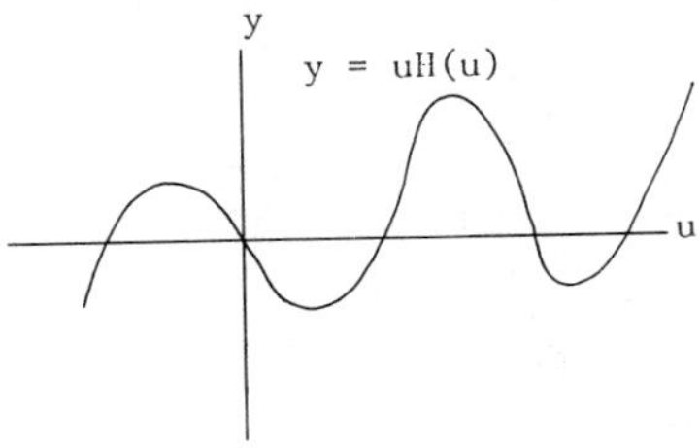

Fig. 7

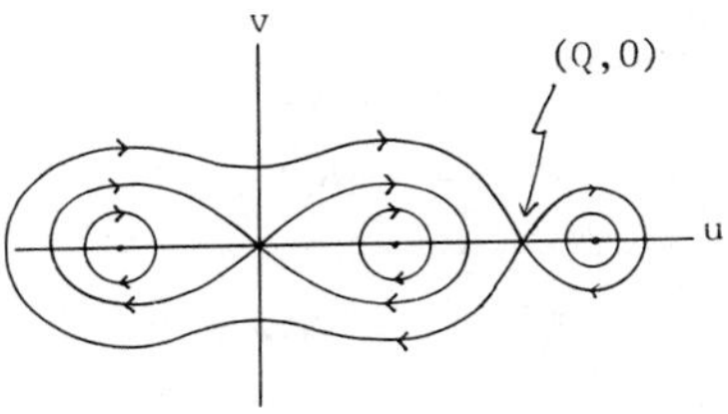

Fig. 8

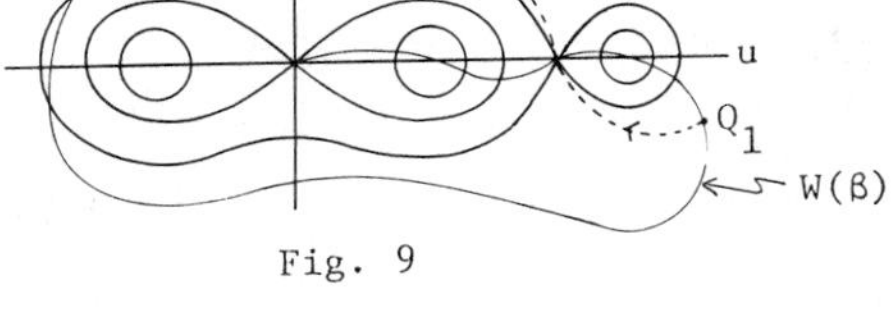

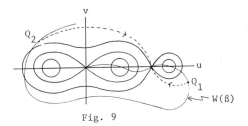

Fig. 9

$$y = r^{\tau}u$$
$$r = \exp(z)$$

where $\tau = 2/s$. This is the Emden-Fowler transformation. It gives a new set of equations that scale the phase space near $r = 0$. One should think of replacing the system (1.9) by the new system resulting from this transformation. On the level of the phase space, the part of fig. 1 near $q = 0$ is changed. There is a new limiting phase portrait at $q= 0$ which turns out to be the one that results from applying the Emden-Fowler transformation to the equation with a pure power, i.e. $a = 0$. It was this case that the transformation was traditionally used to understand, see Joseph-Lundgren [8], and the resulting phase portrait is easily analysed. Using this information, the structure of a transformed version of W can be understood. When this is mapped back to the co-ordinates of eq. (1.9), it can be seen that there are two possible configurations for $W(\gamma)$, where γ is small. The first is shown in fig. 4, this holds if $s > 4/(n-2)$, and the second in fig. 5, which is for $s < 4/(n-2)$. It is too involved for the present paper to explain how these figures are arrived at, suffice it to say that they are determined by perturbation from the limiting case at $r = 0$. The difference between the two situations comes from a global change in structure for the limit problem. At $s = 4/(n-2)$ the limit problem actually has a homoclinic orbit.

I will now explain how a $W(\gamma)$ as in fig. 5 leads to the existence of the solutions described in the theorem. If we apply the time t map for eq. (1.9), to $W(\gamma)$ we get $W(\beta)$ for some β determined by γ and the time t. Apply the flow for a large time t until β is close to 1. It can be shown that the application of the flow does not destroy the winding in W, therefore $W(\beta)$ can be assumed to be qualitatively like $W(\gamma)$. The flow at this point is then well approximated by the one dimensional problem (1.7). To see the fate of points on $W(\beta)$ superimpose the picture of $W(\beta)$ on the phase portrait for (1.9), see fig. 6. As is perhaps plausible from this picture, it can be shown that there is a trajectory starting on $W(\beta)$ for each prescribed amount of winding that goes to the critical point $(0,0,1)$.

This argument supplies a solution, but the winding is in the manifold $W(\beta)$, not in the solution itself. $W(\beta)$ is the manifold of initial conditions in $q = \beta$ that leads to a solution which goes to the line $v = 0$ as $t \to -\infty$. It turns out, however, that this oscillation is related to the number of zeroes of the solution and the above ideas lead to a proof of the theorem.

3. NONUNIQUENESS. The theorem on the existence of standing waves I have discussed is only stated for $F(u) = f(u^2)$ a pure power but it applies to a general class of nonlinearities $F(u)$. In this section I shall discuss an example of an $F(u)$ that is not a pure power. It will lead to a rather curious type of multiplicity. The kind of assumption needed for the existence theorem is that $F(u)$ grow like a power of u that satisfies the usual growth condition. The theorem goes through as the limiting phase portrait obtained by performing the Emden-Fowler transformation only actually depends on the power at large u. The details are in Jones-Kuepper [7].
It will be easier to refer to $H(u)$ where

$$H = F(u) + a. \tag{2.1}$$

Suppose $uH(u)$ has the form depicted in fig. 7. It is clear that $H(u)$ can be written in the form of (2.1).

This nonlinear term leads to a one-dimensional phase portrait eq. (1.7), of the form given in fig. 8. The saddle point in the right half plane will be of particular interest, call it $(Q,0)$. Now repeat the construction of the preceding section. As mentioned above, this will go through so long as the power growth satisfies the usual condition. Superimpose $W(\beta)$ again, for β near 1, on the 1 d. phase portrait and consider the fate of solutions with initial conditions at (x,y,j) where (x,y) is in $W(\beta)$.

The line $\{(Q,0,q): q \text{ in } [0,1]\}$ is invariant and so must lie on $W(\alpha)$ for all α. Take the curve on $W(\beta)$ between $(0,0)$ and $(Q,0)$. I claim that there are infinitely many solutions to the problem emanating from initial conditions on this line. In fact one can prove that there are solutions for any prescribed number of zeroes. To make this believable, consider the image of $W(\beta)$ under a large time t map and superimpose again on the 1d picture, this will look something like fig. 9. The oscillation here is induced by the 1d phase portrait and the shooting argument of the previous section now guarantees the desired solutions.

This construction of solutions has not used the oscillation put into $W(\beta)$ by the behavior near zero. We shall see that this supplies us with another whole set of solutions. As far as the topology of the oscillation is concerned, there is essentially no difference between the invariant lines essentially no difference between the invariant lines $\{(0,0,q): q \text{ in } [0,1]\}$ and $\{(Q,0,q): q \text{ in } [0,1]\}$. One could use the argument of the previous section to guarantee an infinite sequence of solutions which go to $(Q,0,1)$. These would be obtained by taking different amounts of oscillation in $W(\beta)$. Order the corresponding points on the right branch of $W(\beta)$ by their distance along $W(\beta)$ and denote them Q_i, $i = 1,2,\ldots$.

Consider the fate of points on $W(\beta)$ between Q_1 and Q_2. These must all eventually flow into the lefthand bulb which contains the bow-tie. A shooting argument again shows that there must result some solutions although they may have many zeroes. The same argument goes through for the points between Q_{m-1} and Q_m for any m. The resulting solutions will have at least m zeroes. This guarantees the multiplicity of solutions with arbitrarily large numbers of zeroes but something even stranger happens. One can argue that if the section of $W(\beta)$ between Q_{m-1} and Q_m leads to solutions with $i(i > m)$ zeroes then the same section must lead to solutions with k zeroes, for any $k > i$.

One can then prove the following theorem:

THEOREM. There are nonlinearities $f(u)$ with the property that for each m there are at least $k(m)$ solutions to (1.4), (1.6) with exactly m zeroes and $k(m) \to +\infty$ as $m \to +\infty$.

I want to distinguish to types of orbit here. If I linearise the operator:

$$u + uH(u)$$

at a solution $u(r)$ and restrict to spherically symmetric solutions, I obtain the linear operator:

$$v_{rr} + (n-1)/r\, v_r + H'(u)\, v + H(u)\, v$$

The spectrum of this operator can be analysed by Sturm-Liouville theory. The equation $Lv = \lambda v$ will have exactly one solution (up to a scalar multiple) that is bounded at $r = 0$. The number of zeroes of this solution for fixed λ determines the number of eigenvalues to the right of λ. This would be relevant for the stability of the solution relative to the nonlinear heat equation. The corresponding Schroedinger equation is not nearly so simple but this quantity does play a role.

I want to distinguish two different types of solutions:

(I) The number of zeroes of the solution equals the number of zeroes of the above mentioned $\lambda = 0$ solution, (II) These two numbers are not equal.

If there were a unique solution for a given number of zeroes it would have to be type I. I claim that amongst the multiple solutions found above there must be some type II solutions.

4. THE SEARCH FOR REAL EIGENVALUES. Suppose we have a standing wave solution of some nonlinear Schroedinger equation and are interested in the stability of this solution relative to the initial value problem for the full evolution equation. A wave is said to be stable if initial data near to the wave, measured in some norm, lead to solutions that decay to the wave.

A standard technique for deciding such a question is the linearised stability criterion. Linearise the equation at the standing wave and study the spectrum of the resulting linear operator on the right hand side of the equation. I shall give the operator below, but for the moment let me discuss the difficulties in applying this approach to the nonlinear Schroedinger equation. Because of its conservative nature, much of the spectrum will lie on the imaginary axis. The strongest kind of stability that can be hoped for is therefore some kind of marginal stability. There is a symmetry in the problem that forces the spectrum to be symmetric with respect to the real and imaginary axes. If there is any eigenvalue off the imaginary axis, there is therefore one in the right half plane.

Whether the spectrum being entirely on the imaginary axis implies stability is a very difficult question. The proofs of full stability need somewhat more information. In conservative problems such as the one here, one can use the energy to prove stability. The idea is roughly that if the second variation of the energy is definite (either positive or negative) then nearby solutions are trapped by the energy surfaces, see Holm et al. [6] for examples. Cazenave-Lions [4] prove a stability result using the variational approach for the solution which appears as the absolute minimum of the energy functional (this must be positive), this result is for subcritical values of s, in other words for $s < 4/n$. I believe that this is in the same spirit as the second variation method. Weinstein [15], has analysed the one-dimensional problem in somewhat more detail and achieved a similar result while accumulating a good deal of information on the neutral modes. It is most likely that this method breaks down for the solutions with more zeroes, although nothing concrete has been shown either way. I expect that many of these solutions will have spectrum lying entirely on the imaginary axis but indefinite second variations of the energy. Whether such a solution is stable or not is a very deep and interesting question.

Although the relationship between linear and full stability is not understood, I believe it would not be disputed that the existence of eigenvalues off the imaginary axis does entail instability. A natural place to start is to look for real eigenvalues. I shall develop a technique here for finding some real eigenvalues. We shall end up by seeing a mechanism that forces the existence of real eigenvalues.

Write a solution of (1.1) in the form:

$$\exp(-iat)\, u(x,t) = v(x,t) + iw(x,t)$$

The equations then become:

$$\begin{aligned} v_t &= \Delta w + f(v^2 + w^2)w + aw \\ w_t &= -\Delta v - f(v^2 + w^2)v - av \end{aligned} \tag{4.1}$$

The linearisation of the right hand side of these equations at a standing wave, which I shall denote $v(x)$, is:

$$\begin{pmatrix} p \\ q \end{pmatrix}' = M \begin{pmatrix} p \\ q \end{pmatrix} = \begin{pmatrix} 0 & L_- \\ -L_+ & 0 \end{pmatrix} \begin{pmatrix} p \\ q \end{pmatrix} \tag{4.2}$$

where

$$L_+ = \Delta + f(v^2) + 2v^2 f'(v^2) + a$$
$$L_- = \Delta + f(v^2) + a.$$

Consider the case where the underlying standing wave is spherically symmetric, like the ones found above. Restrict also to spherically symmetric perturbations, so assume that p and q are both spherically symmetric. Write

$$M \begin{pmatrix} p \\ q \end{pmatrix} = \lambda \begin{pmatrix} p \\ q \end{pmatrix} \tag{4.3}$$

as a system of ode's:

$$\begin{aligned} p' &= y \\ q' &= z \\ y' &= -((n-1)/r)\, y - g(r)\, p - \lambda q \\ z' &= -((n-1)/r)\, z - h(r)\, q + \lambda p \end{aligned} \tag{4.4}$$

where

$$h(r) = a + f(v^2)$$
$$g(r) = a + f(v^2) + 2v^2 f'(v^2)$$

If λ is real this gives a linear system on R^4. An eigenvalue occurs if (4.4) has a bounded solution on $[0,\infty)$. This requirement determines boundary conditions at $r = 0$ and $r = +\infty$. At $r = 0$, we need that $(y(r), z(r)) \to 0$ as $r \to 0$, for the solution to be regular at the origin. At $r \to +\infty$, the solution must approach the stable subspace of the asymptotic system obtained by letting $r \to +\infty$ in (4.4), note that $g(r)$ and $h(r)$ both approach a as $r \to +\infty$.

The set of solutions satisfying the boundary condition at zero is a 2d. plane, it is therefore natural to view the flow induced by this linear system on the Grassmann manifold of 2-planes in R^4.

Since the nonlinear Schroedinger equation is conservative, one might expect that this flow on the Grassmannian inherits some special structure. To see this, replace q by -q in (4.4), to obtain a system of the form

$$P' = AP \tag{4.5}$$

where $P = (p,q,y,z)$ and A is

$$\begin{pmatrix} 0 & 0 & 1 & 0 \\ 0 & 0 & 0 & -1 \\ -g(r) & \lambda & -(n-1)/r & 0 \\ \lambda & h(r) & 0 & -(n-1)/r \end{pmatrix}$$

If $n = 1$, this system is now Hamiltonian, which is to say it has the form $P' = JS(r)P$ where $S(r)$ is a symmetric matrix and J is the symplectic matrix:

$$\begin{pmatrix} 0 & I \\ -I & 0 \end{pmatrix}$$

I being the 2x2 identity. It is a standard fact that such a system induces a flow on the space of Lagrangian planes. A plane is Langrangian if $(P,JQ) = 0$ for any vectors P,Q in the plane. The space of Lagrangian planes is a submanifold of the Grassmannian described above. There is a flow on the space of Lagrangian planes because the linear system preserves the J-inner product, i.e. if $P(r)$ and $Q(r)$ are solutions then $(P(r), JQ(r))$ is constant.

If $n > 1$, then this is no longer true, however computing $(P,JQ)'$, we get:

$$(P,JQ)' = -((n-1)/r)(P,JQ)$$

which implies that the set $(P,JQ) = 0$ is invariant. Consequently the system still induces a flow on the space of Lagrangian planes. Recall that I am trying to trace a trajectory for each λ which satisfies a boundary condition at $r = 0$ and see what happens to it as $r \to +\infty$. The Lagrangian planes will only be useful if this trajectory lies in that space. However it is easy to see that the boundary condition at 0 is Lagrangian and consequently it does.

The space of 2d. Lagrangian planes in R^4 is well known to be $U(2)/O(2)$, where $U(2)$ = the group of unitary 2x2 matrices and $O(2)$ is the subgroup of real orthogonal matrices. This is a fiber bundle over the circle S^1 with a fiber S^2. It is, however, not a trivial bundle. Think of a line with a sphere attached to each point and the spheres at each end identified via the antipodal map. It can be shown that the total space is equivalent to this by identifying it with the manifold of 2x2 symmetric unitary matrices.

The circle as base space in this bundle means that curves in the total space have some winding by projection down onto the base space. As a function of unitary matrices this projection is the square of the determinant. The index of curves in the total space thus obtained is essentially the Maslov index. To show this in the more general setting of n dimensions is the main theme of Arnol'd [1]. Obviously one has to be a little careful if the projection onto S^1 is not a closed curve, but this is a standard problem that is dealt with easily.

Let $G(r,\lambda)$ be the solution satisfying the boundary condition at 0, for each fixed λ. Denote the Maslov index of the curve $G([0,\infty),\lambda)$ by $I(\lambda)$. I claim it can be shown by a shooting argument that if λ_1, λ_2 are in $[0,\infty)$ with $\lambda_2 > \lambda_1$ and $I(\lambda_2) \neq I(\lambda_1)$, then there is a value of λ in $[\lambda_1, \lambda_2]$ at which the solution G satisfies the boundary condition at $+\infty$. Consequently this value is an eigenvalue. For instance, we can determine fairly easily $I(\lambda)$ when $\lambda >> 1$ and also $I(0)$. If these are different then there must be a real eigenvalue and the solution is unstable.

If $\lambda >> 1$ then λ dominates the equation (4.5) and the flow on the space of Lagrangian planes can be fairly easily analysed. It turns out that the index in this case is zero. Instability will occur in cases when $I(0) = 0$ (so long as it can be shown the eigenvalue found is not actually at 0).

To determine the behavior at $\lambda = 0$, observe that the equations (4.3), or equivalently (4.4) uncouple. The resulting equations are:

$$p_{rr} + ((n-1)/r)\, p_r + g(r)\, p = 0 \tag{4.6}$$

$$q_{rr} + ((n-1)/r)\, q_r + h(r)\, q = 0 \tag{4.7}$$

If we set $H(u) = a + f(u)$, then $g(r) = H'(v(r))$ and $h(r) = H(v(r))$. Notice then that (4.6) is the equation of variations of the ODE (1.4) and (4.7) is satisfied by the solution $v(r)$ itself. Consequently, these are both analysable. As second order equations there are two independent solutions and some characteristic oscillation that measures the winding of the solutions in the projectivised problem. To make sense of this, the winding must be defined correctly. However I claim that this can be done so that it is related to $I(0)$. Let $osc(1)$ be the oscillation in (4.6) and $osc(2)$ that of (4.7), then I claim that the formula:

$$I(0) = osc(1) - osc(2) \tag{4.8}$$

holds,

Let us apply this to the solutions I described in the last section as type II. $osc(1)$ is the oscillation in the equation of variations, which is what I described as not being equal to the number of zeroes of the solution for type II solutions!

osc(2) is the number of zeroes of the solution and so the condition for type II solutions translates into I(0) being non-zero. Type II solutions are therefore unstable, provided that the eigenvalue is not picked up at zero. This is true if the solution is nondegenerate.

It is hoped that this technique will be usable in other instability problems for the nonlinear Schroedinger equation. A particluarly interesting problem is whether different nonlinearities offer different stability characteristics for these solutions with many zeroes. There is an interesting conjecture of Moloney, McLaughlin and Newell [9], on such a phenomenon, which is based on some numerical evidence.

BIBLIOGRAPHY

1. V. I. Arnol'd, "Characteristic class entering in quantization conditions", J. Funct. Anal. and its Applic., 1 (1967), 1-14.

2. H. Berestycki and P. L. Lions, "Nonlinear scalar field equations, II. Existence of infinitely many solutions", Arch. Rat. Mech. Anal., 82 (1983), 347-376.

3. M. S. Berger, "On the existence and structure of stationary states for a nonlinear Klein-Gordon equation", J. Funct. Anal. 9 (1972), 249-261.

4. T. Cazenave and P. L. Lions, "Orbital stability of standing waves for some nonlinear Schroedinger equations", Comm. Math. Phys. 85 (1982), 549-561.

5. C. V. Coffman, "Uniqueness of the ground state solution of $\Delta u - u + u^3 = 0$ and a variational characterisation of other solutions", Arch. Rat. Mech. Anal., 46 (1972), 81-95.

6. D. Holm, J. Marsden, T. Ratiu and A. Weinstein, "Nonlinear stability in fluids and plasmas", to appear in Physical Review.

7. C. K. R. T. Jones and T. Kuepper, "On the infinitely many solutions of a semilinear elliptic equation", submitted to SIAM J. Math. Anal.

8. D. D. Joseph and T. S. Lundgren, "Quasilinear Dirichlet problems driven by positive sources", Arch Rat. Mech. Anal., 49 (1972), 241-269.

9. D. W. McLaughlin, J. Moloney and A. Newell, "Solitary waves as fixed points of infinite-dimensional maps in an optical bistable ring cavity", Phys. Rev. Lett.,. 51 (1983), 75-78.

10. K. McLeod and J. Serrin, "Uniqueness of solutions of semilinear Poisson equations", Proc. Nat. Acad. Sci USA, 78 (1981), 6592-6595.

11. Z. Nehari, "On a nonlinear differential equation arising in nuclear physics", Proc. Roy Irish Acad. 62 (1963), 117-135.

12. S. I. Pohozaev, "Eigenfunctions of the equation $\Delta u + \lambda f(u) = 0$", Sov. Math. Dokl., 5 (1965), 1408-1411.

13. G. H. Ryder, "Boundary value problems for a class of nonlinear differential equations", Pacific J. Math., 22 (1968), 477-503.

14. W. Strauss, "Existence of solitary waves in higher dimensions", Comm. Math. Phys., 55 (1977), 149-162.

15. M. Weinstein, "Modulational stability of ground state of nonlinear Schroedinger equations", SIAM J. Appl. Math., in press.

DEPARTMENT OF MATHEMATICS
UNIVERSITY OF ARIZONA
TUCSON, ARIZONA 85721

Lectures in Applied Mathematics
Volume 23, 1986

ON THE SOLITARY TRAVELING WAVE OF THE GENERALIZED KORTEWEG - DE VRIES EQUATION

Michael I. Weinstein

I. INTRODUCTION

Consider the initial-value problem (IVP) for the generalized Korteweg - de Vries equation (GKdV)

$$(1)\qquad w_t(x,t) + a(w(x,t))w_x(x,t) + w_{xxx}(x,t) = 0,\ (x,t) \in \mathbb{R} \times \mathbb{R}^+$$
$$w(x,0) = w_0(x) \in H^2,$$

where $a \in C^\infty$.
Well-posedness for equations of this type has been studied by Dushane [4], Strauss [8], and Kato [6].

Solitary traveling waves of GKdV can be sought of the form

$$(2)\qquad \tilde{W}(x,t) = \psi(x-ct),\quad c > 0.$$

Substitution of (2) into (1), implies an ODE for ψ. If we integrate the ODE, assuming ψ and its derivatives vanish at $\pm\infty$, we have

$$(3)\qquad \psi'' - c\psi + a_1(\psi) = 0$$

where

$$(4)\qquad a_1(\psi) = \int_0^t a(\mu)d\mu.$$

Integration of (3) leads to

$$\frac{1}{2}(\psi')^2 - \frac{c}{2}\psi^2 + a_2(\psi) = 0 \tag{5}$$

where

$$a_2(t) = \int_0^t (t-\mu)a(\mu)d\mu. \tag{6}$$

We now make the following assumptions on $a(t)$, which guarantee the existence of a unique (up to translation) positive, even, monotonically decreasing solution of (3) that vanishes at $\pm\infty$ (see Berestycki and Lions [2]):

(C1) $t_0 \equiv \min\{t > 0: a_2(t) - \frac{c}{2}t^2 = 0\}$ exists

(C2) $ct_0 - a_1(t_0) < 0$.

For the case $a(w) = w^p$ we have

$$\psi(x-ct) = \left[\frac{(p+1)(p+2)}{2}\right]^{\frac{1}{p}} c^{1/p}\,\mathrm{sech}^{\frac{2}{p}}\left[\frac{p}{2}\, c^{1/2}\,(x-ct)\right]. \tag{7}$$

In this article we present results on the stability of the ground state (Theorem 1) and its role in the global existence theory (Theorem 2) for GKdV.

THEOREM 1. Let $a(\lambda)$ satisfy the "subcritical" growth condition

$$\lim_{|\lambda|\to\infty} \sup|\lambda|^{-4} a(\lambda) \le 0, \quad \text{and} \tag{8}$$

$$0 < \frac{d}{dc} N(\psi(c)) = -c^{-1}\frac{d}{dc} H(\psi(c)), \tag{9}$$

where N and H are defined in (18) and (19). Also, assume that $\int a(\psi(x))[\psi'(x)]^2 dx \ne 0$. Then, the solitary wave of GKdV is orbitally stable, i.e., for any $\varepsilon > 0$, there is a $\delta(\varepsilon) > 0$ such that if

$$\min_{x_0} ||w_0(\cdot + x_0) - \psi(\cdot)||_{H^1} < \delta(\varepsilon) \tag{10}$$

then

$$\min_{x_0} ||w(\cdot + x_0,t) - \psi(\cdot)||_{H^1} < \varepsilon, \quad t > 0. \tag{11}$$

In particular, for the case $a(w) = w^p$, the solitary traveling wave is stable if $p < 4$.

T. B. Benjamin [1] and J. Bona [3] have proved stability for $a(w) = w$ (K-dV). Theorem 1 is a generalization. We remark that our proof is considerably simpler in the following way. Rather than using the explicit spectral representation of the linearized operator that is available in dimension one, we use general spectral properties of the linearized operator which are derivable from a variational principle. Thus, our methods have the benefit of being applicable to higher dimensional problems (see for example stability of ground states for nonlinear Schrödinger equations for dimensions $N \geq 1$ [10]).

For the "critical" case, $p = 4$, we have

THEOREM 2. Let $a(w) = w^4$. A sufficient condition for the existence of a unique global solution to GKdV of class $C([0,\infty);H^2)$ is

$$||u_0||_{L^2} \leq ||\psi||_{L^2} = 15^{1/4}(\tfrac{\pi}{2})^{1/2}. \tag{12}$$

Theorem 2 is analogous to results obtained by the author for the nonlinear Schrödinger equation (NLS)

$$2i\,\frac{\partial\phi}{\partial t} + \Delta\phi + |\phi|^{2\sigma}\phi = 0, \quad x \in R^N, \quad t \in R^+ \tag{13}$$

$$\phi(x,0) = \phi_0(x) \in H^1$$

in the critical case $\sigma = \frac{2}{N}$.

In [9] it was proved that a sufficient condition for the existence of a unique global solution of class $C([0,\infty);H^1)$ to the IVP for NLS is

$$||\phi_0||_2 < ||R||_2. \tag{14}$$

Here R denotes a ground state solution of

$$\Delta R - R + R^{\frac{4}{N}+1} = 0, \tag{15}$$

and $R(x)e^{it/2}$ is an exact solution of NLS.

For NLS in the cases $\frac{2}{N} \leq \sigma < \frac{2}{N-2}$, a class of H^1 initial data has been displayed for which the solution to the initial-value problem blows up in finite time [5]. Specifically, for such data there exists $T < \infty$ such that $\lim_{t \to T^-} ||\nabla\phi(t)||_2 = +\infty$. We believe that if condition (12) is violated, then there exist H^2 initial data w_0 for which the solution $w(x,t)$ of (1) blows up in some sense in finite time. A proof of this conjecture has, thus far, been elusive. We note that such data w_0, by Theorem 2, must satisfy

$$||w_0||_2 \geq ||\psi||_2. \tag{16}$$

We close this section with an existence result, that we require. It is a consequence of a theorem of Kato [6].

THEOREM 3.

(a) Assume the subcritical growth condition (8). Then, there is a unique global solution of (1) of class $C([0,\infty);H^2)$. In particular, if $a(w) = w^p$, then there is global existence for $p < 4$.

(b) For $a(w) = w^p$ and $p = 4$, there exists a positive constant γ, such that if $||w_0||_2 < \delta$, then (1) has a global solution in $C([0,\infty);H^2)$.

(c) The solution of (1) has the following conserved integrals:

$$I(w) = \int w(x,t)dx \tag{17}$$

$$N(w) = \int w^2(x,t)dx \tag{18}$$

(19) $$H(w) = \int \frac{1}{2} w_x^2(x,t) - a_2(w(x,t))dx.$$

2. PROOF OF THEOREM 1 (SKETCH)

The idea is to use the conserved quantity

(20) $$E[w] \equiv H[w] + cN[w]$$

as a Lyapunov function. We write

(21) $$w(x+x_0,t) = \psi(x) + u(x,t)$$

where $x_0 = x_0(t)$ is chosen to minimize the distance to the solitary wave "orbit":

(22) $$||w'(\cdot + x_0,t) - \psi'(\cdot)||^2_{L^2} + c||w(\cdot + x_0) - \psi(\cdot)||^2_{L^2}.$$

The implied linear relation can be used to prove, for perturbations of ψ preserving the L^2 norm, that

(23) $$E[w_0] - E[\psi] \geq g(||u(t)||_{H^1})$$

where $g(s)$ is continuous and nonnegative for $0 < s << 1$ and $g(0) = 0$. The difference on the left hand side of (23) can be made small by choosing w_0 close to ψ in H^1. This implies that $||u(t)||_{H^1}$ is small for all time, i.e., orbital stability. The case of perturbations of arbitrary L^2 norm is handled by constructing solitary waves of arbitrary L^2 norm by varying the speed c in (2).

A more detailed exposition appears in [10].

3. PROOF OF THEOREM 2

We remark that the condition on the "size" of w_0 in Theorems 2 and 3 is to ensure a uniform bound of the form

(24) $$||w(t)||_{H^1} \leq M(N,H).$$

Here, $M(N,H)$ is a constant dependent on the initial data. Bound (24) implies

$$(25)\qquad ||W(t)||_\infty \le C||W(t)||_{H^1} \le C\cdot M(N,H).$$

This uniform bound on $||W(t)||_\infty$ together with an H^2 estimate of the solution, obtained directly from equation (1), provides sufficient a priori control of the local solution for continuation to a solution in $C([0,\infty);H^2)$.

The proof of Theorem 2 requires the following result of Nagy [7]:

LEMMA. Let $f \in H^1$. Then

(i) $f \in L^6$ and

$$(26)\qquad ||f||_6 \le C||f'||_2^{1/3}||f||_2^{2/3}.$$

(ii) The smallest constant C for which the estimate (26) holds is

$$(27)\qquad C = \left(\frac{3}{||f_0||_2^4}\right)^{1/6} = \left(\frac{4}{\pi^2}\right)^{1/6},$$

where

$$(28)\qquad f_0(x) = 3^{1/4}\,\mathrm{sech}^{1/2}(2x).$$

(iii) Equality holds in estimate (26) with constant (27) for

$$(29)\qquad f(x) = \lambda f_0(\mu x + \nu), \quad \lambda\cdot\mu \neq 0.$$

PROOF OF THEOREM 2. By the preceding remark it will suffice to obtain an a priori bound of the form (24). A local solution of equation (1) satisfies the conservation laws (17)-(19). Therefore, using (18,19) and the lemma,

$$(30)\qquad \begin{aligned}\frac{1}{2}||Dw(t)||_2^2 &= H(w_0) + \frac{1}{30}||w(t)||_6^6 \\ &\le H(w_0) + \frac{1}{30}C^6||Dw(t)||_2^2||w_0||_2^4,\end{aligned}$$

where D denotes $\frac{\partial}{\partial x}$.

Hence,

$$\frac{1}{2}\left(1 - \frac{C^6}{15}\,||w_0||_2^4\right)||Dw(t)||_2^2 \leq H(w_0). \tag{31}$$

Using the precise value of C from (27) we get

$$\frac{1}{2}\left(1 - \left(\frac{||w_0||_2}{5^{1/4}||f_0||_2}\right)^4\right)||Dw(t)||_2^2 \leq H(w_0). \tag{32}$$

By expressions (7) and (28) we have

$$\frac{1}{2}\left(1 - \left(\frac{||w_0||_2}{||\psi||_2}\right)^4\right)||Dw(t)||_2^2 \leq H(w_0). \tag{33}$$

Taking $||w_0||_2 < ||\psi||_2$ we obtain the uniform bound (24).

REFERENCES

[1] Benjamin, T. B., "The stability of solitary waves," Proc. Roy. Soc. Lond. A 328 (1972): 153-183.

[2] Berestycki, H., Lions, P. L., "Nonlinear scalar field equations I - Existence of a ground state," Arch. Rat. Mech. Anal. 82 (1983): 313-345.

[3] Bona, J., "On the stability of solitary waves," Proc. Roy. Soc. Lond. A 344 (1975): 363-374.

[4] Dushane, T. E., "Generalization of the Korteweg - de Vries equation," Proc. Symp. Pure Math., Vol. 23, Ann. Math. Soc. 1973: 303-307.

[5] Glassey, R. T., "On the blowing-up of solutions to the Cauchy problem for the nonlinear Schrödinger equation," J. Math. Phys. 18 (1977): 1794-1797.

[6] Kato, T., "On the Cauchy problem for the (generalized) Korteweg - de Vries equation," in Studies in Appl. Math. Advances in Mathematics Supplementary Studies Vol. 8, Academic Press (1983).

[7] Nagy, B. V. Sz., "Über Integralungleichungen zwischen einer Funktion und ihrer Ableitung," Acta Sci. Math. (Szeged) 10 (1941): 64-74.

[8] Strauss, W. A., "Dispersion of low-energy waves to two conservative equations," Arch. Rat. Mech. Anal. 55 (1974): 86-92.

[9] Weinstein, M. I., "Nonlinear Schrödinger equations and sharp interpolation estimates," Comm. Math. Phys. 87 (1983): 567-576.

[10] Weinstein, M. I., "Lyapunov stability of ground states of nonlinear dispersive evolution equations," Comm. Pure Appl. Math., in press.

DEPARTMENT OF MATHEMATICS
STANFORD UNIVERSITY

CURRENT ADDRESS:
DEPARTMENT OF MATHEMATICS
PRINCETON UNIVERSITY
PRINCETON, N. J. 08544

Lectures in Applied Mathematics
Volume 23, 1986

WAVE OPERATORS FOR THE SCATTERING OF SOLITARY WAVES IN MULTIPLE SPATIAL DIMENSIONS

Henry A. Warchall

A definition is proposed for wave operators to describe the scattering of solitary wave solutions of classical nonlinear hyperbolic equations. These operators allow the treatment of asymptotic states whose evolution is not governed by a linear equation. For a particular family of nonlinear Klein-Gordon equations, the convergence of the expressions for the wave operators applied to asymptotic configurations of solitary waves without radiation is examined. The proof of the existence of the wave operators for these states is reduced to the verification of stability criteria for certain solutions.

1. INTRODUCTION. This report outlines work in progress on the development of a scattering theory for solitary waves in multiple spatial dimensions. The goal is to establish, for a family of classical nonlinear wave equations, a time-dependent theory of scattering which is not limited by restrictions of small data or weak nonlinearity. The equations under study have solitary-wave solutions which do not behave asymptotically like solutions of the corresponding linear wave equations, and for which the appropriate "comparative dynamics" are not linear at all, in contrast to the situation for conventional scattering states.

1980 Mathematics Subject Classification. 35L70, 76B25,35P25

This paper focuses on one aspect of this program, the convergence of the formal expressions for the wave operators acting on asymptotic configurations consisting of precisely N solitary waves. We reduce the question of convergence of the wave operators to that of the stability of certain solutions of the nonlinear wave equation.

Section 2 includes the definition of the wave operators and summarizes the theoretical program. Sections 3 through 6 treat the special case of the convergence of the expression for the wave operator Ω_+ on asymptotic states having two solitary waves and no radiation. (Treatment of $\Omega_\pm$ on states with N solitary waves is the same.)

The results presented here are undoubtedly not the best possible; the aim of this report is to summarize the work in progress, and to indicate methods which are applicable to relevant problems. To streamline the discussion, simplifying assumptions are made in several instances, and proofs are merely outlined. Details will appear in a forthcoming paper.

2. GENERAL FRAMEWORK. The development of the mathematical theory of scattering for classical nonlinear wave equations has been hampered in part by the existence of solutions which are not asymptotically free, that is, solutions which do not obey associated linear equations at asymptotically large times. A major problem caused by the existence of such solutions lies in the choice of comparative dynamics for the time-dependent description of scattering; even isolated solitary waves do not obey a linear wave equation, yet there should be a description of the propagation of solitary waves whose interaction with other waves has been "turned off." The framework outlined here is intended to overcome this difficulty and develop a theory of scattering for a class of nonlinear wave equations in multiple space dimensions which have solitary wave solutions.

Consider the (abstract) classical nonlinear wave equation $\dot{\varphi}(t) = -iA\varphi(t) + B(\varphi(t))$ for which solutions φ take values in a Hilbert space $\mathcal{H}$, the operator A is a linear self-adjoint operator on (a dense domain in) $\mathcal{H}$, and B is a suitably nice nonlinear map from $\mathcal{H}$ to itself [7]. The associated linear equation $\varphi(t) = -iA\varphi(t)$ has solution $\dot{\varphi}(t) = U_0(t)\varphi(0)$ with the one-parameter unitary group $U_0(t)$ given by $U_0(t) = e^{-itA}$. Conventional (time-dependent) scattering theory treats solutions $\varphi(t)$ of the nonlinear equation for which there exist vectors $\varphi_\pm$ in $\mathcal{H}$ such that $\lim_{t\to\pm\infty} \|\varphi(t) - U_0(t)\varphi_\pm\| = 0$; this condition reflects the idea that at asymptotically large times the solution $\varphi(t)$ develops according to the "free" dynamics $U_0(t)$. It is well-known, however, that for large classes of nonlinearities B, there exist solutions of the nonlinear equation which do not asymptotically obey the associated linear equation [10]. In particular, there are many nonlinear wave equations in $n \geq 3$ spatial dimensions which have solitary wave solutions whose amplitude does not tend to zero as $t\to\infty$, and which do not develop asymptotically according to $U_0(t)$.

For such solutions, a different comparative "free" dynamics must be used to formulate a time-dependent theory of scattering. To explain this comparison dynamics, we consider for definiteness the nonlinear Klein-Gordon equation $\ddot{u} = \nabla^2 u + g(u)$ where $u:\mathbb{R}\times\mathbb{R}^3\to\mathbb{C}$, and $g:\mathbb{C}\to\mathbb{C}$ is a nonlinear function such that $g(0) = 0$ and $g(re^{i\theta}) = g(r)\, e^{i\theta}$. Suppose that g is one particular function in the large class for which this equation has solitary wave solutions $u_s(x,t)$ which (for fixed t) are exponentially small at spatial infinity [2,4,11]. We can rewrite this equation in the abstract form above in the standard way with $\varphi(t) = (u(t),\dot{u}(t))$ and $\mathcal{H} = \{(u,v) \mid \|u\|_2 + \|\nabla u\|_2 + \|v\|_2 < \infty\}$ with $\|\;\|_2$ the usual norm on $L^2(\mathbb{R}^3)$. If $\varphi(t)$ is a solution of the abstract nonlinear equation, we write $\varphi(t) = U_t(\varphi(0))$ where $U_t: \mathcal{H} \to \mathcal{H}$ is a nonlinear map.

Let $\mathfrak{G}$ be the subset in $\mathfrak{H}$ of initial data corresponding to globally defined solutions. If u_s is a solitary wave solution to the wave equation then the corresponding initial data $\varphi_s(0)$ is in $\mathfrak{G}$. $\mathfrak{G}$ also contains the initial data for all globally-defined solutions which obey the free dynamics at asymptotically large times. Furthermore, because of the time-translation invariance of the wave equation, $U_t \mathfrak{G}$ is contained in $\mathfrak{G}$ for all t.

Note also that the wave equation is invariant under the action $u(x) \to (\Lambda u)(x) = u(Mx + a)$ of the Poincaré group. Here $x \in \mathbb{R}^{3+1}$, $M: \mathbb{R}^{3+1} \to \mathbb{R}^{3+1}$ is an element of SO(3,1), and $a \in \mathbb{R}^{3+1}$. We denote by E[u] the energy functional
$E[u] = \int H(u)\,dx = \frac{1}{2}\|\dot{u}\|^2 + \frac{1}{2}\|\nabla u\|^2 - \int G(|u|)\,dx$ where
$G(s) = \int_0^s g(r)\,dr$.

A solution u(x,t) of the wave equation is said to contain solitary waves at $t = \pm\infty$ if $\sup_{x \in \mathbb{R}^3} |u(x,t)|$ does not tend to zero as $t \to \pm\infty$. A solution u(x,t) of the wave equation which contains solitary waves is said to be an isolated solitary wave if there exists a Poincaré transformation Λ such that for every $\varepsilon > 0$ there exists a ball of radius R around the origin such that $\int_{|x|>R} H(\Lambda u)\,dx < \varepsilon$ for all t. Two solutions u(x,t) and v(x,t) of the wave equation are said to be asymptotically disjoint at $t = \pm\infty$ if $\lim_{t \to \pm\infty} \sup_{x \in \mathbb{R}^3} |u(x,t)v(x,t)| = 0$.

The asymptotic description of solutions which contain solitary waves must of necessity differ from the description of solutions which are asymptotically free. The framework proposed here is based on the observation that isolated solitary waves are in many respects particle-like, while asymptotically free solutions (the usual scattering states) are radiation-like. It is anticipated that a generic solution with initial data in $\mathfrak{G}$ resolves in some sense as $t \to \infty$ into some widely separated (isolated) solitary waves and some asymptotically free "radiation." This is of course precisely the situation known to occur in the completely integrable equations in one space dimension with soliton solutions; it remains to be seen if a class of equations in higher spatial dimensions can be found in which a manageable set of solitary waves "spans" the asymptotic configurations.

These considerations suggest the following tentative framework for the description of asymptotic states. Let Σ be the Banach subspace in $\mathbf{H}$ of initial data for (ordinary) asymptotically free scattering states of the wave equation [10]. Given $\varphi_0 \in \Sigma$ (with $\|\varphi_0\|_{scat}$ small), and given the initial data $\{\varphi_j \in \mathbf{G} \mid j = 1,\ldots,N\}$ for N isolated solitary waves which are pairwise asymptotically disjoint at $t = -\infty$, define the ordered (N+1)-tuple Φ_N in the Cartesian product $\mathbf{G}_N \equiv \Sigma \times \mathbf{G} \times \mathbf{G} \times \ldots \times \mathbf{G}$ (N factors of $\mathbf{G}$; no topology implied) by $\Phi_N \equiv (\varphi_0, \varphi_1, \ldots, \varphi_N)$. Define the nonlinear map $\Gamma_t: \mathbf{G}_N \to \mathbf{G}_N$ by $\Gamma_t \Phi_N \equiv (U_0(t)\varphi_0, U_t(\varphi_1), \ldots, U_t(\varphi_N))$. (We have in mind eventually restricting Γ in the last N slots to work on N copies of a proper subset $\mathbf{S}$ of $\mathbf{G}$, disjoint from Σ, which contains initial data only for isolated solitary waves. For example, $\mathbf{S}$ might consist of the orbit of initial data for a particular solitary wave under the action of the Poincaré group and whatever other internal symmetry group the problem has. For now we leave $\mathbf{S}$ unspecified, noting only that U_t will map $\mathbf{S}$ into $\mathbf{S}$ for all t.) Furthermore, for any $N \geq 0$, let $J: \mathbf{G}_N \to \mathbf{H}$ be defined by $J(\varphi_0, \varphi_1, \ldots, \varphi_N) = \Sigma_0^N \varphi_j$.

The idea here is that Φ_N represents the initial asymptotic configuration of the system, consisting of the incoming radiation φ_0 and the incoming solitary waves $\varphi_1, \ldots, \varphi_N$. The map Γ plays the role of the free time development of the system; each of the isolated solitary waves develops in time independently of the others and of the radiation, while the radiation develops according to the usual free dynamics U_0. By hypothesis the solitary waves are asymptotically disjoint, so for times t farther and farther in the past, the vector $J\,\Gamma_t \Phi_N$ in $\mathbf{H}$ describes a function which should resemble more and more a solution of the nonlinear equation in which some radiation and N solitary waves were incoming in the remote past. That is to say, the analogue to the existence of the wave operator Ω_- is the convergence in some suitable norm of $\lim_{t\to-\infty} U_{-t}\, J\, \Gamma_t\, \Phi_N$. This expression is analogous to that encountered in two-Hilbert-space scattering theory, although here the set $\mathbf{G}_N$ has no relevant linear structure at all.

The aim of the work in progress is to prove results in this framework (for a suitable class of wave equations) analogous to those in "small data" scattering, such as:

(1) Existence and invertibility of nonlinear wave operators $\Omega_{\pm}: \Sigma \times \mathbf{S} \times \ldots \times \mathbf{S} \to \mathbf{G}$ given formally by $\Omega_{\pm} = \lim_{t\to\pm\infty} U_{-t}\, J\, \Gamma_t$.

(2) Weak asymptotic completeness of the wave operators: $\cup_{N=0}^{\infty}\, \Omega_+(\Sigma_N) = \cup_{N=0}^{\infty}\, \Omega_-(\Sigma_N)$ where $\Sigma_N = \Sigma \times \mathbf{S} \times \ldots \times \mathbf{S}$.

(3) Asymptotic completeness: $\mathbf{G} = \cup_{N=0}^{\infty}\, \Omega_+(\Sigma_N)$.

(4) Lippman-Schwinger type integral equations for $\Omega_{\pm}$ to allow perturbative computations of the scattering operator $(\Omega_+)^{-1} \circ \Omega_-$.

The demonstration of weak asymptotic completeness depends on the ability to find a large enough subset $\mathbf{S}$ to encompass all the different types (Poincaré equivalence classes) of solitary waves, on the absence of a solitary-wave "infrared problem" (in which asymptotic states contain infinte numbers of solitary waves with finite total energy), and on the anticipated asymptotic resolution of "in" states in Σ_N to "out" states in Σ_K. The property of full asymptotic completeness is at this point purely conjectural, but is suggested by the observed asymptotic resolution in the completely integrable one-space-dimensional evolution equations. If such an asymptotic resolution of generic initial data into solitary waves and radiation occurs in an interesting three-space-dimensional equation, it is anticipated that a modified Huygens'-principle argument can be employed to separate these components of the asymptotic configuration.

The remainder of this paper treats a special case of the existence problem (1) above.

3. EXISTENCE OF WAVE OPERATORS ON TWO-SOLITARY-WAVE STATES. We now restrict our considerations for simplicity to the investigation of the convergence of the formal expressions for the wave operator Ω_+ acting on asymptotic configurations consisting of two solitary waves. The treatment for $\Omega_\pm$ acting on N solitary waves is the same.

We consider solutions $u: \mathbb{R}^{3+1} \to \mathbb{C}$ of the nonlinear Klein-Gordon equation $\ddot{u} = \nabla^2 u + g(u)$ with $g(0) = 0$ and $g(r\, e^{i\theta}) = g(r)\, e^{i\theta}$, and with $g(s)$ real for real s. We define the energy functional $E[u] \equiv \frac{1}{2}\|\dot{u}\|^2 + \frac{1}{2}\|\nabla u\|^2 - \int G(|u|)\, dx$ where $G(s) = \int_0^s g(r)\, dr$; then if u satisfies $\ddot{u} = \nabla^2 u + g(u)$ and $E[u]$ exists, $E[u]$ is constant in time. If the function g is C^1 and such that $G(s) \leq 0$ for all s, and if $|g(s)| = O(|s|^3)$ as $|s| \to \infty$, then it follows (for example as a special case of more general results in [8]) that for any initial conditions of finite energy, there exists a unique solution of finite energy which is bounded in bounded regions of spacetime. Furthermore this solution is C^1 if the initial conditions are.

Next consider solutions $v: \mathbb{R}^3 \to \mathbb{R}$ of the equation $0 = \nabla^2 v + f(v)$. If $f: \mathbb{R} \to \mathbb{R}$ is C^1, odd, $f'(0) < 0$, $\lim_{s\to\infty} f(s)/s^5 \le 0$, and if $F(s) = \int_0^s f(r)\,dr > 0$ for some $s > 0$, then it follows from [2] (see also [4, 6, 11]) that there exists an infinite sequence of solitary wave solutions v to $0 = \nabla^2 v + f(v)$, each of which has the properties:
(i) v is spherically symmetric,
(ii) v is in $C^2(\mathbb{R}^3)$, and
(iii) v and its derivatives up to second order have exponential decay.

Given such a solution v(x), the function $u_0(x,t) = e^{-i\omega t} v(x)$ is a solution of $\ddot{u} = \nabla^2 u + g(u)$, where f and g are related by $f(s) = g(s) + \omega^2 s$. There are plenty of functions g which guarantee the global existence of unique regular solutions to $\ddot{u} = \nabla^2 u + g(u)$ and for which the corresponding f(s) results in solitary wave solutions to $0 = \nabla^2 v + f(v)$. For example, $g(s) = g_3(s) = -(m + \omega^2)s + \beta|s|s - \alpha|s|^2 s$ with positive m, ω^2, β, and α such that $9\alpha(m+\omega^2) \ge 2\beta^2$ guarantees that $G(s) \le 0$ for all s, and the corresponding $f(s) = f_3(s) = -ms + \beta|s|s - \alpha|s|^2 s$ has $F(\beta/3) > 0$ provided that $2\beta^2 > 9m\alpha$.

Thus if $9\alpha\omega^2 \ge 2\beta^2 - 9m\alpha > 0$ then the equation $\ddot{u} = \nabla^2 u + g_3(u)$ has unique regular global solutions which include ("stationary") isolated solitary wave solutions $u_0(x,t) = e^{-i\omega t} v(x)$ with v a twice differentiable spherically symmetric exponentially decaying function. Henceforth we consider only nonlinear terms g which satisfy all of the criteria above.

Let $\mathcal{H}$ be the Hilbert space $H^1(\mathbb{R}^3) \oplus L^2(\mathbb{R}^3)$; for $\Psi = (v,\dot{v})$ and $\varphi = (u,\dot{u})$ in $\mathcal{H}$ we have $\langle\Psi|\varphi\rangle = \langle v|u\rangle_2 + \langle v|(-\nabla^2)u\rangle_2 + \langle\dot{v}|\dot{u}\rangle_2$ and $\|\varphi\|^2 = \|u\|^2 + \|\nabla u\|^2 + \|\dot{u}\|^2$. For j = 1 or 2, let $\varphi_j(0) = (u_j(0),\dot{u}_j(0))$ be the initial data for an isolated solitary wave solution of $\ddot{u} = \nabla^2 u + g(u)$. (Note that u_1 and u_2 need not be Poincaré equivalent; they may be different "excited states".) Assume that u_1 and u_2 are asymptotically disjoint. Denote by $\Phi(t)$ the scattering state $(0,\varphi_1(t),\varphi_2(t))$ in the space $\mathfrak{A}_2$. To demonstrate the convergence of the expression for the wave operators on $\Phi(0)$, we wish to show that $\Omega_+(t)(\Phi(0)) = U_{-t}(J\,\Gamma_t\,\Phi(0))$ is Cauchy in norm as $t\to\infty$. That is, we want to show that given $\varepsilon > 0$, there exists T so large that $\| \Omega_+(T+S)(\Phi(0)) - \Omega_+(T)(\Phi(0)) \| < \varepsilon$ for all $S > 0$.

It is helpful to rewrite the expression inside the norm. Note that $\Omega_+(T+S)(\Phi(0)) - \Omega_+(T)(\Phi(0)) =$

$$= U_{-T}(U_{-S}(J\,\Phi(S+T)) - U_{-T}(J\,\Gamma_{-S}(\Phi(S+T))).$$

We can separate the question of the norm convergence to zero of this expression into two problems, the "inner problem" and the "outer problem."

<u>Inner problem</u>: Given $\delta > 0$, show that there exists $T(\delta)$ so large that $\| U_{-S}(J\,\Phi(S+T(\delta))) - J\,\Gamma_{-S}(\Phi(S+T(\delta))) \| < \delta$ for all $S \geq 0$.

<u>Outer problem</u>: Given $\varepsilon > 0$, show that there exists $\delta > 0$ such that $\| \Psi - \varphi \| < \delta$ implies $\| U_{-T(\delta)}(\Psi) - U_{-T(\delta)}(\varphi) \| < \varepsilon$, with $T(\delta)$ determined by the inner problem.

The outer problem consists of estimating the growth of the difference between solutions of the nonlinear wave equation with neighboring initial data. (Actually a weaker result than that stated will suffice; φ above can be chosen to be a linear combination of isolated solitary waves.) Since general solutions are not expected to be stable, the veracity of such a bound on the norm of the difference between solutions depends on the rate of growth of $T(\delta)$ with respect to δ. We return to a discussion of the outer problem after discussing the growth rate imposed on $T(\delta)$ by the solution of the inner problem.

The inner problem contains the crux of the matter concerning the existence of the wave operator, because it is the expression $W_{S,T}(t) \equiv U_{-t}(J\ \Phi(S+T)) - J\ \Gamma_{-t}(\Phi(S+T))$ which compares the free and interacting dynamics. To solve it, we need to show that given $\delta > 0$ there is a $T(\delta)$ so large that $\|W_{S,T}(S)\| < \delta$ for all $S \geq 0$.

Suppose T is so large that the (isolated, asymptotically disjoint) solitary waves $u_1(T)$ and $u_2(T)$ are far apart in space and moving away from each other. (Time T is long past their time of closest approach.) The inner problem is to show that if we start with initial data $(\varphi_1(S+T) + \varphi_2(S+T))$ and develop this data (with the interaction switched on) backwards to time T, then the result $U_{-S}(\varphi_1(S+T) + \varphi_2(S+T))$ does not differ much from $(\varphi_1(T) + \varphi_2(T))$.

Clearly this demonstration depends on the stability of the solitary waves. If either is unstable, then the overlap from the other, no matter how small, can cause the interaction to alter the solution substantially from $(\varphi_1(S+T) + \varphi_2(S+T))$ over the time interval of length S. The treatment of the inner problem which follows makes use of a notion of stability inspired by Lyapunov's Direct Method for ordinary differential equations.

4. STABILITY OF SOLITARY WAVES. Let u be an isolated solitary wave solution to $\ddot{u} = \nabla^2 u + g(u)$, and set $\varphi \equiv (u,\dot{u})$. Consider another solution u' with nearby initial conditions, that is, $\|\varphi'(0) - \varphi(0)\|$ is small, where $\varphi' \equiv (u',\dot{u}')$. For general φ' of this sort, we cannot expect that $\|\varphi'-\varphi\|$ stays small for all time. For example, φ' might be the solution φ boosted to a slightly different velocity; then $\|\varphi'(0) - \varphi(0)\|$ is small, but for later times u' drifts away from u at the relative velocity, so that $\|\varphi'-\varphi\|$ tends for large times to $\|\varphi'\|+\|\varphi\|$.

Nevertheless, it is possible to discuss the stability of solitary waves, modulo shifts. For the solitary waves under discussion here, which are spherically symmetric in the rest (center-of-momentum) frame, the following definition is appropriate. Given $\theta \in [0,2\pi)$ and $b \in \mathbb{R}^3$, denote by $\underline{\varphi}$ the vector $\underline{\varphi} \equiv (\underline{u},\underline{\dot{u}})$ where $\underline{u}(x,t) \equiv e^{i\theta}\, u(x-b,t)$ and $\underline{\dot{u}}(x,t) \equiv e^{i\theta}\, \dot{u}(x-b,t)$.

DEFINITION. An isolated solitary wave solution $\varphi \equiv (u,\dot{u})$ to $\ddot{u} = \nabla^2 u + g(u)$ which is spherically symmetric (in its rest frame) will be said to be <u>stable</u> if for every $\varepsilon > 0$ there exists $\delta > 0$ such that for all solutions $\varphi' \equiv (u',\dot{u}')$ to the wave equation, $\|\varphi'(0) - \varphi(0)\| < \delta$ implies $\inf_{\theta,b} \|\varphi'(t) - \underline{\varphi}(t)\| < \varepsilon$ for all $t \geq 0$.

REMARKS:
(1) The definition says that a solution of the wave equation with initial data close to that for a stable solitary wave will stay close to a phase-shifted spatial translate of the solitary wave.
(2) If we were to discuss solitary waves which were not spherically symmetric, this definition would have to be modified by including the possibility of rotation in $\underline{\varphi}$.
(3) J. Shatah has shown [9] (for a class of equations which includes our example g_3 above) that, in the rest frame, an isolated solitary wave of least energy (not an "excited state") is stable with respect to spherically symmetric perturbations, modulo the set of least-energy solutions. That is, if v_0 is a least-energy spherically symmetric solution to $0 = \nabla^2 v + f(v)$, and (for fixed ω) $u(x,t) \equiv e^{i\omega t} v_0(x)$, then the isolated solitary wave solution u to $\ddot{u} = \nabla^2 u + g(u)$ (with $g(u) = f(u) - \omega^2 u$) is such that $\forall\, \varepsilon > 0 \;\; \exists\, \delta > 0$ such that $\rho(u'(0),u(0)) + \|\dot{u}'(0) - i\omega u(0)\|_2 < \delta$ implies $\inf_{v \in M} \{\rho(u'(t),v) + \|\dot{u}'(t) - i\omega v\|_2\} < \varepsilon$ for all $t \geq 0$, for all spherically symmetric solutions u' of the wave equation. Here ρ is a metric given by $\rho(v,w) \equiv \|v-w\|_2 + \|\nabla(v-w)\|_2 + |\int \{G(|v|)-G(|w|)\}dx|$, and the infimum is taken over the set M of least-energy spherically symmetric solutions to the elliptic equation. Note that M contains at

least all phase shifts $e^{i\theta} v_0$ of the existing least-energy solution v_0. An interesting question is whether there exist other functions in M; the definition of stability above anticipates that M consists precisely of the orbit of v_0 under phase shifts. If this is the case, then the term in ρ involving G can be dropped, and the metric replaced by the norm of the difference.

While stability of the solitary waves as defined above is evidently necessary for the convergence of the expression for the wave operators, it is not sufficient for the proof. We next consider a stronger type of stability.

<u>Lyapunov Stability</u>. Rewrite the wave equation $\ddot{u} = \nabla^2 u + g(u)$ as $\dot{\varphi} = Q(\varphi)$ with $\varphi \equiv (u,\dot{u}) \in \mathcal{H}$ and $Q(\varphi) \equiv Q((u,\dot{u})) \equiv (\dot{u}, \nabla^2 u + g(u))$. For $\Psi_0 \in \mathcal{H}$ and $\gamma > 0$, define
$B(\Psi_0,\gamma) \equiv \{ e^{i\theta}\Psi(x-b) \mid \Psi \in \mathcal{H}, \|\Psi-\Psi_0\| < \gamma, \theta \in [0,2\pi), b \in \mathbb{R}^3 \}$,
that is, $B(\Psi_0,\gamma)$ is the set of all phase-shifted spatial translates of the ball of radius γ around Ψ_0.

DEFINITION. A continuous function $V: \mathbb{R} \times B(\Psi_0,\gamma) \times B(\Psi_0,\gamma) \to \mathbb{R}$ will be said to be a Lyapunov function for $\dot{\Psi} = Q(\Psi)$ at Ψ_0 if there exists a $\gamma \dot{>} 0$ such that all of the following conditions hold:
(1) $\inf_{\theta,b} \|\Psi - \underline{\Phi}\| \le V(t;\varphi,\Psi) \le \|\Psi-\varphi\|$ for all $t \in [0,\infty)$ and for all $\Psi, \varphi \in B(\Psi_0,\gamma)$;
(2) V is continuously differentiable in the sense that $(\partial V/\partial t)(t;\varphi,\Psi)$ and the Frechét derivatives $(D_1V)(t;\varphi,\Psi)$ and $(D_2V)(t;\varphi,\Psi)$ in the second and third slots of V exist and are continuous on $[0,\infty) \times B(\Psi_0,\gamma) \times B(\Psi_0,\gamma)$; and
(3) The quantity
$V^*(t;\varphi,\Psi) \equiv (\partial V/\partial t)(t;\varphi,\Psi) + (D_1V)(t;\varphi,\Psi)\cdot Q(\varphi) + (D_2V)(t;\varphi,\Psi)\cdot Q(\Psi)$
is nonpositive on $[0,\infty) \times B(\Psi_0,\gamma) \times B(\Psi_0,\gamma)$.

With this (rather strong) definition as hypothesis, it is not difficult to prove the following generalization of Lyapunov's classical results.

THEOREM. Suppose that there exists a Lyapunov function for $\dot{\varphi} = Q(\varphi)$ around φ_0. If the initial data $\varphi(0)$ is such that $\|\varphi(0)-\varphi_0\| < \gamma$, then the corresponding solution $\varphi(t)$ is stable, that is, given $\varepsilon > 0$ there is a $\delta > 0$ such that $\|\Psi(0)-\varphi(0)\| < \delta$ implies $\inf_{\theta,b} \|\Psi(t)-\underline{\varphi}(t)\| < \varepsilon$ for all $t \geq 0$, where $\Psi(t)$ is the solution with initial data $\Psi(0)$.

Thus the existence of a Lyapunov function at Ψ_0 is a strong guarantee of the stability of solutions to $\dot{\varphi} = Q(\varphi)$ with initial data in the neighborhood of Ψ_0. A spherically symmetric isolated solitary wave with initial data $\varphi(0)$ will be said to be <u>Lyapunov stable</u> if there exists a Lyapunov function for $\dot{\varphi} = Q(\varphi)$ at $\varphi(0)$.

It is interesting to note that Benjamin [1,3] found an explicit Lyapunov function for the soliton solutions to the KdV equation. Because he dealt with real-valued functions in one spatial dimension, Benjamin required an infimum only over translations, not over phase shifts.

The existence of a Lyapunov function implies more than the stability of solitary waves to perturbations of initial values. It also implies a kind of structural stability to time-integrable forcing terms, as the following lemma shows. The reasoning in the proof of this lemma is employed later in the proof of the main result.

LEMMA. Let φ be a Lyapunov stable solution to $\dot{\varphi} = Q(\varphi)$. Suppose that the derivative D_2V of the Lyapunov function V satisfies $\|D_2V(t;\varphi,\eta)\| < (\text{const})$ for all φ and η in $B(\varphi(0),\gamma)$ and all $t \geq 0$. Suppose that for each $S \geq 0$ there exists a continuous $\mathbf{H}$-valued solution Ψ to $\dot{\Psi} = Q(\Psi) + P_{S,T}(t;\Psi)$ with initial data $\Psi(0)$ such that $\|\Psi(0)-\varphi(0)\| < \gamma$, where for all $\eta \in B(\varphi(0),\gamma)$, the perturbation term satisfies $\|P_{S,T}(t;\eta)\|_2 \leq M_{S,T}(t)$ with $M_{S,T}$ a (positive) function such that $\lim_{T\to\infty} \int_0^S M_{S,T}(t)\,dt = 0$ uniformly in S.
Then given $\varepsilon > 0$ there exists $T > 0$ independent of S such that $\inf_{\theta,b} \|\Psi(t)-\Phi(t)\| < \|\Psi(0)-\varphi(0)\| + \varepsilon$ for all $t \in [0,S]$.

SKETCH OF PROOF: By hypothesis, $\varphi(t) \in B(\varphi(0),\gamma)$ for all t. Since $\Psi(0) \in B(\varphi(0),\gamma)$, the continuity of Ψ implies that there exists a $\tau \in (0,S)$ such that $\Psi(t) \in B(\varphi(0),\gamma)$ for $t \in [0,\tau]$. We next compute

$$(d/dt)V(t;\varphi(t),\Psi(t)) = V^*(t;\varphi(t),\Psi(t)) + (D_2V)(t;\varphi(t),\Psi(t))\cdot P_{S,T}(t;\Psi(t))$$
$$\leq (D_2V)(t;\varphi(t),\Psi(t))\cdot P_{S,T}(t;\Psi(t)).$$

So for $t \in [0,\tau]$, we have $\inf_{\theta,b} \|\Psi(t)-\Phi(t)\| \leq V(t;\varphi(t),\Psi(t)) =$

$$= V(0;\varphi(0),\Psi(0)) + \int_0^t ds\,(d/ds)V(s;\varphi(s),\Psi(s))$$
$$\leq \|\Psi(0)-\varphi(0)\| + \int_0^t ds\,(D_2V)(s;\varphi(s),\Psi(s))\cdot P_{S,T}(s;\Psi(s))$$
$$\leq \|\Psi(0)-\varphi(0)\| + (\text{const})\int_0^S ds\,M_{S,T}(s).$$

Because $\lim_{T\to\infty} \int_0^S M_{S,T}(t)\,dt = 0$ uniformly in S, we may choose T (independently of S) so large that $\inf_{\theta,b} \|\Psi(t)-\Phi(t)\|$ is arbitrarily close to $\|\Psi(0)-\varphi(0)\|$. Finally we can extend the results to all $t \leq S$ because the bounds are independent of τ. ///

Note that the lemma shows that a single value of T may be chosen to guarantee that the solutions Ψ_S to the family (parameter S) of initial value problems $\Psi = Q(\Psi) + P_{S,T}(t;\Psi)$ (with $\Psi(0)$ fixed) satisfy $\inf_{\theta,b} \|\Psi_S(S)-\Phi(S)\| < \|\Psi(0)-\varphi(0)\| + \varepsilon$ for all $S \geq 0$.

5. REDUCTION OF INNER PROBLEM TO LYAPUNOV STABILITY. We now indicate how the solution of the inner problem is reduced to the verification of Lyapunov stability of the solitary waves. Recall that we are given two solitary waves u_1 and u_2, and we set $\Phi(t) \equiv (0, \varphi_1(t), \varphi_2(t))$ with $\varphi_j(t) \equiv (u_j(t), \dot{u}_j(t))$. The inner problem is to show that given $\delta > 0$ there exists $T(\delta) > 0$ such that $\|W_{S,T}(S)\| < \delta$ for all $S \geq 0$, where $W_{S,T}(t) \equiv U_{-t}(J\,\Phi(S+T)) - J\,\Gamma_{-t}(\Phi(S+T))$.

THEOREM. Let the function g satisfy the conditions described earlier which guarantee unique regular global solutions to $\ddot{u} = \nabla^2 u + g(u)$ and the existence of isolated solitary waves. Let u_1 and u_2 be asymptotically disjoint isolated spherically symmetric solitary wave solutions, each of which is Lyapunov stable. Assume that the derivative $D_2 V_j(t;\varphi,\psi)$ of each Lyapunov function satisfies $\| D_2 V_j(t;\varphi,\psi) \| < (\text{const})$ for all $\varphi, \psi \in B(\varphi_j(0),\gamma)$ and all $t \geq 0$. Let $W_{S,T}(t)$ be as given above. Then given $\delta > 0$ there is $T > 0$ such that $\|W_{S,T}(S)\| < \delta$ for all $S > 0$.

OUTLINE OF PROOF: First we rewrite $W_{S,T}(t)$ as the difference between the solution of an initial-value problem and a known function of time. Let $v_j(t) \equiv u_j(T+S-t)$; v_j satisfies the initial-value problem $\ddot{v}_j = \nabla^2 v_j + g(v_j)$ with $v_j(0) \equiv u_j(T+S)$ and $\dot{v}_j(0) \equiv -\dot{u}_j(T+S)$. Then the second term in the expression for $W_{S,T}(t)$ is the known function $J\,\Gamma_{-t}(\Phi(S+T)) = (v_1(t)+v_2(t), -\dot{v}_1(t)-\dot{v}_2(t))$. For notational convenience we set $\Upsilon_j(t) \equiv (v_j(t), \dot{v}_j(t))$. Now let y(t) be the solution to the initial-value problem $\ddot{y} = \nabla^2 y + g(y)$ with $y(0) = v_1(0) + v_2(0)$ and $\dot{y}(0) = \dot{v}_1(0) + \dot{v}_2(0)$. In terms of y, the first term in $W_{S,T}(t)$ is $U_{-t}(J\,\Phi(S+T)) = (y(t), -\dot{y}(t))$. Thus $W_{S,T}(t) = (y(t)-v_1(t)-v_2(t), -\dot{y}(t)+\dot{v}_1(t)+\dot{v}_2(t))$.

The idea of the proof is the following. For large T, $v_1(t)$ and $v_2(t)$ overlap only slightly for $t \in [0,S]$ owing to the exponential falloff of u_1 and u_2. So $y(t)$ should stay close to $v_1(t)+v_2(t)$. To show this, we will split y into two pieces y_1 and y_2 whose overlap is small, and then employ the Lyapunov stability of u_j to keep y_j near a phase-shifted translate $\underline{v}_j$ of v_j. Finally we will show that for large T, the phase shift θ_j and the translation b_j remain small during the time interval [0,S].

Choose T_0 so large that the ("free", asymptotically disjoint) solitary waves $u_1(s)$ and $u_2(s)$ are headed away from each other for all $s \geq T_0$. Henceforth we consider only $T \geq T_0$. Since each u_j is the Poincaré transform of a "stationary" isolated solitary wave, the two transformations determine a relative (three-vector) velocity β between u_1 and u_2. Because u_1 and u_2 are asymptotically disjoint, $\beta \neq 0$. There is thus a plane which divides $\mathbb{R}^3$ into two half-spaces R^1 and R^2 such that given $\varepsilon > 0$, $\int_{(R^j)} |u_j(x,s)|\, dx < \varepsilon$ for sufficiently large s, that is, u_j overlaps R^j only slightly at large times. Perform a Euclidean transformation to make $R^1 = \{ (x_1,x_2,x_3) \mid x_3 < 0 \}$ and $R^2 = \{ (x_1,x_2,x_3) \mid x_3 > 0 \}$. Let $\sigma > 0$, and let $\{\chi_1,\chi_2\}$ be a two-member C^∞ partition of unity such that $\chi_1(x) + \chi_2(x) = 1$ for all x in $\mathbb{R}^3$, $\chi_1(x) = 1$ if $x_3 \geq \sigma$, and $\chi_1(x) = 0$ if $x_3 \leq -\sigma$.

For notational convenience, we introduce the function $h(r,s) \equiv g(r+s) - g(r) - g(s)$. Note that $h(r,0) = h(0,s) = 0$ and $(\partial h/\partial r)(r,0) = (\partial h/\partial s)(0,s) = 0$. If $g(s) = g_3(s)$ as in our earlier example, then $h_3(r,s)$ consists of a sum of terms of orders 2 and 3 involving products of r and s, of order at most 2 in either.

For $j = 1$ or 2, let y_j satisfy $\ddot{y} = \nabla^2 y + g(y) + \chi_j h(y_1, y_2)$ with $y_j(0) = v_j(0)$ and $\dot{y}_j(0) = \dot{v}_j(0)$. If y_1 and y_2 are solutions to these equations, then $y = y_1 + y_2$ satisfies $\ddot{y} = \nabla^2 y + g(y)$ with $y(0) = v_1(0) + v_2(0)$ and $\dot{y}(0) = \dot{v}_1(0) + \dot{v}_2(0)$. Thus if we set $\Psi_j(t) \equiv (y_j(t), \dot{y}_j(t))$, we have $\|W_{S,T}(t)\| \leq \|\Psi_1(t) - \Upsilon_1(t)\| + \|\Psi_2(t) - \Upsilon_2(t)\|$, so the inner problem becomes that of keeping $\|\Psi_j(S) - \Upsilon_j(S)\|$ small.

By hypothesis u_j is Lyapunov stable, and from the time reversal invariance of the wave equation it follows that v_j is Lyapunov stable. If we can show that the term $\chi_j h(y_1, y_2)$ in the equation for y_j satisfies the conditions on $P_{S,T}$ in the lemma above, we will be able to conclude that T may be chosen so large that $\inf_{b,\theta} \|\Psi_j(S) - \underline{\Upsilon}_j(S)\|$ stays less than ε.

Suppose for simplicity that the minimum $\inf_{b,\theta} \|\Psi_j(t) - \underline{\Upsilon}_j(t)\|$ is attained for each t at the unique values $b_j(t)$ and $\theta_j(t)$. That is, given $y_j(t)$ and $v_j(t)$, $b_j(t)$ and $\theta_j(t)$ are the unique shifts which minimize the quantity $\| y_j(x,t) - e^{i\theta_j(t)} v_j(x - b_j(t), t) \|_{(H^1)} +$

$$+ \| \dot{y}_j(x,t) - e^{i\theta_j(t)} \dot{v}_j(x - b_j(t), t) \|_2 \text{ at each } t.$$

Note that we may take $b_j(0) = \theta_j(0) = 0$. Set $\xi_j(x,t) \equiv y_j(x,t) - \underline{v}_j(x,t) \equiv y_j(x,t) - e^{i\theta_j(t)} v_j(x - b_j(t), t)$ and $\zeta_j(x,t) \equiv \dot{y}_j(x,t) - \underline{\dot{v}}_j(x,t) \equiv \dot{y}_j(x,t) - e^{i\theta_j(t)} \dot{v}_j(x - b_j(t), t)$, and put $\eta_j(t) \equiv (\xi_j(t), \zeta_j(t))$. Then the statement above that $\inf_{b,\theta} \|\Psi_j(S) - \underline{\Upsilon}_j(S)\|$ is small can be rephrased as $\|\eta_j(S)\|$ is small. Henceforth $\underline{\Upsilon}_j$ will denote Υ_j phase-shifted and translated by the values of θ_j and b_j which minimize $\|\Psi_j - \underline{\Upsilon}_j\|$ at each time t. Thus $\eta_j = \Psi_j - \underline{\Upsilon}_j$.

Now, $\|\Psi_j(S)-\Upsilon_j(S)\| \le \|\eta_j(S)\| + \|\underline{\Upsilon}_j(S)-\Upsilon_j(S)\|$, so if we can also show that the minimizing shifts $b_j(S)$ and $\theta_j(S)$ can be kept arbitrarily small with a choice of large T, then we can keep $\|\underline{\Upsilon}_j(S)-\Upsilon_j(S)\|$ arbitrarily small, and we will have solved the inner problem.

This is in essence how the proof proceeds; we must, however, handle both the estimates on $P_{S,T}$ and on θ_j and b_j at the same time. To do this, let Ψ_j, Υ_j, and $\eta_j = \Psi_j - \underline{\Upsilon}_j$ be as above. Let γ be the size of the neighborhood on which there exists a Lyapunov function for Υ_j. Assume, without loss of generality, that $\varepsilon/4 < \gamma$. Let $\tau \in (0,S)$ be such that for $t \in [0,\tau]$, $\Psi_j(t) \in B(\varphi_j(0), \varepsilon/4)$, that is, $\|\eta_j(t)\| < \varepsilon/4$, <u>and</u> such that $b_j(t)$ and $\theta_j(t)$ remain so small that $\|\underline{\Upsilon}_j(t)-\Upsilon_j(t)\| < \varepsilon/4$.

<u>Bounding Shape Differences</u>. We will use the Lyapunov function $V(t;\Upsilon_j,\Psi_j)$ to bound $\|\eta_j(t)\|$. To this end, note that $\dot{\Psi}_j = Q(\Psi_j) + (0, \chi_j h(y_1,y_2))$ and $\dot{\Upsilon}_j = Q(\Upsilon_j)$. Thus we must inspect the time integral of $\|P_{S,T}(t;\Psi_j)\| = \|\chi_j h(\underline{v}_1+\xi_1,\underline{v}_2+\xi_2)\|_2$. For purposes of illustration we take $g(s) = g_3(s)$ with g_3 our earlier example function. We can estimate $\|h_3(y_1,y_2)\|_2$ as follows. First, we easily get $|h_3(y_1,y_2)| \le 2\beta|y_1y_2| + 3\alpha|y_1{}^2y_2| + 3\alpha|y_2{}^2y_1|$. Hence we obtain the estimate $\|h_3(y_1,y_2)\|_2 \le 2\beta\|y_1y_2\|_2 +$
$+ (\text{const})(\|\nabla y_1\|_2 + \|\nabla y_1\|_2)(\|y_1y_2\|_2 + \|\nabla(y_1y_2)\|_2)$, and so because $\|\Psi_j(t)\| \le \|\underline{\Upsilon}_j(t)\| + \|\eta_j(t)\| \le \|\Upsilon_j(0)\| + \varepsilon/4$ for $t \in [0,\tau]$, we have $\|h_3(y_1(t),y_2(t))\|_2 \le (\text{const})\|y_1(t)y_2(t)\|_{(H^1)}$ for $t \in [0,\tau]$.

It is not difficult to see that the exponential (spatial) decay of the solitary waves implies $\|v_1(t)v_2(t)\|_{(H^1)} \leq (\text{const})\, e^{-\ell(T+S-t)}$ where ℓ is a constant depending on the relative velocity of the two solitary waves. Making use of the bound on $|b_j(t)|$ for $t \in [0,\tau]$, we can obtain the same form of estimate for $\|\underline{v}_1(t)\underline{v}_2(t)\|_{(H^1)}$ on $[0,\tau]$. We thus have
$\int_0^s \|\underline{v}_1(t)\underline{v}_2(t)\|_{(H^1)}\, dt \leq (\text{const})\, (e^{-\ell(S-s)} - e^{-\ell S})\, e^{-\ell T}$ for $s \in [0,\tau]$. So this time integral can be made arbitrarily small, uniformly in s, by choosing T large. Note that by virtue of the properties of the u_j, exactly the same statements are true of the integrals $\int_0^s \|X_1\underline{v}_2(t)\|_{(H^1)}\, dt$ and $\int_0^s \|X_2\underline{v}_1(t)\|_{(H^1)}\, dt$, as well as for repeated integrals $\int_0^s dt_1 \int_0^{t_1} dt_2\, \|\underline{v}_1(t_2)\underline{v}_2(t_2)\|_{(H^1)}$, etc.

Now, exactly as in the lemma above, the hypothesized Lyapunov stability of the solitary waves yields the estimate
$\|\eta_j(t)\| \leq (\text{const}) \int_0^t ds\, \|X_j h_3(y_1(s),y_2(s))\|_2 \leq$

$\leq (\text{const}) \int_0^t ds\, \|X_j y_1(s)y_2(s)\|_{(H^1)} \leq (\text{const}) \int_0^t ds\, \|\underline{v}_1(s)\underline{v}_2(s)\|_2 +$

$+ (\text{const}) \int_0^t ds\, \|X_j[\xi_1(s)\underline{v}_2(s) + \xi_2(s)\underline{v}_1(s) + \xi_1(s)\xi_2(s)]\|_2$ for $t \in [0,\tau]$. Combining these coupled estimates for $\|\eta_1\|$ and $\|\eta_2\|$, the fact that $\xi_1(0) = \xi_2(0) = 0$, and the uniformly small nature of the repeated time integrals involving $\underline{v}_1\underline{v}_2$, $X_1\underline{v}_2$, and $X_2\underline{v}_1$, we can show by an iterative argument that $\|\eta_j(t)\| \leq (\text{const})\, e^{-kT}$ uniformly for $t \in [0,\tau]$.

<u>Bounding Phase Shifts and Translations</u>. In order to extend the upper time limit from τ to S, we must show that $b_j(t)$ and $\theta_j(t)$ can also be made arbitrarily small uniformly in t by a choice of sufficiently large T. For brevity in the following summary, we assume all required differentiability and continuity. Since we deal with only one pair $\{b_1,\theta_1\}$ or $\{b_2,\theta_2\}$ at a time, we often drop the subscript j in the following.

Because $b(t)$ and $\theta(t)$ are chosen for each t to minimize $\|\Psi - \underline{\Upsilon}\|$, they maximize $\mathrm{Re}\langle\Psi|\underline{\Upsilon}\rangle$ at each t. Setting $(\partial/\partial\theta)\mathrm{Re}\langle\Psi|\underline{\Upsilon}\rangle = 0$ yields $e^{2i\theta} = \langle\Upsilon^b|\Psi\rangle/\langle\Psi|\Upsilon^b\rangle$ where Υ^b indicates the spatial translate of Υ by b (that is, Υ^b is $\underline{\Upsilon}$ with $\theta=0$). We note that $\theta(t)$ is chosen at each t to make $\langle\Psi|\underline{\Upsilon}\rangle$ real. Furthermore, we have an explicit formula for $e^{2i\theta}$ in terms of b.

Setting $(\partial/\partial b^k)\mathrm{Re}\langle\Psi|\underline{\Upsilon}\rangle = 0$ yields the equation $\langle\nabla_k\Upsilon^b|\Psi\rangle\langle\Psi|\Upsilon^b\rangle + \langle\Upsilon^b|\Psi\rangle\langle\Psi|\nabla_k\Upsilon^b\rangle = 0$. We note that maximizing $\mathrm{Re}\langle\Psi|\underline{\Upsilon}\rangle$ over θ and b is equivalent to maximizing $|\langle\Upsilon^b|\Psi\rangle|^2$ over b (with the explicit formula above for θ in terms of b). Taking the derivative of the equation for b with respect to time gives $\sum_n M_{kn}(b)\,\dot{b}^n = J_k(b)$ where

$$M_{kn}(t;b) \equiv -\langle\nabla_k\underline{\Upsilon}|\nabla_n\Psi\rangle\langle\Psi|\underline{\Upsilon}\rangle + \langle\nabla_k\underline{\Upsilon}|\Psi\rangle\langle\Psi|\nabla_n\underline{\Upsilon}\rangle + \mathrm{cc} \quad \text{and}$$

$$J_k(t;b) \equiv \langle\nabla_k\dot{\underline{\Upsilon}}|\Psi\rangle\langle\Psi|\underline{\Upsilon}\rangle + \langle\nabla_k\underline{\Upsilon}|\dot{\Psi}\rangle\langle\Psi|\underline{\Upsilon}\rangle + {}$$
$$+ \langle\nabla_k\underline{\Upsilon}|\Psi\rangle\langle\dot{\Psi}|\underline{\Upsilon}\rangle + \langle\nabla_k\underline{\Upsilon}|\Psi\rangle\langle\Psi|\dot{\underline{\Upsilon}}\rangle + \mathrm{cc},$$

where the phase shift has been (trivially) reintroduced into the formulas for J and M for notational convenience. Here $\dot{\underline{\Upsilon}}$ means the phase-shifted translate of the time derivative of Υ.

We now use the facts that $\dot{\underline{\Upsilon}}_j = Q(\underline{\Upsilon}_j)$, $\Psi_j = \underline{\Upsilon}_j + \eta_j$, and $\dot{\Psi}_j = Q(\underline{\Upsilon}_j) + \Xi_j$, where $\Xi_j \equiv Q(\eta_j) + (0, h(\underline{v}_j,\xi_j) + \chi_j h(y_1,y_2))$. From our estimates above we conclude, in view of the bound on $|b_j|$ during the time interval $[0,\tau]$, that the integral $\int_0^t ds\,\|\Xi_j(s)\|$ can be made arbitrarily small by choice of large T, uniformly in t.

Using these formulas, we can write the 3×3 matrix M as

$$M_{kn}(t;b) = -\mathrm{tr}[(|\Upsilon\rangle\langle\nabla_k\Upsilon| + |\nabla_k\Upsilon\rangle\langle\Upsilon|)(|\Upsilon\rangle\langle\nabla_n\Upsilon| + |\nabla_n\Upsilon\rangle\langle\Upsilon|)] +$$
$$+(\text{expression involving } \Xi \text{ which can be made small uniformly}).$$

That is, M is the sum of a constant (independent of t and b) manifestly negative definite matrix and a matrix which can be made uniformly small for $t \in [0,\tau]$. Thus M is invertible for $t \in [0,\tau]$, and M^{-1} is uniformly bounded on $[0,\tau]$.

Similarly, use of the formulas for $\dot{\underline{\Upsilon}}$ and $\dot{\Psi}$ generates many pairwise cancelling terms in the expression for the 3-vector J, and we can conclude from the bounds on Ξ that $\int_0^t ds\, |\mu_k(t;b)|$ can be bounded uniformly. Writing $\dot{b} = M^{-1}J$ and combining these estimates with the fact that $b_j(0) = 0$, we conclude that $|b_j(t)|$ can be made arbitrarily small by choice of large T, uniformly for $t \in [0,\tau]$. Finally, we may make use of these estimates and the explicit formula for θ_j to show that $e^{2i\theta_j(t)}$ can be made arbitrarily close to unity by choice of large T, uniformly for $t \in [0,\tau]$.

We have therefore shown that the postulated upper bounds on $\|\eta_j\|$, $|\theta_j|$, and $|b_j|$ which determined the upper limit τ are superfluous, because we may choose T sufficiently large to ensure that each of these quantities is arbitrarily small uniformly in time. Since all estimates are valid for $t \in [0,S]$, we have solved the inner problem by bounding each of the terms on the right hand side of $\|W_{S,T}(S)\| \le \|\eta_1(S)\| + \|\eta_2(S)\| + \|\Upsilon_1(S)-\underline{\Upsilon}_1(S)\| + \|\Upsilon_2(S)-\underline{\Upsilon}_2(S)\|$, and $\|W_{S,T}(S)\|$ can thus be made arbitrarily small by choosing T sufficiently large.

6. THE OUTER PROBLEM. Because the outer problem has significantly less structure, we are limited to estimates for the outer problem which are rougher than those for the inner problem. Recall that given $\varepsilon > 0$, we seek to show that there exists $\delta > 0$ such that $\|\Psi-\Phi\| < \delta$ implies $\|U_{-T(\delta)}(\Psi) - U_{-T(\delta)}(\Phi)\| < \varepsilon$. Here $T(\delta)$ is the large time T which guarantees that $\|W_{S,T}(S)\| < \delta$ for all $S \geq 0$ in the inner problem.

By virtue of the time-reversal invariance of the equation, the outer problem as posed above is equivalent to showing that $\|U_{T(\delta)}(\Psi) - U_{T(\delta)}(\varphi)\| < \varepsilon$. Let u and v be two solutions to the wave equation $\ddot{u} = \nabla^2 u + g(u)$, and set $w \equiv u - v$. Then $(w(t), \dot{w}(t)) = U_t((u(0), \dot{u}(0))) - U_t((v(0), \dot{v}(0)))$. The outer problem is thus equivalent to showing that $\|(w(T(\delta)), w(T(\delta)))\| < \varepsilon$, where w satisfies $\ddot{w} = \nabla^2 w + g(w) + h(v,w)$ with $(w(0), \dot{w}(0)) = \Psi - \varphi$, and where v satisfies $\ddot{v} = \nabla^2 v + g(v)$ with $(v(0), \dot{v}(0)) = \varphi$.

To get a bound on the size of $\|(w(t), \dot{w}(t))\|$, we employ the energy functional $E[(y_1,y_2)] \equiv \frac{1}{2}\|y_2\|^2 + \frac{1}{2}\|\nabla y_1\|^2 - \int G(|y_1|)\, dx$. Assume that the function G, in addition to being nonpositive, is such that $-G(s) \geq \mu s^2$ for all $s \geq 0$ with μ a positive constant. (The example function G_3 can certainly satisfy this requirement.) Then the functional Y defined by $Y[(y_1,y_2)] \equiv [\, 2\, E[(y_1,y_2)]\,]^{1/2}$ satisfies $\|(y_1,y_2)\| \leq (\text{const})\, Y[(y_1,y_2)]$. It is straightforward to compute that

$$(d/dt)Y[(w(t),\dot{w}(t))] = \{\, (d/dt)E[(w(t),\dot{w}(t))]\, \} \,/\, Y[(w(t),\dot{w}(t))] =$$
$$= \mathrm{Re}\langle \dot{w}(t) \mid h(v(t),w(t))\rangle \,/\, Y[(w(t),\dot{w}(t))].$$

So we have $|(d/dt)Y[(w(t),\dot{w}(t))]| \leq \| h(v(t),w(t)) \|_2$. Therefore $Y[(w(t),\dot{w}(t))] \leq Y[(w(0),\dot{w}(0))] + \int_0^t ds\, \| h(v(t),w(t)) \|_2$.

Now, $E[(v(t),\dot{v}(t))]$ is constant in time because v is a solution of the wave equation, hence $\|v\|$ is bounded uniformly in time. Sobolev estimates of the same type as for the inner problem give $\| h(v(t),w(t)) \|_2 \leq (\text{const})\{\, \|w(s)\|_{(H^1)} + [\|w(s)\|_{(H^1)}]^2\, \}$.

Thus setting $Z(t) \equiv Y[(w(t),\dot{w}(t))]$, we have $Z(t) \leq Z(0) + L\int_0^t ds\,\{ Z(s) + Z(s)^2 \}$ with L a positive constant. Hence we may employ a Gronwall-type argument to obtain the inequality $Z(t) \leq Z(0)\, e^{Lt} / (1 + Z(0)(1 - e^{Lt}))$. Since $\|(w(t), \dot{w}(t))\| \leq Z(t)$, it is straightforward to verify that if $e^{LT(\delta)} = o(1/\delta)$ as $\delta \to 0$, then given $\varepsilon > 0$, there exists a $\delta > 0$ such that $\|\Psi - \varphi\| < \delta$ implies $\|U_{-T(\delta)}(\Psi) - U_{-T(\delta)}(\varphi)\| < \varepsilon$.

Comparing these growth restrictions (as $\delta \to 0$) on $T(\delta)$ from the outer problem with the growth requirements imposed on $T(\delta)$ by the inner problem requires a detailed analysis, which we do not attempt here, of the rates involved in the estimates made in the solution of the inner problem. We note, however, that logarithmic growth of $T(\delta)$ as $\delta \to 0$ is consistent with the bounds we obtained in the inner problem.

BIBLIOGRAPHY

[1] T. B. Benjamin, The stability of solitary waves. Proc. R. Soc. Lond. A 328(1972) 153-183.

[2] H. Berestycki and P.-L. Lions, Nonlinear scalar field equations, I and II. Arch. Rat. Mech. Anal. 4(1983) 313-375.

[3] J. Bona, On the stability theory of solitary waves. Proc. R. Soc. Lond. A 344(1975) 363-374.

[4] H. Brezis and E. Lieb, Minimum action solutions of some vector field equations. Comm. Math. Phys. (1984, to appear).

[5] T. Cazenave and P. Lions, Orbital stability of standing waves for some nonlinear Schrödinger equations. Comm. Math. Phys. 85(1982) 549-561.

[6] C. Jones and T. Küpper, On the infinitely many solutions of a semilinear elliptic equation. University of Arizona preprint (1984).

[7] M. Reed, Abstract nonlinear wave equations. Springer Lecture Notes in Math. 507(1976).

[8] M. Reed, Unsolved problems in the theory of nonlinear wave equations, in: Many Degrees of Freedom in Field Theory, ed. L. Streit (1978).

[9] J. Shatah, Stable standing waves of nonlinear Klein-Gordon equations. Comm. Math. Phys. 91(1983) 313-327.

[10] W.A. Strauss, Nonlinear invariant wave equations, in: Springer Lecture Notes in Physics 73, ed. Velo and Wightman (1977).

[11] W.A. Strauss, Existence of Solitary waves in higher dimensions. Comm. Math. Phys. 55(1977) 149-162.

DEPARTMENT OF MATHEMATICS
UNIVERSITY OF TEXAS
AUSTIN, TX 78712

Lectures in Applied Mathematics
Volume 23, 1986

GLOBAL SOLUTIONS OF A DISPERSIVE EQUATION IN HIGHER DIMENSIONS

Joel AVRIN and Jerome A. GOLDSTEIN[1]

ABSTRACT. Global existence theorems are established for the initial-boundary value problem for the partial differential equation

$$\partial u/\partial t + \mathrm{div}(\phi(u)) - \partial \Delta u/\partial t = 0.$$

Growth rates are imposed on ϕ' and depend on the spatial dimension.

Long waves in a nonlinear dispersive system have been modelled by the Korteweg-de Vries (KdV) equation

(1) $$u_t - u_{xxx} + uu_x = 0$$

for $x \in \mathbb{R}$ and $t \geq 0$. There are various ways to "derive" this equation formally. Some of them lead equally well to the regularized long wave equation or Benjamin-Bona-Mahony (BBM) equation

(2) $$u_t - u_{xxt} + uu_x = 0.$$

(See Benjamin, Bona and Mahony [3].) Equation (2) is easier to treat conceptually than is (1) because of the following observation. In operator form (2) becomes

$$(I - d^2/dx^2)u_t = N(u)$$

1980 Mathematics Subject Classification. 35Q20, 47H17, 34G20.
[1]Partially supported by an NSF grant.

0075-8485/86 $1.00 + $.25 per page

where

$$N(u) = -uu_x = -\frac{1}{2}(u^2)_x.$$

In integrated form this becomes

$$(3) \qquad u(t) = u(0) + \int_0^t (I - d^2/dx^2)^{-1} N(u(s))ds.$$

Here x has been suppressed and we regard $u(t) = u(t,\cdot)$ as belonging to some Banach space X of functions of x. Since N is a (nonlinear) first order differential operator and since applying $(I - d^2/dx^2)^{-1}$ results in a gain of two derivatives, (2) or (3) can be thought of as an integral equation for which the initial value problem can be solved by iteration in the standard way.

Our purpose here is to indicate how (2) can be embedded in a class of partial differential equations involving a spatial variable of arbitrary dimension. For these equations we can establish global (in time) existence. This is the good news. The bad news is that more technical assumptions are required for solvability of the problem as the space dimension increases.

The generalized BBM equation that we treat is

$$(4) \qquad \begin{cases} u_t - \Delta u_t + \operatorname{div}(\phi(u)) = 0 & \text{for } x \in \Omega,\ t \geq 0, \\ u(0,x) = u_0(x) & \text{for } x \in \Omega, \\ u(t,x) = 0 & \text{for } x \in \partial\Omega,\ t \geq 0. \end{cases}$$

Here Ω is a smooth bounded domain in $\mathbb{R}^n$. Letting $\psi = \phi'$, so that

$$\operatorname{div}(\phi(u)) = \phi'(u) \cdot \nabla u = \psi(u) \cdot \nabla u,$$

we assume that

$$\psi \in C^1(\mathbb{R}\,;\mathbb{R}^n)$$

with no further restriction if $n = 1$; while if $n \geq 2$ we assume in addition that $\psi \in C^2$ and

$$(5) \qquad |\psi(s)| \leq \text{Const}(1 + |s|^\alpha), \quad s \in \mathbb{R},$$

where

(6) $$\alpha < \infty \quad \text{if} \quad n = 2 ,$$

(7) $$\alpha \leqslant \frac{n+2}{2(n-2)} \quad \text{if} \quad n \geqslant 3 ,$$

(8) $$\alpha \leqslant \frac{2}{n-4} \quad \text{if} \quad n \geqslant 6 .$$

Then, if $u_0 \in W_0^{2,p}(\Omega)$ is given and $p > n$, there is a unique (global) solution u of (4). As a function of the time t, the solution u belongs to the space

$$C([0,\infty) ; W^{2,p}(\Omega) \cap W_0^{1,p}(\Omega)).$$

(For the Sobolev spaces see e.g. Adams [1] or Friedman [5]. Note that $W^{2,p}(\Omega) \cap W_0^{1,p}(\Omega)$ is the domain of the Dirichlet Laplacian acting on $L^p(\Omega)$ for $1 < p < \infty$.)

These results are due to Goldstein and Wichnoski [6] and Avrin and Goldstein [2]. See also Calvert [4], which overlaps [6]. Here is a brief sketch of the proof.

The local existence is standard. We let

$$L = I - \Delta, \ N(u) = - \operatorname{div}(\phi(u)) ,$$

$$S(v)(t) = u_0 + \int_0^t L^{-1} N(v(s))ds .$$

Choosing X to be a subspace of $W^{m,p}(\Omega)$ and viewing S as a mapping on

$$Y = \{v \in C([0,T];X) : v(0) = u_0 , \ \|v(t) - u_0\| \leqslant a \text{ for } 0 \leqslant t \leqslant T\}$$

for suitable $T > 0$, $a > 0$, S becomes a strict contraction from Y into Y. The resulting unique fixed point of S is the solution of (4). The solution u can be as regular as one likes; simply assume u_0 and ϕ to be smooth enough and choose m accordingly. For details see [6].

If $0 < T_{max} \leqslant \infty$ defines the maximal interval of existence $[0,T_{max})$, then either $T_{max} = \infty$ or else

$$\limsup_{t \uparrow T_{max}} \|u(t)\| = \infty .$$

Thus the solution is global (i.e. $T_{max} = \infty$) if it can be shown that the norm of the solution doesn't blow up in a finite time, i.e. is locally bounded on $[0,\infty)$. Here the $W^{1,p}(\Omega)$ norm is understood for $p > n$.

The first step is to multiply $u_t - \Delta u_t + \mathrm{div}(\phi(u)) = 0$ by $\bar{u}$ and integrate over Ω; an application of the divergence theorem then gives

$$\frac{d}{dt}\left(\int_\Omega |u(t,x)|^2 dx + \int_\Omega |\nabla u(t,x)|^2 dx\right) = 0,$$

whence the $W^{1,2}(\Omega)$ norm of $u(t)$ is constant in t. In one space dimension global existence follows.

For $n \geqslant 2$ we find that the differential equation satisfied by $v = \partial u/\partial x_j$ is

$$v_t - \Delta v_t + \mathrm{div}(v\phi'(u)) = 0 .$$

The divergence theorem then reduces matters to the inequality

$$\frac{1}{2}\frac{d}{dt}\|v(t)\|^2_{W^{1,2}} \leqslant \|\nabla_x v(t)\|_{L^2}\|v(t)\phi(u(t))\|_{L^2}.$$

This in turn implies, using Hölder's inequality and suppressing the t,

$$\frac{d}{dt}\|v\|_{W^{1,2}} \leqslant \|v\|_{L^{2q}}\|\phi(u)\|_{L^{2r}}$$

where $q^{-1} + r^{-1} = 1$. If ϕ satisfies the growth condition (6) or (7), according as $n = 2$ or $n \geqslant 3$, then we get

$$\|\phi(u)\|_{L^{2r}} \leqslant \mathrm{Const}\left(1 + \int_\Omega |u|^{2r\alpha}\, dx\right)^{1/2r}.$$

Choosing $q = 2n(n-2)^{-1}$ determines both q and r. Since $\|u\|_{W^{1,2}}$ is locally bounded in time, so is $\|u\|_{L^{2r\alpha}}$ by Sobolev's inequality. This gives

$$\frac{d}{dt}\|v\|_{W^{1,2}} \leqslant C\|v\|_{W^{1,2}}$$

which implies that $\|u\|_{W^{2,2}}$ is locally bounded. This gives the desired L^p bound for $p > n$ provided that $n \leq 5$. For details see [6].

For $n \geq 6$ we follow the argument of Avrin-Goldstein [2] for global existence. Rather than seek a local $W^{m,2}$ bound with $m \geq 2$ we seek directly a $W^{1,p}$ bound with $p > n$. Apply ∇_x to (4) and taking L^p norms we obtain, for the known local solution u,

$$\|\nabla_x u(t)\|_{L^p} \leq \|u_0\|_{W^{1,p}} + \int_0^t \|\nabla_x (I - \Delta)^{-1} \phi(u(s)) \cdot \nabla_x u(s)\|_{L^p} \, ds.$$

By Sobolev's inequality $\nabla_x (I - \Delta)^{-1}$ is bounded from $L^q(\Omega)$ to $L^p(\Omega)$ provided that $q = np(n + p)^{-1}$; whence we deduce, suppressing the s,

$$\|\nabla_x (I - \Delta)^{-1} \phi(u) \cdot \nabla u\|_{L^p}$$

$$\leq C\|\phi(u) \cdot \nabla u\|_{L^q} \leq C\|\phi(u)\|_{L^{rq}} \|\nabla u\|_{L^{sq}}$$

for $r^{-1} + s^{-1} = 1$. By (7) and the above reasoning we have a local $W^{2,2}$ bound on u. If r is chosen so that $rq = n$, then $sq = p$ and

$$\|u\|_{L^{\alpha n}} \leq C\|u\|_{W^{2,2}}$$

holds, and so $\|\phi(u)\|_{L^{rq}}$ is locally bounded, by assumption (8). An application of Gronwall's inequality then gives a local bound on $\|u\|_{W^{1,p}}$, and this implies global existence.

Note that (8) implies (7) for $n \geq 6$.

Conditions (7) and (8) allow us to take $\alpha = 1$ when $n \leq 6$. The case of $\alpha = 1$ is of interest since $\alpha = 1$ is that power occurring in the original BBM equation. If α_n is, as a function of n, the maximum value of α for which global existence holds, then it is an open question as to whether $\alpha_n \to 0$ as $n \to \infty$ or not.

BIBLIOGRAPHY

1. Adams, R. A., Sobolev Spaces, Academic, New York 1975.

2. Avrin, J. and J. A. Goldstein, "Global existence for the Benjamin-Bona-Mahony equation in arbitrary dimensions", Nonlinear Analysis, TMA, to appear.

3. Benjamin, T. B., J. L. Bona, and J. J. Mahony, "Model equations for long waves in nonlinear dispersive systems", Phil. Trans. Roy. Loc. London A272 (1972), 47-78.

4. Calvert, B., "The equation $A(t,u(t))' + B(t,u(t)) = 0$", Math. Proc. Camb. Phil. Soc. 79 (1976), 545-561.

5. Friedman, A., Partial Differential Equations, Holt, Rinehart & Winston, New York, 1969.

6. Goldstein, J. A. and B. Wichnoski, "On the Benjamin-Bona-Mahony equation in higher dimensions", Nonlinear Analysis, TMA 4 (1980), 665-675.

Department of Mathematics
Tulane University
New Orleans, LA 70118

Current address of J. Avrin:
Department of Mathematics
University of North Carolina
at Charlotte
Charlotte, NC 28223

Lectures in Applied Mathematics
Volume 23, 1986

ENERGY DECAY OF THE DAMPED NONLINEAR KLEIN - GORDON EQUATION

Joel Avrin

ABSTRACT. By the use of energy functional techniques we show that if a damping term bu_t with $b > 0$ is added to the nonlinear Klein-Gordon equation

$$u_{tt} + (-\Delta + m^2)u + \lambda|u|^{p-1}u = 0$$

then the energy decays like $O(e^{-bt})$ if $b < 2m$. We are assuming that the usual dimension-dependent restrictions on p apply, or the initial data are small enough, so that global existence holds.

We briefly indicate how the results extend to an abstract setting and discuss the importance of addressing the relationship of b to m.

1. INTRODUCTION. Energy functional techniques have for several years played a key role in discussions of global existence and stability for nonlinear wave equations. In this note we illustrate the ease with which these methods yield fairly accurate decay rates for a class of equations which include the damped nonlinear Klein - Gordon equation:

$$u_{tt}(x,t)+bu_t(x,t)+B^2u(x,t)=-\lambda|u(x,t)|^{p-1}u(x,t) \quad (1.1)$$

$$u(x,0)=f(x) \ , \ u_t(x,0)=g(x), \ x \in \mathbb{R}^n.$$

0075-8485/86 $1.00 + $.25 per page

Here $B = (-\Delta + m^2)^{\frac{1}{2}}$ and b, m, and λ are positive real numbers.

The sense in which we mean accurate is described in Theorem 1.1 below. Meanwhile, it is straightforward to establish global existence and uniqueness of (1.1) using the standard reduction-of-order method (see [4]) when $f \in D(B^2)$, $g \in D(B)$, and the usual dimension-dependent restrictions on p are imposed (see p.21 of [4]); for more general n, and p large enough, global existence also holds (as in [4] with b = 0) when the initial data are small enough.

We define the energy functional E(t) in the usual manner:

$$E(t) = \tfrac{1}{2} \{\|u'(t)\|_2^2 + \|Bu\|_2^2 + \frac{2\lambda}{p+1}\int |u(x,t)|^{p+1} dx\}.$$

Here $u'(t) = \frac{d}{dt} u(t)$ and $\|\cdot\|_2$ is the norm on $L^2(\mathbb{R}^n)$. Our main result is the following:

<u>THEOREM 1.1</u>. If $b < 2m$ then

$$E(t) \leqslant \gamma E(0)e^{-bt}, t\varepsilon[0,+\infty),$$

where $\gamma = (1 + b/2m)\,(1-b/2m)^{-1}$.

From the proof of Theorem 1.1 in the next section, it will become clear that the decay estimate holds in a broad abstract setting. Replace $L^2(\mathbb{R}^m)$ by an abstract Hilbert space H with norm $\|\cdot\|$ and $B^2u + \lambda|u|^{p-1}u$ by a map A(u) from a dense domain of H into H. If $(A(u),u) = \overline{(A(u),u)} \geqslant m^2\|u\|^2$ for some $m > 0$, $u \in D(A)$, and there exists a positive convex functional G(u) such that

i) $\frac{d}{dt} G(u(t)) = (u'(t),A(u(t)) + (A(u(t),u'(t))$

whenever $u \in C^1([0,+\infty); D(A))$ and

ii) $(u, A(u)) \geq G(u)$, $u \in D(A)$

then the conclusion of Theorem 1.1 applies with $E(t) = \frac{1}{2}\{\|u'(t)\|^2 + G(u(t))\}$.

Note that a special case of the abstract setting is the underdamped case of linear second-order constant-coefficient ordinary differential equations. Also note that we do not expect energy decay like $O(e^{-bt})$ in the overdamped or critically damped ODE case; hence all abstract extensions of Theorem 1.1 must address the relationship between b and m in order to obtain comparatively accurate decay bounds.

2. PROOF OF THEOREM 1.1. For notational simplicity we write $u(x,t)$ as u. Set

$$\tilde{E}(t) = E(t) + \operatorname{Re}(\tfrac{b}{2}\,(u,u_t)). \tag{2.1}$$

Using the fact that $E'(t) = -b\,\|u_t\|_2^2$ we have

$$\begin{aligned}
\tilde{E}'(t) &= E'(t) + \operatorname{Re}\,(\tfrac{b}{2}[(u_t,u_t) + u,u_{tt})]) \\
&= -b\,\|u_t\|_2^2 + \tfrac{b}{2}\,\|u_t\|_2^2 \\
&\quad -\operatorname{Re}(\tfrac{b}{2}(u,\, \mathbf{b}\mathbf{u}_t + B^2u + \lambda|u|^{p-1}u)) \\
&\leq -\tfrac{b}{2}\,\|u_t\|_2^2 - b\,\operatorname{Re}(\tfrac{b}{2}\,(u,u_t)) \\
&\quad -\tfrac{b}{2}\,[\|B^2u\|_2^2 + \tfrac{2\lambda}{p+1}\int|u|^{p+1}dx] \\
&= -b\,[E(t) + \operatorname{Re}[(\tfrac{b}{2}\,(u,u_t))] \\
&= -b\,\tilde{E}(t)
\end{aligned}$$

so that

$$\tilde{E}(t) \leq \tilde{E}(0)e^{-bt}. \tag{2.2}$$

Now by Cauchy - Schwartz

$$\frac{b}{2}\,|(u,u_t)| \leqslant \frac{b}{2}\|u\|_2\|u_t\|_2$$

$$\leqslant \frac{b}{2m}\,(m\|u\|_2\|u_t\|_2)$$

$$\leqslant \frac{b}{4m}\,(m^2\|u\|_2^2 + \|u_t\|_2^2)$$

$$\leqslant \frac{b}{2m}\,E(t);$$ combining this with (2.1)

and (2.2) we obtain $(1 - \frac{b}{2m})E(t) \leqslant [E(0) + \frac{b}{2}(f,g)]e^{-bt}$

but again by Cauchy - Schwartz $E(0)+\frac{b}{2}(f,g) \leqslant (1+\frac{b}{2m})E(0)$

so that if $b < 2m$

$$E(t) \leqslant \gamma\, E(0)e^{-bt}$$

where $\gamma = (1 + b/2m)(1 - b/2m)^{-1}$.

REMARKS. The linear homogeneous case $\lambda \equiv 0$ has been handled before; see e.g. [2]. Here we have modified the arguments therein so that we may take a purely inner-product approach. We note that in our case as well, it is easy to obtain the lower bound $E(0)e^{-2bt} \leqslant E(t)$ directly from the relationship $E'(t) = -b\,\|u_t\|_2^2$. If we now take the limit as $b \to 0$ formally above and in Theorem 1.1, we recover the result that $E(t) = E(0)$ in the undamped case.

In fact, it is easy to establish that the solutions of the damped equation converge to the solution of the undamped equation as $b \to 0$ in a suitably strong sense. This will be briefly discussed in a later paper, along with a more complete discussion of the abstract case. Also in the future we will deal with the case when $G(u)$ need not be positive, noting that some results may be expected if the data are small enough.

Meanwhile, some examples of recent energy functional techniques can be found in [1], [3], [5] and the references contained therein.

BIBLIOGRAPHY

1. P. Aviles and J. Sandefur, "Non-Linear Second-Order Equations with Applications to Partial Differential Equations", to appear in J. Differential Equations.

2. J. A. Goldstein and S. I. Rosencrans, "Energy Decay and Partition for Dissipative Wave Equations", J. Differential Equations, 36 (1980), pp. 66 - 73.

3. P. Holmes and J. Marsden "Bifurcation to Divergence and Flutter in Flow Induced Oscillations: an Infinite-Dimensional Analysis", Automatica, 14 (1978), 367 - 384.

4. M. Reed, "Abstract Non-Linear Wave Equations", Lecture Notes in Mathematics, 507, Springer-Verlag, Berlin-Heidelberg - New York, 1975.

5. G. F. Webb, "Existence and Asymptotic Behavior for a Strongly Damped Non-Linear Wave Equation", Can. J. Math., 32 (1980), pp. 631 - 643.

DEPARTMENT OF MATHEMATICS
UNIVERSITY OF NORTH CAROLINA AT CHARLOTTE
CHARLOTTE, NORTH CAROLINA 28223

Lectures in Applied Mathematics
Volume 23, 1986

NONLINEAR FOCUSING OF DISPERSIVE WAVES

Richard Haberman
Ren-ji Sun

ABSTRACT. Perturbation analyses for slowly varying weakly nonlinear dispersive waves have difficulties since the wave amplitude is predicted to be infinite when the wave number becomes triple-valued. Energy first focuses along a cusped caustic (the envelope of the rays or characteristics). The nonlinear Schrödinger equation (NSE) describes a thin focusing region with relatively large wave amplitudes. Connection formulas (before and after focusing) for the NSE are derived from an equivalent linear singular integral equation related to a Riemann-Hilbert problem. A slowly varying nearly monochromatic wave evolves after focusing into a triple-phased weakly nonlinear wave with nonlinear phase shifts, which reduce to the well-known linear theory determined from a Pearcey integral.

1. CHARACTERISTICS AND THEIR CUSPED CAUSTICS. We summarize our results which appear elsewhere [7] concerning a model weakly nonlinear dispersive wave equation,

$$u_t = L(u) + \varepsilon^{\frac{1}{2}}\gamma u u_x \quad , \tag{1.1}$$

where $L(u)$ is a linear dispersive operator (such as u_{xxx}). We assume linear waves ($e^{i(kx-\omega t)}$ with $\gamma = 0$) exist, such that ω satisfies a real dispersion relation, $\omega = \omega(k)$. We consider slowly varying waves ($\bar{x} = \varepsilon x$, $\bar{t} = \varepsilon t$), in which case the method of multiple scales shows that to leading order ($u = u_0 + \varepsilon^{\frac{1}{2}}u_1 + \ldots$)

1980 Mathematics Subject Classification. 35Q20, 76B15, 35B25.

$$u_0 = A(\bar{x},\bar{t})e^{i\theta(\bar{x},\bar{t})/\varepsilon} + (*) \quad , \tag{1.2}$$

where (*) represents the complex conjugate. Here $A(\bar{x},\bar{t})$ is the slowly varying amplitude. The slowly varying wave number $k(\bar{x},\bar{t})$ and frequency $\omega(\bar{x},\bar{t})$ are defined from the phase

$$k(\bar{x},\bar{t}) = \theta_{\bar{x}} \quad \text{and} \quad \omega(\bar{x},\bar{t}) = -\theta_{\bar{t}} \quad , \tag{1.3}$$

yielding conservation of waves,

$$k_{\bar{t}} + \omega_{\bar{x}} = 0 \quad . \tag{1.4}$$

Using a perturbation expansion ($k = k_0 + \ldots$, $\omega = \omega_0 + \ldots$), it is well known (Whitham [14]) that the leading order frequency satisfies the dispersion relation $[\omega_0(\bar{x},\bar{t}) = \omega(k_0)]$. Thus, the wave number propagates with the group velocity $\omega'(k_0)$:

$$k_{0_{\bar{t}}} + \omega'(k_0)k_{0_{\bar{x}}} = 0 \quad . \tag{1.5}$$

The method of characteristics shows that

$$k_0(\bar{x},\bar{t}) = k_0(\xi,0) \quad , \tag{1.6a}$$

where the characteristics satisfy

$$\bar{x} = F(\xi)\bar{t} + \xi \tag{1.6b}$$

$$F(\xi) = \frac{d\omega}{dk}(k_0(\xi,0)) \quad . \tag{1.6c}$$

The wave number may fold when $\xi = \xi_f$ (as illustrated in Figure 1) and become triple-valued within the region described by a cusped caustic, the envelope of rays (characteristics). Consideration of the amplitude $A(\bar{x},\bar{t})$ shows that a singularity first develops in time $\bar{t} = \bar{t}_f$ at the cusp $\bar{x} = \bar{x}_f$ of the caustic, described by the generic cubic equation,

$$\bar{x} - \bar{x}_f - \omega'(k_f)(\bar{t} - \bar{t}_f) = -(\xi - \xi_f)\frac{\bar{t} - \bar{t}_f}{\bar{t}_f} + (\xi - \xi_f)^3 \frac{\bar{t}_f}{3!}F'''(\xi_f). \tag{1.7}$$

At the cusp the energy focuses with wave number $k = k_f$ resulting in an infinite wave amplitude there, according to the slowly

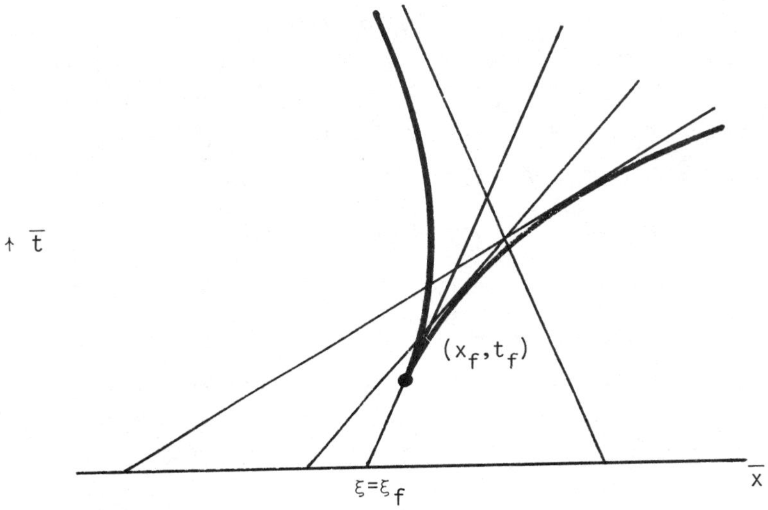

Figure 1a. Cusped caustic (envelope of characteristics).

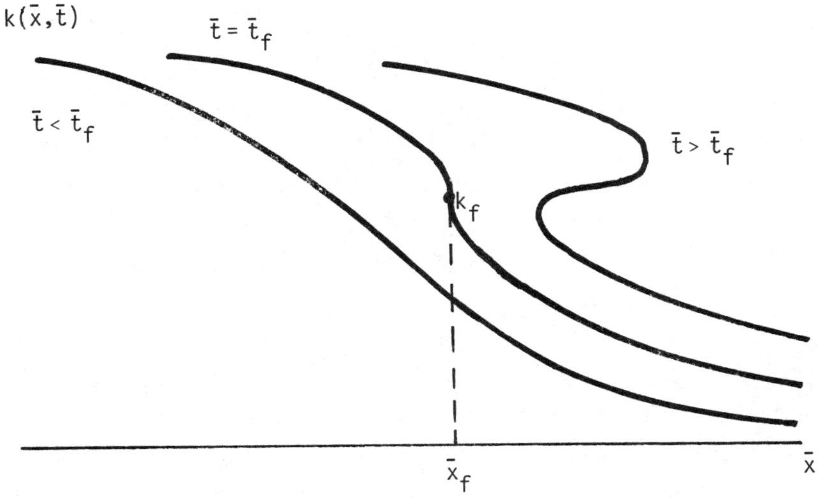

Figure 1b. Triple-valued wave number (generic cubic equation).

varying theory.

2. FOCUSING REGION: THE NONLINEAR SCHRÖDINGER EQUATION. Higher order terms in the perturbation expansion (see [7]) are more singular at the cusp. Thus, (1.1) needs to be rescaled in a focusing region near the cusp. The method of matched asymptotic expansions shows that the appropriate variables are

$$\bar{t} - \bar{t}_f = \varepsilon^{\frac{1}{2}}\tau \tag{2.1a}$$

$$\bar{x} - \bar{x}_f - \omega'(k_f)(\bar{t} - \bar{t}_f) = \varepsilon^{\frac{3}{4}} z \ , \tag{2.1b}$$

the same scales as needed for linear problems ($\gamma = 0$). The focusing region is of intermediate size between the fast and slow multiple scales. In [7] the solution is shown to be a relatively large carrier wave,

$$u = \varepsilon^{-\frac{1}{4}} e^{i[k_f x - \omega(k_f)t]} q(z,\tau) + (*) \ , \tag{2.2}$$

where the complex envelope $q(z,\tau)$ satisfies the nonlinear Schrödinger equation (NSE):

$$q_\tau = i\frac{\omega''}{2} q_{zz} + i\delta q|q|^2 \ . \tag{2.3}$$

Here δ is a real parameter defined in [7]. Peregrine and Smith [12] derived the NSE to describe nonlinear effects near caustics. This should be anticipated, since Benney and Newell [1] showed that weakly nonlinear dispersive waves satisfy the NSE if the solution is nearly monochromatic (one wave number as implied by (2.2)) and is viewed from a coordinate system moving with the group velocity.

3. LINEAR CUSPED CAUSTIC: A PEARCEY INTEGRAL. Well-known results for the linear problem ($\gamma = 0$ and hence $\delta = 0$ in (2.3)) will be briefly reviewed. The linear Schrödinger equation has a solution which is appropriate for the cusped caustic. It has a quartic phase and is known as a generalized Airy integral or

Pearcey integral [2,3,4,8,11]

$$q = E \int e^{i(\xi z - \frac{\omega''}{2}\xi^2\tau + \gamma\xi^4)} d\xi \quad , \tag{3.1}$$

where E and γ are determined either by matching or by the analysis of coalescing stationary points. The asymptotic expansion for large τ and z of q given by (3.1) is obtained using the method of stationary phase. The stationary points of the quartic phase satisfy the generic cubic equation (1.7). It can be shown that each stationary point corresponds to a slowly varying wave propagating along the characteristics given by (1.6). Before focusing, there is one root of cubic, one stationary point. The solution is a slowly varying wave given by (1.2) - (1.6). After focusing, the cubic has three roots, representing three stationary points. The solution is a linear superposition of three slowly varying waves, each satisfying (1.2) - (1.6). The three waves are the ones predicted by the method of characteristics. It turns out that only the middle wave is phase shifted (by $\pi/2$). The Pearcey integral (3.1) describes the birth of a triple-phased linear wave at a cusped caustic.

4. NONLINEAR SCHRÖDINGER EQUATION: EQUIVALENT SINGULAR INTEGRAL EQUATIONS. The corresponding nonlinear problem for a cusped caustic, described by the NSE (2.3), is more difficult. Important work on the NSE include multiply-interacting envelope solitons obtained using inverse scattering by Zakharov and Shabat [15] and slowly varying similarity solutions developed by Segur and Ablowitz [13]. These require that the solution decay sufficiently fast as $x \to \pm\infty$. For our problem, it can be shown that $q = O(x^{-\frac{1}{3}})$ as $x \to \pm\infty$, such that the usual reflection and transmission coefficients do not exist. Thus, these earlier results on the NSE cannot be applied.

More recent work by Fokas and Ablowitz [5] shows that there are singular integral equations which are equivalent to the

Korteweg-deVries equation. The derivation uses a Riemann-Hilbert problem which requires certain analytic properties that follow from sufficiently rapid decay as $x \to \pm\infty$. However, they also showed that the singular integral equations could be independently established. Thus, these singular integral equations are capable of describing a larger family of solutions than the inverse scattering transform. We follow the procedure of Fokas and Ablowitz [5], but we apply it to the NSE. We claim that

$$q(z,\tau) = -\frac{1}{\pi}\int_{-\infty}^{\infty} \psi_1 e^{-i\xi z}\, d\xi \tag{4.1}$$

satisfies a normalized version of the NSE,

$$q_\tau = iq_{zz} + 2i\sigma q|q|^2 , \tag{4.2}$$

where $\sigma = \pm 1$, if $\psi_1(\xi;z,\tau)$ and $\psi_2(\xi;z,\tau)$ satisfy the following coupled system of linear nonhomogeneous singular integral equations:

$$\sigma\psi_1 = -\frac{1}{2\pi i}\int_{-\infty}^{\infty} \frac{\psi_2^*}{\xi - \bar{\xi} + i0} e^{i\xi z} e^{-ig} r_0^* \, d\bar{\xi} \tag{4.3a}$$

$$\psi_2^* = e^{-i\xi z} - \frac{1}{2\pi i}\int_{-\infty}^{\infty} \frac{\psi_1}{\xi - \bar{\xi} - i0} e^{-i\xi z} e^{ig} r_0 d\bar{\xi} , \tag{4.3b}$$

where $g = \bar{\xi}z + 4\bar{\xi}^2\tau$. It can be shown that ψ_1 and ψ_2 also satisfy the system of differential equations [see (7.1)], corresponding to the direct scattering problem of Zakharov and Shabat [15], as well as linear time evolution equations.

There is a unique solution for ψ_1 and ψ_2 (and hence q) for each weight $r_0(\xi)$. We can show that q matches to the focusing outer solution only if

$$r_0(\xi) = ce^{i\beta\xi^4} , \tag{4.4}$$

where c is a complex constant. We will show that this yields the quartic phase. First, we introduce a change of variables motivated by the time dependence of the direct scattering problem:

$$\psi_1 = u_1 \exp[-i(\xi z + 4\xi^2\tau + \beta\xi^4)] \tag{4.5a}$$

$$\psi_2 = (u_2 + 1)\exp(i\xi z) \tag{4.5b}$$

The system of singular integral equations becomes

$$\sigma u_1 \exp[i\Phi_0(\xi)] = \frac{1}{2\pi i}\int_{-\infty}^{\infty} \frac{u_2^*+1}{\bar{\xi}-\xi-i0} \exp[i\Phi_0(\bar{\xi})]c^* d\bar{\xi} \tag{4.6a}$$

$$u_2^* = \frac{1}{2\pi i}\int_{-\infty}^{\infty} \frac{u_1}{\bar{\xi}-\xi+i0}\, c\, d\bar{\xi} \quad , \tag{4.6b}$$

where $\Phi_0(\xi)$ is the quartic phase

$$\Phi_0(\xi) = -(2\xi z + 4\xi^2\tau + \beta\xi^4) \quad . \tag{4.7}$$

5. RELATED RIEMANN-HILBERT PROBLEMS. The system of singular integral equations (4.6) does not have a closed form solution (that we are aware of). However, we will obtain asymptotic solutions as $\tau \to \pm\infty$, providing connection formulas across the focusing region. The integral in (4.6a) may be approximated as $\tau \to \pm\infty$ [with $z = O(|\tau|^{\frac{3}{2}})$ and $\xi = O(|\tau|^{\frac{1}{2}}]$ using the ideas of stationary phase for oscillatory singular integrals (correcting a sign error in Manakov [9]). In this way, we obtain

$$\sigma u_1 \sim \begin{cases} c^*(1+u_2^*) & \text{if } \Phi_0' > 0 \\ 0 & \text{if } \Phi_0' < 0 \quad , \end{cases} \tag{5.1}$$

in which case (4.6b) becomes

$$u_2 \sim -\frac{\sigma}{\pi} \int_{\Phi_0'>0} \frac{1+u_2}{\bar{\xi}-\xi-i0} |c|^2 d\bar{\xi} \tag{5.2}$$

and the solution of the NSE is

$$q \sim -\frac{c^*}{\pi} \int_{\Phi_0'>0} (1+u_2^*) \exp[i\Phi_0(\xi)]d\xi \quad . \qquad (5.3)$$

The stationary points of the quartic phase [in either (4.6a) or (5.3)] satisfy $d\Phi_0/d\xi = 0$, which yields the fundamental cubic associated with the cusped caustic. The cubic has one root (one stationary point) for $\tau < 0$. The solution (5.3) is a wave with a single phase. However, inside the cusped caustic the cubic has three roots (three stationary points). The corresponding solution (5.3) is a triple-phased wave. We must determine u_2 from (5.2).

The simpler singular integral equation (5.2) may be solved by introducing the function $F(z)$ of the complex variable $z = \xi + i\eta$. $F(z)$ is an analytic function on the plane cut on the real axis wherever $\Phi_0' > 0$:

$$F(z) = \frac{1}{2\pi i} \int_{\Phi_0'>0} \frac{|c|^2(1+u_2)}{\bar{\xi} - z} d\bar{\xi} \quad . \qquad (5.4)$$

The Plemelj formulas

$$F_+ - F_- = |c|^2(1+u_2) \qquad (5.5a)$$

$$F_+ + F_- = 2F_p \quad , \qquad (5.5b)$$

where F_p is the Cauchy principle value integral, shows that the singular integral equation (5.2) is equivalent to a non-homogeneous Riemann-Hilbert problem on the cut (different for $\tau < 0$ and $\tau > 0$):

$$\frac{F_+ - F_-}{|c|^2} - 1 = -\sigma F_+ \quad . \qquad (5.6)$$

Equation (5.6) is solved in [7] using the ideas of Muskhelishvili [10] and Gakhov [6]. Finite cuts must be introduced. The solutions for u_2 [7] require a somewhat unusual regularization of divergent integrals. In this way, we have obtained connection formulas for ψ_1 and ψ_2.

The solutions u_2 of the Riemann-Hilbert problems have discontinuities at the roots (stationary points) of the cubic. We have not succeeded in determining q from (5.3) this way, since the solutions u_2 actually have a boundary layer behavior in the neighborhood of the stationary points. Instead, we obtain connection formulas using slowly varying similarity solutions.

6. SLOWLY VARYING SIMILARITY SOLUTIONS. The appropriate asymptotic matching condition for the NSE (4.2) involves fixing the similarity variable $s = z/|\tau|^{\frac{3}{2}}$ (obtained from the generic cubic (1.7)) and letting $\tau \to \pm\infty$ corresponding to pre- and post-focusing. Using the method of multiple scales, we show in [7] that the leading order asymptotic solution for the NSE is single-phased before focusing and triple-phased after focusing (within the cusped caustic):

$$q_0 = \sum_m A_m e^{i\theta^m} \quad , \tag{6.1}$$

where $\sum$ represents a sum over the different phases (one before focusing and three after). Here, the phase θ^m satisfies a non-linear first-order partial differential equation corresponding to the linearized dispersion relation for the NSE:

$$\theta^m_\tau = -(\theta^m_z)^2 \quad . \tag{6.2}$$

Each phase is one of the quartic phases [see (4.7)]

$$\theta^m = \Phi_0(\xi) = -(2\xi z + 4\xi^2\tau + \beta\xi^4) \quad , \tag{6.3}$$

where ξ satisfies the generic cubic

$$0 = 2z + 8\xi\tau + 4\beta\xi^3 \quad . \tag{6.4}$$

The cubic has one root before focusing and three after

$$\xi = |\tau|^{\frac{1}{2}} f_m(s) \qquad m = 1 \text{ or } m = 1,2,3 \ . \qquad (6.5)$$

Thus, $f_m(s)$ satisfies the scaled cubic.

The complex amplitudes $A_m(z,\tau)$ satisfy certain partial differential equations obtained by eliminating higher order secular terms in the multiple scale expansion. If real amplitudes and phases are introduced,

$$A_m = B_m e^{i\phi^m} \ , \qquad (6.6)$$

then we show [7] that the real amplitude is the same as occurs in linear theory (equivalently obtained by stationary phase or linear transport theory):

$$B_m = d_m |f_m'(s)/\tau|^{\frac{1}{2}} \ . \qquad (6.7)$$

The only nonlinear affect is the phase ϕ^m, which is propagated with the group velocity $2\theta_z^m$:

$$\phi_\tau^m + 2\theta_z^m \phi_z^m = \frac{2\sigma}{|\tau|} [-d_m^2 |f_m'(s)| + 2 \sum_n d_n^2 |f_n'(s)|] \ . \qquad (6.8)$$

After focusing within the cusped caustic, (6.8) determines the phase modulations caused by the interaction of three weakly nonlinear coupled dispersive waves (without resonances). By solving (6.8) in [7], we obtain the phases (to within unknown phase shifts — homogeneous solutions of (6.8)). By matching to the initial conditions, we obtain (before focusing, $\tau \to -\infty$)

$$q \sim d|f'(s)/\tau|^{\frac{1}{2}} e^{i[\Phi_0(\xi) + \sigma d^2 \ell n |f'/\tau|]} \ , \qquad (6.9)$$

while in the triple-phased region after focusing

$$q \sim d \sum_m |f_m'(s)/\tau|^{\frac{1}{2}} e^{i[\Phi_0(\xi_m) + {}^m\Phi_1]} \ , \qquad (6.10)$$

where ${}^{m}\Phi_1$ represents the three unknown phase shifts (to be determined).

7. DIRECT SCATTERING. We determine the phase shifts using the differential equations of direct scattering:

$$\psi_{1_z} = -i\xi\psi_1 + q\psi_2 \tag{7.1a}$$

$$\psi_{2_z} = i\xi\psi_2 - \sigma q^{*}\psi_1 \ . \tag{7.1b}$$

We substitute the asymptotic expansions of q for $\tau \to \pm\infty$, given by (6.9) and (6.10), into (7.1). We use the method of multiple scales. Unfortunately, a type of resonance occurs at a particular value of z, requiring a complicated boundary layer analysis there. The details [7] show that parabolic cylinder functions occur in the boundary layer. After considerable algebra, we obtain ψ_1 and ψ_2, depending on the yet unknown nonlinear phase shifts ${}^{m}\Phi_1$. This calculation agrees with the earlier asymptotic analysis of ψ_1 and ψ_2 that we obtained from the singular integral equations only if the nonlinear phase shifts ${}^{m}\Phi_1$ are chosen correctly. The interested reader is referred to [7] for the detailed derivation and result.

BIBLIOGRAPHY

1. Benney, D. J. and A. C. Newell, "The propagation of nonlinear wave envelopes", J. Math. & Phys., 46(1967), 133-139.

2. Bleistein, N., "Uniform asymptotic expansions of integrals with many nearby stationary points and algebraic singularities", J. Math. Mech., 17(1967), 533-559.

3. Brekhovskikh, L. M., Waves in Layered Media, 2nd Ed., Academic, New York, 1980.

4. Brillouin, L., "Sur une méthode de calcul approachée de certaines intégrales dite méthode de col", Ann. Sci. l'Ecole Norm. Sup., 33(1916), 17-69.

5. Fokas, A. S. and M. J. Ablowitz, "Linearization of the Korteweg-deVries and Painlevé II equations", Phys. Rev. Lett., 47(1981), 1096-1100.

6. Gakhov, F. D., Boundary Value Problems, Pergamon, New York, 1966.

7. Haberman, R. and R. Sun, "Nonlinear cusped caustics for dispersive waves", Stud. Appl. Math., to appear.

8. Ludwig, D., "Uniform asymptotic expansions at a caustic", Comm. Pure Appl. Math., 19(1966), 215-250.

9. Manakov, S. V., "Nonlinear Fraunhofer diffraction", Soviet Phys. JETP, 38(1974, 693-696.

10. Muskhelishvili, N. I., Singular Integral Equations, Noordhoff, Groningen, 1953.

11. Pearcey, T., "The structure of an electromagnetic field in the neighborhood of a cusp of a caustic", Philos. Mag., 37(1946), 311-317.

12. Peregrine, D. H. and R. Smith, "Nonlinear effects upon waves near caustics", Philos. Trans. Roy. Soc. Lond. A, 292(1979), 341-370.

13. Segur, H. and M. J. Ablowitz, "Asymptotic solutions and conservation laws for the nonlinear Schrödinger equation, I.", J. Math. Phys., 17(1976), 710-713.

14. Whitham, G. B., Linear and Nonlinear Waves, Wiley-Interscience, New York, 1974.

15. Zakharov, V. E. and A. B. Shabat, "Exact theory of two-dimensional self-focusing and one-dimensional self-modulation of waves in nonlinear media", Soviet Phys. JETP, 34(1972), 62-69.

DEPARTMENT OF MATHEMATICS
SOUTHERN METHODIST UNIVERSITY
DALLAS, TEXAS 75275

Variational Problems

Lectures in Applied Mathematics
Volume 23, 1986

NOETHER'S THEOREMS AND SYSTEMS OF CAUCHY-KOVALEVSKAYA TYPE

Peter J. Olver[1]

ABSTRACT. Noether's Theorem relating symmetries and conservation laws is refined to provide a one-to-one correspondence between nontrivial variational symmetry groups and nontrivial conservation laws for "normal" systems of Euler-Lagrange equations. Besides these, "underdetermined" systems admit nontrivial depencies among the equations, and thus by Noether's Second Theorem admit infinite-dimensional groups of nontrivial variational symmetries with corresponding trivial conservation laws, while "overdetermined" systems have nontrivial integrability conditions, adding further complications.

1. INTRODUCTION AND HISTORY. In 1916, inspired by recent developments in classical mechanics and relativity, E. Noether, [14], formulated and proved two remarkable theorems relating symmetry groups and conservation laws for conservative systems arising from variational principles. The first of these results, justly famous as Noether's Theorem, provides an effective general means of computing conservation laws when used in conjunction with Lie's theory of symmetry groups of differential equations, [19]. With the refinement of Bessel-Hagen, [4], then, all the tools were available to conduct a systematic investigation into the symmetry properties and corresponding conservation laws of the equations of mathematical physics, but, amazingly, such did not occur. On the

1980 Mathematics Subject Classification 22E70, 35A10, 35A30, 58G35.

1 Supported in part by NSF Grant MCS 81-00786.

contrary, this important result went unappreciated for 30 years until Hill, [10], popularized a limited, special version of Noether's general theorem among the physics community. As a result, a sizable proportion of subsequent theoretical work in this area has unfortunately been devoted to proving various special cases of Noether's Theorem, followed by rediscoveries of more or less general versions, despite the fact that Noether's original paper provides the most general connection between (generalized) symmetries+ and conservation laws. In contrast, although many authors have looked at the conservation laws associated with geometrical symmetries, only very recently have there been attempts to completely classify conservation laws for some equations of importance in mathematical physics, [3],[15], [16],[23],[25]. Much work remains to be done in this practical direction, and it can be safely said that Noether's Theorem remains the most widely quoted but most under-utilized result in the entire mathematical physics literature.

On the theoretical side, far less attention has been payed to the role played by trivial conservation laws and trivial symmetries in the Noether correspondence, triviality in each case referring to the fact that no new information on the equations or their solutions is provided by the relevant object. Indeed, Noether's Theorem will provide a truly effective means for computing and completely classifying conservation laws only when nontrivial symmetries give rise to nontrivial conservation laws and conversely. For conservation laws, there are, in fact, two distinct kinds of triviality, a fact that causes much of the complication in this aspect of the theory. In the first kind, the conserved density itself vanishes for all solutions of the

+ As far as I can determine, Noether herself was the first to introduce and study generalized symmetries (also mis-named "Lie-Bäcklund transformations", [2]), which have since been rediscovered many times.

system in question; in the second kind the law holds not just for solutions but for all functions. In either case, from the standpoint of the solution set of the system, no new information results. Similarly a trivial (generalized) symmetry is one whose infinitesimal generator vanishes on all solutions, and hence has correspondingly trivial group action. In both cases, symmetries and conservation laws, one is really only interested in equivalence classes of such objects, two of them being equivalent if they differ by a trivial one. The most desirable and effective form of Noether's Theorem, then, would determine a one-to-one correspondence between equivalence classes of conservation laws and equivalence classes of variational symmetries. (As is well known, not every symmetry of the Euler-Lagrange equations gives rise to a conservation law-only those leaving the variational problem itself invariant, called "variational symmetries", are relevant.)

For such a result to hold, one must improve certain non-degeneracy conditions on the system in question, including a "local solvability" criterion, which naturally leads one to consider systems for which the Cauchy-Kovalevskaya existence theorem is applicable. The requisite class of differential equations is the normal systems, which are characterized by the existence of at least one noncharacteristic direction at each point, and include practically every system of importance in physical applications. For such systems, one can indeed prove the above refined version of Noether's Theorem. The first person to recognize the importance of such a normality condition is Vinogradov, who in very recent work, [27], [28], [29] uses a closely related condition to prove a similar correspondence using complicated cohomological machinery. The proofs in the present case are entirely elementary and will appear in [18]. (Incidentally, the correct concept of triviality in the case of conservation laws apparently first appeared in [22], but no attempt to incorporate this into the Noether correspondence was made until

Vinogradov's work.)

From this new vantage point, a natural question to appear is the range of validity of this refined version of Noether's theorem, or, to put it another way, how does one characterize normal systems, meaning systems, which under a change of variables are in Kovalevskaya form. Bourlet, [6], was the first to ascertain the existence of "un-normal" systems, but it was not until the under-appreciated work of Finzi, [7] (see also [9]) that the true nature of these systems was revealed. Finzi proved the striking result that a system has the property that every direction is characteristic if and only if it has some kind of "integrability condition". The un-normal systems split naturally into two further distinct classes; the over-determined case, where this integrability condition prescribes further relations amongst lower order derivatives, and the under-determined case, in which there is a nontrivial differential relation among the Euler-Lagrange equations. This latter case is precisely that dealt with by Noether's Second Theorem, which states that such a relation exists if and only if there is a infinite-dimensional group of variational symmetries depending on an arbitrary function. These nontrivial variational symmetries give rise to conservation laws using the original version of Noether's Theorem, but the resulting laws are readily seen to be trivial. Thus under-determined systems of Euler-Lagrange equations are uniquely prescribed by the property that they have nontrivial symmetries giving rise to trivial conservation laws.

The overdetermined case is harder to fathom, and, as yet I know of no counterexample to the refined version of Noether's Theorem relating nontrivial symmetries with nontrivial conservation laws. In particular, does there exist a system of Euler-Lagrange equations which has a nontrivial conservation law coming from a trivial variational symmetry? The answer to this question remains unclear, but as indicated at the end of section 5, if such an example exists it must be quite complicated. (In [8], it is remarked that there is such an example, but the paper referred to, [11], does not actually contain one.)

Lack of space precludes the inclusion of proofs in this paper; these will appear in the forthcoming book, [18]. For the same reason, indications of the vast range of applications of Noether's Theorem for producing new and interesting conservation laws for equations of mathematical, physical and engineering applications must be foregone in this brief summary. Suffice it to say that the methods are completely constructive, to the extent that one could envision symbol-manipulating programs systematically computing conservation laws directly by these techniques. The interested reader can refer to [2], [18], [19] and the other papers referred to in the bibliography for some indication of the range of possibilities.

It is a pleasure to thank the organizing committee of this conference for a most enjoyable and productive stay in Santa Fe.

2. SOLVABILITY FOR SYSTEMS OF PARTIAL DIFFERENTIAL EQUATIONS.

We will be concerned with systems of partial differential equations

$$\Delta_\nu(x,u^{(n)}) = 0 \quad , \quad \nu = 1,\dots,q \quad , \tag{1}$$

involving p independent variables $x = (x^1,\dots,x^p) \in X \approx \mathbb{R}^p$ and q dependent variables $u = (u^1,\dots,u^q) \in U \approx \mathbb{R}^q$ defined over an open subset $M \subset X \times U$. Here $u^{(n)}$ denotes all the partial derivatives $u^\alpha_J = \partial^k u^\alpha / \partial x^{j_1} \dots \partial x^{j_k}$ of the u's up to order n, and the functions Δ_ν are smooth or even analytic in their arguments. All our considerations are local, justifying restrictions to Euclidean space, but extensions to vector bundles and smooth manifolds are immediate. By <u>solution</u> $u = f(x)$ we mean a smooth (C^∞) solution for convenience, although with care the differentiability requirements on both the equations and their solutions can be considerably weakened. Since, in our final analysis, the system (1) appears as the Euler-Lagrange equations of some variational problem, we can justifiably restrict our attention to systems having the same number of equations as unknowns, although extensions can, in some instances, be easily envisioned.

In the study of algebraic properties of such a system, including symmetries and conservation laws, a persistent question that arises is the characterization of all differential functions, meaning a function $P(x,u^{(m)})$ depending on x,u and derivatives of u, which vanish for all solutions of the system. The answer to this question rests on certain nondegeneracy hypotheses.

DEFINITION. A system of differential equations (1) is of maximal rank if the Jacobian matrix $(\partial\Delta_\nu/\partial u_J^\alpha)$ with respect to all variables $u^{(n)}$ has rank q whenever $(x,u^{(n)})$ is a solution. The system is locally solvable if for every point $(x_o,u_o^{(n)})$ solving (1) there exists a solution $u=f(x)$ defined in a neighborhood of x_o whose derivatives have the prescribed values $u_o^{(n)}=f^{(n)}(x_o)$.

The maximal rank condition is purely algebraic in nature, which reflects the fact that (1) determines a smooth submanifold of the "jet space" $M^{(n)}$, with coordinates $(x,u^{(n)})$. The local solvability condition addresses the differential properties of system, reflecting the discovery of H. Lewy, [12], of systems of differential equations which have no solutions. For systems of ordinary differential equations, the local solvability problem is the same as the usual initial value problem, whereas for partial differential equations it is of a quite different character from the usual Cauchy or boundary - value problems since the "initial data" $(x_o,u_o^{(n)})$ is prescribed merely at one point x_o rather than on a whole submanifold of the space X. It is closely related to the Riquier existence theory discussed in

Ritt, [21]; also Nirenberg, [13], considers this problem in the context of elliptic systems.

Beside the Lewy-type counter-examples having no solutions, the principal source of systems of differential equations which fail to be locally solvable are those with integrability conditions. For example, the system

$$u_{xx} + v_{xy} + v_x = 0 \quad , \quad u_{xy} + v_{yy} - u_x = 0 \quad , \tag{2}$$

which forms the Euler-Lagrange equations for the variational problem

$$\iint [\tfrac{1}{2}(u_x + v_y)^2 - uv_x]\,dxdy$$

is not locally solvable. Indeed, differentiating the first equation with respect to x, the second with respect to y and subtracting, we find $u_{xx} + v_{xy} = 0$, hence $v_x = 0$, etc. Thus any assignation of initial data $(x^o, y^o, u^o, v^o, u_x^o, u_y^o, v_x^o, v_y^o, u_{xx}^o, u_{xy}^o, u_{yy}^o, v_{xx}^o, v_{xy}^o, v_{yy}^o)$ satisfying (2), but with $v_x^o \neq 0$, will fail the local solvability test.

The appearance of integrability conditions suggests that we should not only look at the system (1) itself, but also all prolongations of it obtained by differentiation. To discuss these we introduce the <u>total derivative</u> operators $D_1, \ldots, D_p$, which differentiate differential functions $P(x, u^{(m)})$ with respect to $x^1, \ldots, x^p$, treating u as a function of x; for instance $D_x(u_{xy} + u^2) = u_{xxy} + 2uu_x$. The m-th prolonged system $\Delta^{(m)}$ corresponding to (1) is the system of differential equations

$$D_J \Delta_\nu = 0 \quad , \quad \nu = 1, \ldots, n \ , \ \#J \leq m$$

obtained by differentiating (1) in all possible ways up to order m, so $D_J = D_{j_1} \cdots D_{j_k}$, where $1 \leq j_\kappa \leq p$, $k = \#J \leq m$.

DEFINITION. A system of differential equations is <u>non-degenerate</u> if it and all its prolongations $\Delta^{(m)}$ are both of maximal rank and locally solvable.

THEOREM 1. Let Δ be a nondegenerate system of differential equations. Let $P(x,u^{(m)})$ be a differential function. Then $P(x,u^{(m)})=0$ for all solutions $u=f(x)$ to Δ if and only if there exist differential operators $\mathcal{D}_\nu=\Sigma Q_J^\nu D_J$ with

$$P=\mathcal{D}_1\Delta_1+\dots+\mathcal{D}_q\Delta_q .$$

The principal tool available to prove the nondegeneracy of a given system is the Cauchy-Kovalevskaya theorem, which is concerned with analytic systems in <u>Kovalevskaya form</u>, [20],

$$\frac{\partial^{n_\nu} u^\nu}{\partial t^{n_\nu}}=\Gamma_\nu(y,t,\widetilde{u^{(n)}}) \ , \quad \nu=1,\dots,q \tag{3}$$

in which $(y,t)=(y^1,\dots,y^{p-1},t)$ are the independent variables, each dependent variable u^α appears up to some order n_α in each of the equations, with the particular derivatives $\partial^{n_\alpha}u^\alpha/\partial t^{n_\alpha}$ appearing only on the left-hand side of the α-th equation. (This is the meaning of the symbol $\widetilde{u^{(n)}}$ in (3).) More generally, an arbitrary system of differential equations (1) can be transformed into one in Kovalevskaya form by a suitable change of variable provided we can find a noncharacteristic direction.

DEFINITION. Let the point $(x_0,u_0^{(n)})\in M^{(n)}$ be a solution to (1). The system of differential equations Δ is <u>normal</u> at $(x_0,u_0^{(n)})$ if it has at least one non-characteristic direction there. The system is normal if it is normal at every such point.

Recall that a p-tuple $\omega=(\omega^1,\dots,\omega^p)$ determines a <u>characteristic direction</u> (and gives the normal direction to a characteristic surface) if the $q\times q$ matrix $M(\omega)=M(\omega;x_0,u_0^{(n)})$ with polynomial entries

$$M^{\nu}_{\alpha}(\omega) = \sum_{\#J = n_\alpha} \omega^J \frac{\partial \Delta_\nu}{\partial u^\alpha_J} (x_o, u_o^{(n)}) \tag{4}$$

is singular, i.e. $\det M(\omega) = 0$. In (4), the sum is over all multi-indices $J = (j_1, \ldots, j_k)$, $1 \leq j_\kappa \leq p$, of order $k = \#J$ equal to the maximal order of derivatives of u^α which appear in (1), and $\omega^J = \omega^{j_1}\omega^{j_2}\ldots\omega^{j_k}$. Otherwise, if $\det M(\omega) \neq 0$, ω determines a noncharacteristic direction, and we can apply the Cauchy-Kovalevskaya existence theorem to any noncharacteristic surface through x_o with ω as its normal direction there. Thus the only way for a system to fail to be normal at a point $(x_o, u_o^{(n)})$, whereby there is no way to apply the Cauchy-Kovalevskaya theorem there in any direction, is for the matrix $M(\omega)$ to be singular for all directions ω. Such systems exist; for instance the matrix for (2) is

$$M(\omega) = M(\xi, \eta) = \begin{pmatrix} \xi^2 & \xi\eta \\ \xi\eta & \eta^2 \end{pmatrix}$$

which is singular for all values of $\omega = (\xi, \eta)$, so every direction for (2) is characteristic.

In the case of analytic systems, the Cauchy-Kovalevskaya theorem immediately implies that an analytic, normal system is nondegenerate in the sense of the above definition. Surprisingly, the converse of this statement is also true - an analytic system which is not normal either fails the maximal rank condition or the local solvability condition. (The C^∞ case is more delicate owing to the appearance of Lewy-type examples there, and little is known in general.)

THEOREM 2. Let Δ be an analytic system of differential equations. Then Δ is nondegenerate if and only if it is normal.

The proof rests on a remarkable result due to Finzi, [7], culminating the historical investigations into the algebraic nature of characteristics. In essence, Finzi's theorem says that a system of differential equations is not normal if and only if it has some kind of integrability condition such as (2).

THEOREM 3. Let Δ be a system of differential equations. Then Δ is not normal if and only if there exist homogeneous differential operators. $\mathcal{D}_1,\dots,\mathcal{D}_q$ of some order k such that the linear combination

$$\mathcal{D}_1\Delta_1 + \dots + \mathcal{D}_q\Delta_q = R \tag{5}$$

depends on derivatives of u^α up to order $n_\alpha + k - 1$ only, for $\alpha = 1,\dots,q$.

Since u^α appears in $\Delta_1,\dots,\Delta_q$ to order n_α, if $\mathcal{D}_1,\dots,\mathcal{D}_q$ were any old k-th order differential operators, one would expect u^α to appear in (5) up to order $n_\alpha + k$. Finzi's theorem says that for unnormal system, one can find special operators such that (5) depends on derivatives of order at least one less than might otherwise be expected. (Such indeed was the case with (2) where $k = 1$.)

The operators $\mathcal{D}_\nu$ in (5) are k-th order, so the combination R would appear as a consequence of the equations in the k-th (and higher) order prolongations of Δ. On the other hand, the derivatives of u^α up to order $n_\alpha + k - 1$ already appear in the previous prolongation $\Delta^{(k-1)}$. At this stage there are two possibilities: a) either R vanishes as a consequence of the equations in the previous prolongation, in which case the integrability condition (5) is illusory, or b) $R = 0$ introduces new relations among the $(n_\alpha+k-1)$st order derivatives not appearing in $\Delta^{(k-1)}$, thereby introducing new integrability conditions into the system. These are called respectively under-determined and over-determined systems, and have the formal definition:

DEFINITION. A system of differential equations (1), in which u^α appears up to order n_α, is <u>under-determined</u> if there exist homogeneous k-th order differential operators $\mathcal{D}_1,\ldots,\mathcal{D}_q$, not all zero, such that the combination (5) vanishes as a consequence of the (k-1)st prolongation $\Delta^{(k-1)}$. In the contrary case that the combination (5) does not vanish as a result of $\Delta^{(k-1)}$ the system is called <u>over-determined</u>. (Note that a system can be both under- and over-determined if it has several relations of the form (5) holding.)

For example, the system (2) is over-determined, whereas the closely related system

$$\Delta_1 = u_{xx} + v_{xy} = 0 \quad , \quad \Delta_2 = u_{xy} + v_{yy} = 0 \tag{6}$$

is under-determined since $D_y\Delta_1 - D_x\Delta_2 = 0$ for all solutions. Finzi's result thus gives a complete trichotomy for analytic systems of differential equations; either the system is normal, in which case it satisfies both nondegeneracy criteria and is in a well-defined sense precisely determined; or it is under-determined and some prolongation violates the maximal rank condition; or it is over-determined and some prolongation fails to be locally solvable. For a purely under-determined system, one can go further and prove that there is at least one arbitrary function in the general solution to the system. Conversely, for an over-determined system, one can prove that it is not possible to prescribe Cauchy data or boundary data arbitrarily and expect to have a solution. Only for normal systems are the natural Cauchy and boundary-value problems well-posed.

3. CONSERVATION LAWS. Consider a system of partial differential equations (1). By a <u>conservation law</u> we mean a divergence expression

$$\operatorname{Div} P = D_1 P_1 + \cdots + D_p P_p = 0 \tag{7}$$

with the p-tuple $P = (P_1,\ldots,P_p)$ depending on x,u and derivatives of u , which vanishes for all solutions $u = f(x)$ of the given system. (If one of the independent variables is time t , so that (7) takes the form $D_t T + \mathrm{Div}\, X = 0$, the corresponding entry of P is called the conserved density and has the property that $\int T\, dx$ is constant for all solutions $u = f(x,t)$ which decay sufficiently rapidly as $|x| \to \infty$.) There are two types of trivial conservation laws which hold for any system.

1) If $P = 0$ for all solutions to Δ , then its divergence (7) also vanishes on solutions.

2) If $\mathrm{Div}\, P \equiv 0$ for all functions $u = f(x)$ then (7) automatically holds for all solutions. For example,

$$D_x(u_y) + D_y(-u_x) = 0$$

is a conservation law for any system involving $u = f(x,y)$. Such trivial conservation laws, known as null divergences, [17], have been characterized as "total curls" using the variational complex that arises in the global theory of the calculus of variations on manifolds, [27], [25], [1].

THEOREM 4. A p-tuple $(P_1,\dots,P_p)$ is a null divergence if and only if there exist differential functions Q_{ij} , $i,j = 1,\dots,p$ so that

$$\begin{aligned} &\text{i)} \quad Q_{ij} = -Q_{ji} , \\ &\text{ii)} \quad P_i = \sum_{j=1}^{p} D_j Q_{ij} . \end{aligned} \tag{8}$$

A conservation law is trivial if it is the sum of trivial laws of the above two types, i.e. (8) holds for all solutions of the system of differential equations in question. Two conservation laws are equivalent if they differ by a trivial conversation law, $P - \tilde{P} = P_0$, where P_0 is trivial.

Now suppose the system of differential equations is nondegenerate, so that by theorem 1 (7) vanishes for all solutions if and only if there exist function $Q_\nu^J(x,u^{(n)})$ so that

$$\text{Div } P = \Sigma \, Q_\nu^J D_J \Delta_\nu \ . \tag{9}$$

A simple integration by parts shows that there is an equivalent conservation law $\tilde{P}$ in <u>characteristic form</u>

$$\text{Div } P = Q \cdot \Delta = Q_1 \Delta_1 + \ldots + Q_q \Delta_q \ , \tag{10}$$

where the characteristic $Q = (Q_1, \ldots, Q_q)$ is given by $Q_\nu = \Sigma (-D)_J Q_\nu^J$. For example,

$$Q D_x \Delta = (-D_x Q) \cdot \Delta + D_x (Q \cdot \Delta)$$

and the second term is a trivial conservation law of the first kind. The characteristic Q is uniquely determined only up to the addition of a <u>trivial characteristic</u>, meaning one which vanishes for all solutions $u = f(x)$, owing to an elementary algebraic lemma.

LEMMA 5. If Δ is nondegenerate, and $Q \cdot \Delta = \tilde{Q} \cdot \Delta$, then $Q_\nu - \tilde{Q}_\nu = 0$ for all solutions $u = f(x)$ to Δ .

Two characteristics are equivalent if they differ by a trivial characteristic. For normal systems of differential equations, each conservation laws is, up to equivalence, uniquely determined by its characteristic and vice versa. This result is fundamental to the systematic study of conservation law and their ultimate connection with symmetries in the case of variational problems.

THEOREM 6. Let Δ be a normal, nondegenerate system of partial differential equations. The conservation laws $\text{Div } P = 0$, $\text{Div } \tilde{P} = 0$ are equivalent if and only if their corresponding characteristics Q and $\tilde{Q}$ are equivalent.

In other words, there is a one-to-one correspondence between (equivalence classes of) conservation laws and (equivalence classes of) characteristics provided the underlying system is normal. The case of "unnormal" systems will be taken up in section 5.

The direct proof of theorem 6 is quite tricky owing to the

two types of triviality for conservation laws. Details will appear in [18]; see also [28].

4. SYMMETRIES AND NOETHERS THEOREM. By a geometrical symmetry group of a system of differential equations we mean a local group of transformations acting on the space $M \subset X \times U$ of independent and dependent variables which transforms solutions of the system to other solution. The group transformations $g:(x,u) \to (\tilde{x},\tilde{u})$ act on functions $u = f(x)$ by a point-wise transformation of their graphs. There is thus an induced action on the derivatives $u^{(n)}$ of such functions, called the <u>prolonged group action</u> and denoted $pr^{(n)}g:(x,u^{(n)}) \to (\tilde{x},\tilde{u}^{(n)})$ determined so that if g transforms $u = f(x)$ to $\tilde{u} = \tilde{f}(\tilde{x})$, then it takes the derivatives $u^{(n)} = f^{(n)}(x)$ of u at the point x to the corresponding derivatives $\tilde{u}^{(n)} = \tilde{f}^{(n)}(\tilde{x})$ at the image point $\tilde{x}$.

For a connected, local Lie group of transformations, we can explicitly determine symmetries by looking at their <u>infinitesimal generators</u>, which are vector fields

$$\underline{v} = \sum_{i=1}^{p} \xi^i(x,u) \frac{\partial}{\partial x^i} + \sum_{\alpha=1}^{q} \varphi_\alpha(x,u) \frac{\partial}{\partial u^\alpha}$$

on M, the corresponding one-parameter group being found by integrating the system of ordinary differential equations

$$\frac{dx^i}{d\varepsilon} = \xi^i(x,u) , \quad \frac{du^\alpha}{d\varepsilon} = \varphi_\alpha(x,u) \tag{11}$$

determining the flow for $\underline{v}$. There is a corresponding prolonged vector field

$$pr^{(n)}\underline{v} = \underline{v} + \sum_{\alpha,J} \varphi_\alpha^J(x,u^{(n)}) \frac{\partial}{\partial u_J^\alpha} \tag{12}$$

on the jet space $M^{(n)}$ generating the one parameter group of prolonged transformations. The coefficient functions of $pr^{(n)}\underline{v}$ have the explicit form

$$\varphi_\alpha^J = D_J Q_\alpha + \sum_{i=1}^{p} \xi^i u_{J,i}^\alpha ,$$

in which $u_{J,i}^\alpha = \partial u_J^\alpha / \partial x^i$, and $Q = (Q_1, \dots, Q_q)$,

$$Q_\alpha = \varphi_\alpha - \sum_{i=1}^{p} \xi^i \cdot u_i^\alpha , \quad u_i^\alpha = \partial u^\alpha / \partial x^i , \tag{13}$$

is the characteristic of the given symmetry. The basic Lie-Ovsiannikov technique for computing symmetry groups of differential equations hinges on the basic result:

THEOREM 7. A connected Lie group of transformations G forms a symmetry group of the nondegenerate system of differential equations (1) if and only if

$$\text{pr } \underline{v}(\Delta_\nu) = 0 , \quad \nu = 1, \dots, q \tag{14}$$

for all solutions $u = f(x)$ and all infinitesimal generators $\underline{v}$ of the group G.

In practice, (14) forms a large system of elementary differential equations for the coefficients ξ^i, φ_α of $\underline{v}$ whose general solution gives the most general symmetry group of the given system of differential equations. See [5], [19], [18] for examples and applications thereof.

Noether, [14], generalized the notion of symmetry by allowing the coefficients ξ^i, φ_α of the infinitesimal generator to depend on derivatives of u also. The resulting generalized symmetries have the same formula (12) for their prolongations and criterion (14) to be symmetries of some system of differential equations; however the group transformations themselves no longer have an elementary geometric interpretation.

An easy computation shows that any (generalized) vector field $\underline{v}$ can always be replaced by one in evolutionary form

$$\underline{v}_Q = \sum_{\alpha=1}^{q} Q_\alpha \cdot \partial / \partial u^\alpha ,$$

Q being the characteristic (13), which is a symmetry if and only if $\underline{v}$ itself is. In this case, the group transformations are recovered by solving a system of evolution equations

$$\frac{\partial u^\alpha}{\partial \epsilon} = Q_\alpha(x, u^{(n)}) \ , \ \alpha = 1, \ldots, q \ , \tag{15}$$

which replaces the flow equations (11) in the geometrical case; here the symmetries are "non-local". Again generalized symmetries can be systematically determined through an analysis of the symmetry criterion (14), [2].

If the characteristic Q of the vector field $\underline{v}$ vanishes on all solutions of the system Δ, then (14) trivially holds and we obtain a <u>trivial symmetry</u>. The corresponding group transformations do not change solutions at all, and hence shed no new light on the system. Two symmetry groups are <u>equivalent</u> if their infinitesimal generators $\underline{v}$ and $\tilde{\underline{v}}$ differ by a trivial symmetry $\underline{v}_0 = \underline{v} - \tilde{\underline{v}}$, and, as with conservation laws, we really need only be interested in nontrivial inequivalent groups.

The connection between symmetry groups and conservation laws holds only for systems with some form of variational structure. In the present discussion, we presume the existence of a variational principle

$$\mathcal{L}[u] = \int_\Omega L(x, u^{(n)})\, dx$$

for which our system of differential equations are the Euler-Lagrange equations

$$\Delta_\nu = E_\nu(L) \equiv \sum_J (-D)_J (\partial L / \partial u_J^\nu) = 0 \ , \ \nu = 1, \ldots, q \ . \tag{16}$$

The variational problem is normal or nondegenerate insofar as its Euler-Lagrange equations are. As for symmetries, those of the variational problem are of principal importance, where we define a connected local Lie group of symmetries or generalized symmetries to be a <u>variational symmetry group</u> if for every infinitesimal generator $\underline{v}$,

$$\mathrm{pr}\, \underline{v}(L) + L \cdot \mathrm{Div}\, \xi = \mathrm{Div}\, B \tag{17}$$

for some p-tuple $B = (B_1, \dots, B_p)$ of differential functions, and where $\mathrm{Div}\, \xi = \sum_i D_i \xi^i$. In essence, the criterion (17), which is due to Bessel-Hagen, [4], says that the variational integral is unchanged under the group action, except for the addition of boundary terms due to the integral of B over $\partial\Omega$. Again, we can replace a generalized vector field $\underline{v}$ by its evolutionary representative without loss of generality, for which (17) simplifies to

$$\mathrm{pr}\, \underline{v}_Q(L) = \sum_{\alpha, J} D_J Q_\alpha \frac{\partial L}{\partial u_J^\alpha} = \mathrm{Div}\, \tilde{B} \tag{18}$$

for some p-tuple $\tilde{B}$. The connection between variational symmetries and symmetries of the Euler-Lagrange equations is as follows:

THEOREM 8. If $\underline{v}$ generates a one-parameter group of variational symmetries for $\mathcal{L} = \int L\,dx$, then it generates a symmetry group of the Euler-Lagrange equations $E(L) = 0$.

The converse is <u>not</u> true, the principle source of counter-examples being groups of scaling transformations. The easiest means of computing variational symmetries is usually to first determine symmetries of the Euler-Lagrange equations and then determine which groups satisfy the additional variational criterion (17). In particular, a variational symmetry group is trivial if it generates a trivial symmetry group of the Euler-Lagrange equations, and two variational symmetries $\underline{v}$ and $\tilde{\underline{v}}$ are equivalent if their characteristics agree on all solutions of the Euler-Lagrange equations.

Noether's theorem relating symmetry groups and conservation laws arises through an simple integration by parts on (18), which shows that

$$\mathrm{pr}\, \underline{v}_Q(L) = Q \cdot E(L) + \mathrm{Div}\, A \tag{19}$$

for some well-defined p-tuple A depending on Q and L. Combining (18), (19) and (10) we see that the characteristic Q of a variational symmetry is the characteristic of a conservation law and vice-versa.

THEOREM 9. Let $\Delta = E(L) = 0$ be the system of Euler-Lagrange equations for a variational problem. A q-tuple Q is the characteristic of a conservation law for this system if and only if Q is the characteristic of a variational symmetry. In particular, if the Euler-Lagrange equation are normal, non-degenerate, there is a one-to-one correspondence between (equivalence classes of) conservation laws and (equivalence classes of) one-parameter groups of generalized variational symmetries.

In other words, to each nontrivial variational symmetry there corresponds a nontrivial conservation law and conversely. Thus an effective and systematic means of computing conservation laws for a system of Euler-Lagrange equations is to first determine all variational symmetry groups by checking which symmetries of the system satisfy (17) (actually, this can be done directly, [18]) and then computing the resulting conservation laws using (19). Explicit formulae are available to this latter task, but are not particularly enlightening. This is then, for normal systems, our refined version of Noether's Theorem.

5. NOETHER'S SECOND THEOREM. The connection between variational symmetries and conservation laws for unnnormal systems is less transparent. Although theorem 9 still yields a variational symmetry for each conservation law and vice versa, there is no guarantee that nontrivial symmetries will result in only trivial conservation laws, or trivial symmetries might give rise to non-trivial laws. For over-determined systems, the complete answer to this problem has yet to be found. Underdetermined systems, however, fall within the scope of Noether's Second Theorem,

which deals with infinite-dimensional groups of variational symmetries depending on arbitrary functions.

THEOREM 10. Let $\mathcal{L} = \int L dx$ be a variational problem with Euler-Lagrange equations $E(L) = 0$. This problem admits an infinite-dimensional group of variational symmetries depending on an arbitrary function $h(x)$ if and only if there is a nontrivial dependency between the Euler-Lagrange equation of the form

$$\mathcal{D}_1 E_1(L) + \dots + \mathcal{D}_q E_q(L) \equiv 0 \tag{20}$$

holding identically, the $\mathcal{D}_\nu$'s being differential operators, not all zero.

Note that if the differential operators $\mathcal{D}_\nu$ in (20) are homogeneous, we recover (5) with $R = 0$, meaning that the system of Euler-Lagrange equations is under-determined. In the general under-determined case, presuming the (k-1)st prolongation is of maximal rank, we necessarily have $R = \Sigma \tilde{\mathcal{D}}_\nu E_\nu(L)$ for certain (k-1)st order differential operators $\tilde{\mathcal{D}}_\nu$, so (5) changes into (20) by replacing $\mathcal{D}_\nu$ by $\mathcal{D}_\nu - \tilde{\mathcal{D}}_\nu$. In other words, Noether's Second Theorem says that a system of Euler-Lagrange equations is under-determined if and only if the associated variational problem admits an infinite dimensional group of symmetries depending on an arbitrary function.

The proof of the theorem proceeds in outline as follows. One multiplies (20) by an arbitrary function $h(x)$ and integrates by parts, yielding

$$\mathcal{D}_1^*(h) \cdot E_1(L) + \dots + \mathcal{D}_q^*(h) E_q(L) = \operatorname{Div} P , \tag{21}$$

where $\mathcal{D}_\nu^*$ is the (formal) L^2-adjoint of the differential operator $\mathcal{D}_\nu$, and P is some well-determined p-tuple of differential functions depending linearly on h and $E(L)$, whose precise form is unimportant. If we set $Q_\nu = \mathcal{D}_\nu^*(f)$, and use (19) we find that $Q = (Q_1, \dots, Q_q)$ is the characteristic of a variational symmetry of $\mathcal{L}$ depending linearly on an arbitrary function $h(x)$. The calculation clearly works in reverse provided the characteristic Q of the group of variational symmetries depends linearly on the arbitrary function; otherwise we can replace it by its Frechet derivative

$$Q_\nu'[u;h_o,h] = \sum_J \frac{\partial Q_\nu}{\partial h_J}[u;h_o]D_J h \qquad (h_J = D_J h)$$

(evaluated at any convenient $h_o(x)$) without losing the symmetry property. The proof is thereby completed - see Noether, [14], for the details.

Now return to the key intermediary relation (21). Defining Q as above, we see that (21) is precisely of the form of a conservation law with characteristic Q, (10). However, P depends linearly on $E(L)$, and hence vanishes whenever $u = f(x)$ is a solution to the Euler-Lagrange equations. In other words $\operatorname{Div} P = 0$ is a trivial conservation law (of the first kind), but whose characteristic Q is, in general, nontrivial and hence corresponds to a nontrivial symmetry group. Since the arbitrary function $h(x)$ apears in both P and Q, we have, in fact, an entire infinite family of trivial conservation laws which arise from nontrivial symmetries whenever the system of Euler-Lagrange equations is underdetermined. In fact, more than this is ture.

THEOREM 11. Let $\mathcal{L}$ be a variational problem. Suppose there exists a nontrivial variational symmetry of $\mathcal{L}$ such that the corresponding conservation law obtained via Noether's Theorem is trivial. Then the Euler-Lagrange equations for $\mathcal{L}$ are under-determined and there in fact exists an infinite dimensional family of such conservation laws depending on an arbitrary function.

As an example, the system (6) arises from the variational problem $\mathcal{L} = \iint \frac{1}{2}(u_x + v_y)^2 dxdy$, which admits the infinite dimensional group of variational symmetries generated by

$$\underline{v} = h_y(x,y)\partial_u - h_x(x,y)\partial_v$$

for h an arbitrary function, i.e. the one-parameter group

$$(u,v) \to (u+\epsilon h_y, v-\epsilon h_x)$$

leaves $\mathcal{L}$ unchanged. The corresponding family of trivial conservation laws are

$$D_x[h(u_{xy}+v_{yy})] - D_y[h(u_{xx}+v_{xy})] = 0 ,$$

with characteristics $(h_x,-h_y)$.

As for overdetermined systems, one might conjecture the possibility of there being systems with trivial symmetry groups corresponding to nontrivial conservation laws. If such an example exists, it must be quite complicated, and I have been unable to produce it. One reason for the complication is the following result, proved with the aid of the homotopy operator for the variational complex, [1], [18].

THEOREM 12. Suppose Δ is a homogeneous system of differential equations, meaning

$$\Delta_\nu(x,\lambda u^{(n)}) = \lambda^\alpha \Delta_\nu(x,u^{(n)}) , \quad \nu = 1,\dots,q$$

for all x,u, and all $\lambda \in \mathbb{R}$, where α is some nonzero constant. Then every trivial characteristic of a conservation law corresponds to a trivial conservation law.

Thus for homogeneous systems of differential equations, in particular linear systems, nontrivial conservation law necessarily have nontrivial characteristics. Then, by Theorem 12, if a homogeneous system is not under-determined, Noether's Theorem 9 holds as stated for normal systems. In particular, any example exemplifying the above phenomenon must be at least a nonhomogeneous polynomial system!

BIBLIOGRAPHY

1. I.M. Anderson & T. Duchamp, On the existence of global variational principles, Amer. J. Math. 102 (1980) 781-868.

2. R.L. Anderson & N.H. Ibragimov, Lie-Bäcklund Transformations in Applications, SIAM Studies in Applied Mathematics, Philadelphia, 1979.

3. T.B. Benjamin & P.J. Olver, Hamiltonian structure, symmetries and conservation laws for water waves, J. Fluid Mech. 125 (1982) 137-185.

4. E. Bessel-Hagen, Über die Erhaltungssätze der Elektrodynamik, Math. Ann. 84 (1921) 258-276.

5. G.W. Bluman & J.D. Cole, Similarity Methods for Differential Equations, Applied Mathematical Sciences 13, Springer-Verlag, New York, 1974.

6. M.C. Bourlet, Sur les équations aux dérivées partielles simultanées, Ann. Sci Ecole Norm. Sup. 8 (3) (1891) Suppl S.3-S.63.

7. A. Finzi, Sur les systèmes d'équations aux dérivées partielles qui, comme les systemes normaux, comportent autant d'équations que de fonctions inconnues, Proc. Kon. Neder. Akad. v. Wetenschappen 50 (1947), 136-142; 143-150; 288-297; 351-356.

8. A.S. Fokas, Generalized symmetries and constants of motion of evolution equations, Lett. Math. Phys. 3 (1979) 467-473.

9. J.S. Hadamard, La Theorie des Equations aux Derivées Partielles, Editions Scientifiques, Peking, 1964.

10. E.L. Hill, Hamilton's principle and the conservation theorems of mathematical physics, Rev. Mod. Physics 23 (1951) 253-260.

11. N.H. Ibragimov, Group theoretical nature of conservation laws, Lett. Math. Phys. 1 (1977) 423-428.

12. H. Lewy, An example of a smooth linear partial differential equation without solution, Ann. of Math. 64 (1956) 514-522.

13. L. Nirenberg, Lectures on partial differential equations, CBMS Regional Conference Series in Mathematics, No. 17, American Mathematical Society, Providence, R.I., 1973.

14. E. Noether, Invariante Variations probleme, Nachr. Königl Wissen. Göttingen Math.-Phys. Kl. (1918) 235-257. (See Transport Theory and Stat. Phys. 1 (1971) 186-207 for an English translation.)

15. P.J. Olver, Euler operators and conservation laws of the BBM equation, Math. Proc. Camb. Phil. Soc. 85 (1979) 143-160.

16. P.J. Olver, Conservation laws of free boundary problems and the classification of conservation laws for water waves. Trans. Amer. Math. Soc. 277 (1983) 353-380.

17. P.J. Olver, Conservation laws and null divergences, Math. Proc. Camb. Phil. Soc. 94 (1983) 529-540.

18. P.J. Olver, Applications of Lie Groups to Differential Equations, Springer-Verlag, New York (to appear).

19. L.V. Ovsiannikov, Group Analysis of Differential Equations, Academic Press, New York, 1982.

20. I.G. Petrovsky, Lectures on Partial Differential Equations, Interscience Publ., Inc., New York, 1954.

21. J.F. Ritt, Differential Algebra, Dover, New York, 1966.

22. H. Steudel, Über die Zuordnung zwischen Invarianzeigenschaften und Erhaltungsätzen, Zeit. fur Naturforsch, 17A (1962) 129-133.

23. G.-Z. Tu, The Lie algebra of invariant group of the KdV, MKdV or Burgers equation, Lett. Math. Phys. 3 (1979) 387-393.

24. W.M. Tulczyjew, The Lagrange complex, Bull. Soc. Math. France 105 (1977) 419-431.

25. T. Tsujishita, Conservation laws for free Klein-Gordon fields, Lett. Math. Phys. 3 (1979) 445-450.

26. A.M. Vinogradov, On the algebro-geometric foundations of Lagrangian field theory, Sov. Math. Dokl. 18 (1977) 1200-1204.

27. A.M. Vinogradov, The $\mathcal{C}$-Spectral Sequence, Lagrangian formalism and conservation laws. I. The linear theory. J. Math. Anal. Appl. 100 (1984) 1-40.

28. A.M. Vinogradov, The $\mathcal{C}$-spectral sequence, Lagrangian formalism, and conservation laws. II. The nonlinear theory. J. Math. Anal. Appl. 100 (1984) 41-129.

29. A.M. Vinogradov, Local symmetries and conservation laws, Acta. Appl. Math. 2 (1984) 21-78.

School of Mathematics
University of Minnesota
Minneapolis, MN 55455

Lectures in Applied Mathematics
Volume 23, 1986

SOME GINZBURG-LANDAU TYPE VECTOR FIELD EQUATIONS

Elliott H. Lieb[1]

The lecture reported joint work with Haïm Brezis.

The system of equations studied is

$$-\Delta u_i = g^i(u) \text{ in } \mathcal{D}', \quad i=1,\ldots,n \qquad (*)$$

on $\mathbb{R}^d$, $d \geq 2$ and with $u: \mathbb{R}^d \to \mathbb{R}^n$, $n \geq 1$. (The case $n=1$ is called the scalar field case.) The functions g^i are defined by

$$g^i(u) = \partial G(u)/\partial u_i, \quad u \neq 0$$
$$g^i(0) = 0$$

for some $G: \mathbb{R}^n \to \mathbb{R}$ with

$$G \in C^1(\mathbb{R}^n \setminus \{0\})$$
$$G \in C(\mathbb{R}^n)$$
$$G(0) = 0$$

and statisfying appropriate additional conditions given below.

1980 Mathematical Subject Classifications 35J20, 35J60.
[1]Partially supported by U.S. National Science Foundation grant PHY-8116101-A02.

Associated with (*) is the action

$$S(u) = \frac{1}{2}\int|\nabla u|^2\,dx - \int G(u)dx$$

(which is not, in general, bounded below).

We prove that (*) has a non-zero solution in $\mathcal{D}'$ in the following class of functions:

$\mathcal{C} = \{u \mid u \in L^1_{loc}(\mathbb{R}^d),\ \nabla u \in L^2(\mathbb{R}^d),\ G(u) \in L^1(\mathbb{R}^d),\ u \to 0$ at infinity$\}$.

(The meaning of $u \to 0$ at infinity is that meas $\{x \mid |u(x)| > a\} < \infty$ for all $a > 0$.) This solution is also shown to have certain regularity properties.

Among these non-zero solutions there is one (call it $\bar{u}$) with the property that

$$S(\bar{u}) \leq S(v)$$

for any $v \in \mathcal{C}$ such that $v \not\equiv 0$, $g(v) \in L^1_{loc}(\mathbb{R}^d)$, v satisfies (*) in $\mathcal{D}'$. Thus, even though S is not bounded below, it has a minimum within the class of non-zero solutions to (*).

The conditions on G are the following

$d \geq 3$: With $p = 2^* = 2d/(d-2)$,

(i) $\limsup_{|u|\to\infty} |u|^{-p}\, G(u) \leq 0$.

(ii) $\limsup_{|u|\to 0} |u|^{-p}\, G(u) \leq 0$.

(iii) $G(u_o) > 0$ for some $u_o \in \mathbb{R}^n$.

(iv) $|g(u)| \leq C + C|u|^{p-1}$, for some C and all $u \in \mathbb{R}^n$. [Condition (iv) can be relaxed somewhat.]

<u>d = 2</u>: (i) $G(u) < 0$ for $0 < |u| < \varepsilon$ for some $\varepsilon > 0$.

(ii) $G(u_o) > 0$ for some $u_o \in \mathbb{R}^n$.

(iii) $|g(u)| \leq C + C|u|^{p-1}$, for some C, $1 < p < \infty$ and all $u \in \mathbb{R}^n$.

These results were first announced in ref. 1. The full details, together with related results and a summary of earlier work, appear in ref. 2.

BIBLIOGRAPHY

1. Lieb, E. H.: Some vector field equations. In: Differential equations. Proc. of the Conference held at the University of Alabama in Birmingham, USA, March 1983, Knowles, I., Lewis, R. (eds.) Math. Studies Series, Vol. 92 Amsterdam: North-Holland 1984.

2. Brezis, H. and Lieb, E. H.: Minimum action solutions of some vector field equations, Commun. Math. Phys. (to be published, 1984).

DEPARTMENTS OF MATHEMATICS AND PHYSICS
PRINCETON UNIVERSITY
P.O. BOX 708
PRINCETON, NJ 08544

Lectures in Applied Mathematics
Volume 23, 1986

EXISTENCE AND REGULARITY THEORY FOR ISOPERIMETRIC VARIATIONAL PROBLEMS ON ORLICZ-SOBOLEV SPACES: A REVIEW

Pierre A. Vuillermot

1. INTRODUCTION AND OUTLINE. In this review article, we outline and discuss our most recent results regarding the existence and the regularity theory for a class of strongly nonlinear eigenvalue problems on Orlicz-Sobolev spaces, with a glance at other contemporary attempts to understand the structure of some strongly nonlinear variational boundary-value problems defined on certain nonreflexive Banach spaces. The class of eigenvalue problems recently investigated can best be defined as follows. With $2 \leq n \in \mathbb{N}^+$, let $\Omega \subset \mathbb{R}^n \ni (x_1;\ldots;x_n) = x$ be an open bounded domain with closure $\bar{\Omega}$ and smooth $(C^{2,\alpha} -)$ boundary $\partial\Omega$; let $\hat{Y} \equiv \{Y_i\}_{i=1}^n$, Y be a family of $C^{(2)}$-Young functions, which means that Y_i, $Y \in C^2(\mathbb{R};\mathbb{R}^+)$ are even and convex for each $i \in \{1,\ldots,n\}$, with $\lim_{t\to\pm\infty} t^{-1}Y_{(i)}(t) = \pm\infty$ and $\lim_{t\to 0} t^{-1}Y_{(i)}(t) = 0$. Pick $\rho \in C(\bar{\Omega};\mathbb{R}^+)$ and let $F\colon \Omega\times\mathbb{R}\ni(x;\tau)\to\mathbb{R}$ be a Carathéodory function odd in τ. With $\lambda \in \mathbb{R}$, we then consider the class of real-valued, elliptic boundary-value problems

$$\left\{\begin{array}{c} \sum_{i=1}^{n} \{Y_i'(z_{x_i}(x))\}_{x_i} + \lambda\rho(x)Y'(z(x)) = F(x;z(x)) \quad \text{in} \quad \Omega \\ z = 0 \quad \text{on} \quad \partial\Omega \end{array}\right\} \tag{1.1}$$

where we have used the standard notation $\{z_{x_i}\}_{i=1}^n$ for the partial derivatives of z. Our forthcoming discussion of problem (1.1) will be entirely centered around a theorem stated in Section 2, which represents a blending of the main results from [1] and [2]; in that theorem, we exhibit nearly optimal growth conditions regarding Y_i, Y and F, which ensure the existence of countably many eigensolutions to problem (1.1). The existence proof for these eigensolutions is briefly sketched in Section 3; it rests on a new principle of restored compactness, which allows one to bypass the lack of reflexivity of the Banach spaces of distributions over which the isoperimetric variational problems associated with (1.1) are defined. We also exhibit nearly optimal regularity properties for those eigensolutions; the corresponding method of proof, also sketched in Section 3, rests on a combination of Schauder's inversion method with new convexity inequalities which characterize the shape of the given nonlinearities in (1.1). This new technique allows one to bypass the use of Nirenberg's translation method and of the related bootstrap procedures [3], which cannot be applied to our case to get Hölder-continuity estimates for the second derivatives, because of the stiffness or lack of good homogeneity properties in the nonlinear term of the principal part of (1.1). Our studies were motivated in the first place by certain problems in elasticity theory and in combustion theory ([4], [5], [6]), and by some questions in the theory of diffusion and reaction of gases ([7], [8]). They also represent a first attempt at elaborating a complete existence and regularity theory for isoperimetric variational problems defined on nonreflexive Banach spaces. For alternate formulations of the results, see [9]; for the corresponding one-dimensional Sturm-Liouville case, see [10].

2. BEYOND POHOŽAEV'S RESULT: STATEMENT OF THE MAIN THEOREM.

Regarding the eigensolutions to second-order nonlinear Dirichlet boundary-value problems without nonlinear terms in their principal part, one can go back to the celebrated result proved by Pohožaev in 1965 to find optimal growth conditions which ensure their existence [11]. Pohožaev considered nonlinear boundary-value problems of the form

$$\left\{ \begin{array}{c} \Delta z(x) + \lambda\Phi(z(x)) = 0 \quad \text{in} \quad \Omega \\ z = 0 \quad \text{on} \quad \partial\Omega \end{array} \right\} \tag{2.1}$$

with Φ odd and sufficiently regular (typically $\Phi \in C^{0,\alpha}(\mathbb{R};\mathbb{R})$ for $\alpha \in (0;1]$). Among other things, he proved that if $n \geq 3$ and Ω is starshaped, problem (2.1) has at least countably many nontrivial eigensolutions in a Banach space W of Sobolev type <u>if, and only if</u>, Φ grows more slowly than $|t|^{n+2/n-2}$. Using classic methods of potential theory, he also proved that with $\Phi \in C^{0,\alpha}(\mathbb{R};\mathbb{R})$ and $\alpha \in (0;1]$, those eigensolutions are in fact $C^{(2)}(\Omega) \cap C(\overline{\Omega})$-regular, a non-optimal result since no Hölder-continuity estimates for the second derivatives were given (see for example [12] for the derivation of such estimates). It is worth noting at this point that with $\Phi = Y'$, problem (2.1) is a particular case of problem (1.1) with $2Y_i(t) = t^2$ for each $i \in \{1,\dots,n\}$, $\rho = 1$ and $F = 0$. Recently, we proved the following generalization of Pohožaev's results (in the next statement and in all subsequent examples, Ω denotes an open, bounded <u>convex</u> domain in $\mathbb{R}^n$ with $C^{2,\alpha}$-boundary $\partial\Omega$):

THEOREM 2.1. Pick $2 \leq n \in \mathbb{N}^+$, ε, $\mu \in (0;+\infty)$, $\gamma \in (0;1)$ and consider problem (1.1) with $\rho \in C^{0,\gamma}(\overline{\Omega};\mathbb{R}^+)$. Assume moreover that the following hypotheses hold:

(H_1) For each $i \in \{1,\dots,n\}$, the Y_i's are <u>strictly</u> convex $C^{(3)}$-Young functions.

(H_2) The Young functions $\{Y_i\}_{i=1}^n$ and Y are convex in t^2 in the sense of [13].

(H_3) There exist $\kappa > 0$ and $t_0 \geq 0$ such that $|t|^{n+\varepsilon} \leq Y(\kappa t)$ for each $t \in \mathbb{R}$ with $|t| \geq t_0$. Moreover for each $i \in \{1,\ldots,n\}$, there exist $\kappa_i > 0$, $t_{0,i} \geq 0$ such that $|t|^{n+\varepsilon} \leq Y_i(\kappa_i t)$ for each $t \in \mathbb{R}$ such that $|t| \geq t_{0,i}$.

(H_4) For each $\kappa > 0$, we have

$$\lim_{t\to+\infty} \frac{e^{|t|^{n/n-1}} - 1}{Y(\kappa t)} = +\infty .$$

(H_5) For each $i \in \{1,\ldots,n\}$, there exist $\nu_i > 2$ and $t_{0,i} \geq 0$ such that

$$Y_i(2t) \leq \nu_i Y_i(t) .$$

Assume finally that F is Hölder continuous in x with Hölder constant uniform in τ, Hölder continuous in τ with Hölder constant uniform in x and satisfies the following hypothesis:

(H_6) There exists $\eta > 0$ such that

$$|F(x;\tau)| \leq \eta \tilde{Y}^{-1}(Y(\tau))$$

uniformly in x, where $\tilde{Y}^{-1}$ is the reciprocal inverse on $\mathbb{R}^+$ of the Legendre transform of Y, namely $\tilde{Y}(s) = \max_{t\geq 0} \{|s|t - Y(t)\}$, $s \in \mathbb{R}$.

Then problem (1.1) possesses at least a countable infinity $\{\pm z_\mu^{(m)}\}_{m\in\mathbb{N}^+}$ of $C^{2,\alpha}(\Omega) \cap C^{0,\beta}(\bar{\Omega})$ - eigensolutions with appropriate eigenvalues $\lambda_\mu^{\tilde{m}}$, $\beta \in (0; \frac{\varepsilon}{n+\varepsilon}]$, some $\alpha \in (0;1)$ and $\int_\Omega dx\rho(x)Y(\pm z_\mu^{(m)}(x)) = \mu$. We have written $\tilde{m}$ instead of m in $\lambda_\mu^{\tilde{m}}$ because of possible degeneracy.

To illustrate the significance of Theorem 2.1 and allow comparison with Pohožaev's result, we now discuss some examples.

EXAMPLE 2.1. Consider the nonlinear eigenvalue problem in $\mathbb{R}^3$

$$\left\{\begin{aligned} 3\sum_{i=1}^{3} z_{x_i}^2(x) z_{x_i,x_i}(x) + \Delta z(x) + Y'(z(x)) = 0 \quad &\text{in} \quad \Omega \\ z = 0 \quad &\text{on} \quad \partial\Omega \end{aligned}\right\} \tag{2.2}$$

where Y denotes any $C^{(2)}$-Young function which grows essentially more slowly than $e^{|t|^{3/2}} - 1$ (Hypothesis (H_4)), satisfies Hypothesis (H_2) and grows at least as fast as $|t|^{3+\varepsilon}$, $\varepsilon > 0$ (Hypothesis (H_3)). Problem (2.2) is of the form (1.1) with $\rho = 1$, $Y_i(t) = \frac{t^4}{4} + \frac{t^2}{2}$ for each $i \in \{1,2,3\}$ and $F \equiv 0$. From Theorem 2.1, we infer that problem (2.2) has a countable infinity of distinct $C^{2,\alpha}(\Omega) \cap C^{0,\beta}(\overline{\Omega})$-eigensolutions for some $\alpha \in (0;1)$ and $\beta \in (0;1/4]$ (we may take $\varepsilon = 1$ in Hypothesis (H_3)). For instance, we may take $Y(t) = t^8/8$.

The above example contrasts sharply with the following:

EXAMPLE 2.2. Consider the eigenvalue problem

$$\left\{\begin{aligned} \Delta z(x) + \lambda Y'(z(x)) = 0 \quad &\text{in} \quad \Omega \\ z = 0 \quad &\text{in} \quad \partial\Omega \end{aligned}\right\} \tag{2.3}$$

where Y is as in Example 2.1, but grows at least as fast as $|t|^6$ (for instance, we may still choose $Y(t) = t^8/8$); in this case, we have $\rho = 1$, $F = 0$ but $Y_i(t) = t^2/2$ in problem (1.1): Hypothesis (H_3) fails to apply and in fact we know from Pohožaev's results that problem (2.3) has <u>no</u> classical eigensolutions.

REMARKS. (1) Comparison of Examples 2.1 and 2.2 shows that Theorem 2.1 goes, in a certain sense, beyond Pohožaev's results since it means that one can restore the existence of countably many eigensolutions to problem (1.1) with an appropriate growth in the principal part of the equation; more specifically, one can restore the existence of such eigensolutions even with an exponential growth in Y (controlled by the cutoff function

$t \to e^{|t|^{n/n-1}} - 1$, see Hypothesis (H_4)), provided the Young functions Y_i governing the nonlinear terms in the principal part of (1.1) grow faster than $|t|^n$ (Hypothesis (H_3)), but no faster than polynomially (because of Hypothesis (H_5)). Whether Hypothesis (H_3) defines an optimal borderline between existence and non existence results of the above type, and whether Theorem 2.1 holds without Hypothesis (H_5) (that is with possible exponential growth in the principal part of (1.1)), are open questions at this time.

(2) The strict convexity assumption (H_1) makes problem (1.1) strictly elliptic, hence the $C^{2,\alpha}(\Omega) \cap C^{0,\beta}(\bar{\Omega})$ - regularity statement of Theorem 1.1 (see [2]). Without Hypothesis (H_1), problem (1.1) becomes degenerate since we may have $Y_i''(0) = 0$ for each $i \in \{1,\dots,n\}$ (for instance, with $Y_i(t) = \frac{t^4}{4}$); easy counter-examples then show that only weaker, though often optimal, regularity statements hold. For complete details regarding the transition mechanism from C^2 - regularity to $C^{1,\alpha}_{\ell oc}$ - regularity when degeneracy occurs, see [14], which goes beyond the recent results of [15], [16] and [17]; for the corresponding one-dimensional Sturm-Liouville case, see [18] and [19].

In the next section, we sketch the proof of Theorem 2.1; we deliberately avoid all technical details, and we restrict ourselves to the explanation of the main ideas.

3. SKETCH OF THE PROOF OF THEOREM 2.1. Let $\overset{\circ}{B}_{\hat{Y},Y,\rho}$ be the real Banach space obtained by closure of $C_0^\infty(\Omega)$ with respect to the norm $\|u\|^2_{\hat{Y},Y,\rho} = \|u\|^2_{Y,\rho} + \sum_{i=1}^n \|u_{x_i}\|^2_{Y_i}$, where

$$\|u\|_{Y,(\rho)} = \inf\left\{k > 0: \int_\Omega dx(\rho(x))Y(\frac{u(x)}{k}) \leq 1\right\}.$$

The main idea inherent in the proof of Theorem 2.1 amounts to transforming problem (1.1) into the following isoperimetric variational problem on $\overset{\circ}{B}_{\hat{Y},Y,\rho}$. Define the two functionals

$$\left\{\begin{aligned} V(z) &= \sum_{i=1}^{n} \int_\Omega dx Y_i(z_{x_i}(x)) + \int_\Omega dx \int_0^{z(x)} d\tau F(x;\tau) \\ C_\mu(z) &= \int_\Omega dx \rho(x) Y(z(x)) - \mu \end{aligned}\right\} \tag{3.1}$$

along with the cylinder $K_\mu = \{z \in \overset{\circ}{B}_{\hat{Y},Y,\rho} : C_\mu(z) = 0\}$. We then look for critical points, z_μ, of V on K_μ; indeed, such critical points do solve problem (1.1) with appropriate eigenvalues λ_μ, which emerge as Lagrange's multipliers of the functional V with respect to K_μ. These critical points either minimize V, or maximize V, or represent saddle-points of V. In so doing, we meet three major difficulties: first, the nonreflexivity of $\overset{\circ}{B}_{\hat{Y},Y,\rho}$; second, the unboundedness of K_μ with respect to the $\overset{\circ}{B}_{\hat{Y},Y,\rho}$-topology; third, the lack of good homogeneity properties (stiffness) of the Young functions Y_i, $i \in \{1,\dots,n\}$.

(A) Bypass around the lack of reflexivity of $\overset{\circ}{B}_{\hat{Y},Y,\rho}$.

The first difficulty is bypassed using a new principle of restored compactness: we first embed $\overset{\circ}{B}_{\hat{Y},Y,\rho}$ into a reflexive Sobolev space $W^{1,n}_\rho$ using Hypothesis (H_3); we then proceed in looking for a compact embedding of $W^{1,n}_\rho$ into a subspace of $\overset{\circ}{B}_{\hat{Y},Y,\rho}$ (in this particular instance, the small Orlicz space $E_{Y,\rho}$); this can be done using a classic result by Trudinger [20] and Hypothesis (H_4). In $\overset{\circ}{B}_{\hat{Y},Y,\rho}$, we finally restore the compactness of the critical levels of V with respect to K_μ, using Hypothesis (H_5) and new convexity inequalities valid for all C^2-Young functions which satisfy Hypothesis (H_2); these inequalities characterize the shape of the nonlinearities in (1.1).

(B) Bypass around the unboundedness of the cylinder K_μ.

The second difficulty is bypassed in exploring the geometric structure of K_μ, and in exhibiting a simple infinite family of nested and bounded subsets of K_μ with finite genus, with respect to which the minimax principles of critical point theory can be defined. The existence proof of Theorem 2.1 is then completed following an adaptation of the traditional Ljusternik-Schnirelmann scheme ([21], [22]).

(C) Bypass around the stiffness of the Y_i's.

In fact, steps (A) and (B) lead to the existence proof of countably many $C^{0,\beta}(\overline{\Omega})$-eigensolutions to problem (1.1), with $\beta \in (0;\frac{\varepsilon}{n+\varepsilon}]$; however, the stiffness of the Y_i's prevents one to apply the usual bootstrap procedures based on Nirenberg translation method [3], which would allow one to improve the regularity properties of z_μ, get Hölder-continuity estimates for the second derivatives and therefore prove the $C^{2,\alpha}(\Omega) \cap C^{0,\beta}(\overline{\Omega})$-regularity of the eigensolutions. To bypass this difficulty, we devised in [2] a new method which can be summarized as follows: with every eigensolution $z_\mu \in C^{0,\beta}(\overline{\Omega})$ which solves problem (1.1) in the distributional sense, we associate a unique $\hat{z}_\mu \in C^{2,\alpha}(\overline{\Omega})$ with $\alpha \in (0;1)$; this function $\hat{z}_\mu$ solves the auxiliary Dirichlet boundary-value problem

$$\left\{\begin{array}{c} \sum_{i=1}^{n} \{Y_i'(\hat{z}_{\mu,x_i}(x)\}_{x_i} = f(z_\mu(x)) \\ \hat{z}_\mu = 0 \end{array}\right\} \tag{3.2}$$

where $x \to f(z_\mu(x)) = F(x;z_\mu(x)) - \lambda_\mu \rho(x) Y'(z_\mu(x))$ is in $C^{0,\gamma}(\overline{\Omega};\mathbb{R})$ for some $\gamma \in (0;1)$. The unique solvability of problem (3.2) in $C^{2,\alpha}(\overline{\Omega})$ follows from a routine application of Schauder's inversion method [12], through Hypotheses (H_1), (H_2) and (H_6). Using then new convexity inequalities (different from those mentioned under (A)), we finally prove that $\nabla z_\mu = \nabla \hat{z}_\mu \in C^{1,\alpha}(\Omega;\mathbb{R}^n)$, which gives the desired $C^{2,\alpha}(\Omega) \cap C^{0,\beta}(\overline{\Omega})$-regularity. This indirect procedure can be summarized in the following diagram:

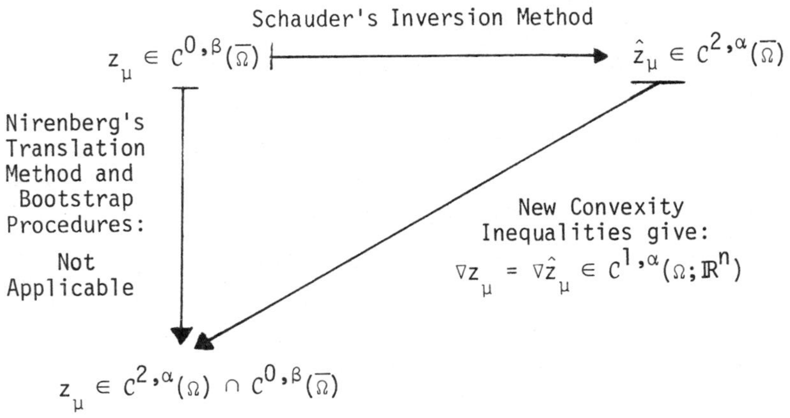

This completes the sketch of the proof of Theorem 2.1.

In the next section, we briefly describe other recent attempts at understanding the structure of nonlinear boundary-value problems defined on some nonreflexive Banach spaces.

4. VARIATIONAL BOUNDARY-VALUE PROBLEMS ON NONREFLEXIVE BANACH SPACES: A GLANCE AT CONTEMPORARY RESEARCH. The general ideas outlined in the preceding sections are complementary to other comtemporary attempts at understanding the structure of certain strongly nonlinear problems for integro-differential equations and for variational boundary-value problems. In particular, we refer the reader to the classic monograph by Krasnosel'skii and Rutickii [23] and to the numerous references therein for an account of the research done in eastern european schools. We also mention Visik's work in [24], which is more specifically concerned with boundary-value problems having coefficients taking on values in Orlicz classes. In the western hemisphere, one can trace things down to [25], in which Browder initiated a study of nonlinear elliptic functional equations in nonreflexive Banach spaces; Orlicz spaces and related spaces of distributions are also briefly mentioned in the paper by Leray and Lions [26], as being potentially relevant to the study of variational boundary-

value problems with nonlinear terms which grow faster than polynomially. This class of problems was later systematically studied by Donaldson in [27], and by Donaldson and Trudinger in [28], who considerably developed the functional analysis of Orlicz-Sobolev spaces, in particular the embedding theorems relevant to problems in nonlinear analysis (see also [29] for an account of their work). Further progress was accomplished by Gossez in [30], [31], [32] and most recently in [33]. Regarding nonlinear Dirichlet problems on unbounded regions of $\mathbb{R}^n$, see [34]; we also refer the reader to [35] for most recent results concerning variational inequalities on Orlicz-Sobolev spaces. Regarding regularity theory, we refer the reader to the series of papers by de Thélin ([36], [37], [38], [39]) and to the thesis by Wardi-Lamrini [17]. For discussions of critical point theories on Orlicz-Sobolev spaces, their applications, their limitations and their extensions see the aforementioned references to the author's work and also [40]. See [41] for an elementary summary. The challenge of all this is the elaboration of a complete existence and regularity theory for strongly nonlinear eigenvalue problems on nonreflexive Banach spaces.

ACKNOWLEDGEMENTS. The author would like to thank Drs.B. Nicolaenko D. D. Holm and J. M. Hyman from the Los Alamos National Laboratory for their invitation to this conference and their financial support. He also thanks Mrs. M. Pruiett for her excellent typing of the manuscript.

BIBLIOGRAPHY

[1] P. A. Vuillermot, "A Class of Elliptic Partial Differential Equations with Exponential Nonlinearities," Mathematische Annalen, 268, 4 (1984), 497-518.

[2] P. A. Vuillermot, "$C^{2,\alpha}(\Omega) \cap C^{0,\beta}(\overline{\Omega})$-Regularity for the Solutions of Strongly Nonlinear Eigenvalue Problems on Orlicz-Sobolev Spaces," for the Proceedings of the American Mathematical

Society (1985).

[3] L. Nirenberg, "Remarks on Strongly Nonlinear Partial Differential Equations," Commun. Pure and Appl. Math., 8 (1955), 648-674.

[4] J. Moseley, "A Nonlinear Eigenvalue Problem with an Exponential Nonlinearity," Contemporary Mathematics (V. Komkov, Editor), 4 (1981), 11-24.

[5] J. Moseley, "Asymptotic Solutions for a Dirichlet Problem with an Exponential Nonlinearity," SIAM Journal Math. Analysis, 14, 4 (1983), 719-735.

[6] J. Moseley, "A Two-Dimensional Dirichlet Problem with an Exponential Nonlinearity," SIAM Journal Math. Analysis, 14, 5 (1983), 934-946.

[7] R. Aris, The Mathematical Theory of Diffusion and Reaction of Permeable Catalysts, Clarendon Press, Oxford (1975).

[8] R. Aris, "Some Characteristic Nonlinearities of Chemical Reaction Engineering," North-Holland Math. Studies (A. R. Bishop, D. K. Campbell, B. Nicolaenko, editors), 61 (1982), 247-265.

[9] P. A. Vuillermot, "On a Class of Strongly Nonlinear Dirichlet Boundary-Value Problems: Beyond Pohožaev's Results," Proceedings of Symposia in Pure Mathematics (F. E. Browder, editor), in press (1985).

[10] P. A. Vuillermot, "A Class of Sturm-Liouville Eigenvalue Problems with Polynomial and Exponential Nonlinearities," Nonlinear Analysis: Theory, Methods, and Applications, 8, 7 (1984), 775-796.

[11] S. I. Pohožaev, "Eigenfunctions of the Equation $\Delta u + \lambda f(u) = 0$," Doklady, 165, 1 (1965), 1408-1411.

[12] D. Gilbarg and N. S. Trudinger, Elliptic Partial Differential Equations of Second-Order, Springer Verlag, Berlin, Heidelberg, New York (1977).

[13] P. A. Vuillermot, "A Class of Orlicz-Sobolev Spaces with Applications to Variational Problems involving Nonlinear Hill's Equations," J. Math. Anal. Appl., 89, 1 (1982), 327-349.

[14] P. A. Vuillermot, "Regularity Theory for Strongly Nonlinear Degenerate Dirichlet Boundary-Value Problems on Orlicz-Sobolev Spaces," to be submitted to Inventiones Mathematicae (1984).

[15] L. C. Evans, "A New Proof of $C^{1,\alpha}$-Regularity of Solutions of Certain Degenerate Elliptic Partial Differential Equations," J. Diff. Equations, 45, 3 (1982), 356-373.

[16] P. Tolksdorf, "Regularity of a More General Class of Quasilinear Elliptic Equations," J. Diff. Equations, 51, 1 (1984), 126-150.

[17] S. Wardi-Lamrini, "Régularité Höldérienne de la Solution d'une Equation aux Dérivées Partielles fortement Nonlinéaire," Thése de 3ème cycle, Publication Number 2868, Université Paul Sabatier, Toulouse, France (1983).

[18] P. A. Vuillermot, "Remarks on Some Strongly Nonlinear Degenerate Sturm-Liouville Eigenvalue Problems," for the Proceedings of the 6th UTA-International Conference on Trends in Theory and Practice of Nonlinear Analysis, North-Holland Math. Studies (1985).

[19] P. A. Vuillermot, "An Optimal Regularity Result for a Class of Strongly Nonlinear Degenerate Sturm-Liouville Eigenvalue Problems," to be submitted to the Journal of Differential Equations (1984).

[20] N. S. Trudinger, "On Embeddings into Orlicz Spaces and Some Applications," Journ. Math. and Mech., 17, 5 (1967), 473-483.

[21] R. Palais, "Ljusternik-Schnirelmann Theory on Banach Manifolds," Topology, 5 (1966), 115-132.

[22] P. Rabinowitz, "Variational Methods for Nonlinear Eigenvalue Problems," Course of Lectures CIME, Varenna, Italy (1974).

[23] M. A. Krasnosels'kii and Ya. B. Rutickii, Convex Functions and Orlicz Spaces, Noordhoff, Groningen, The Netherlands (1961).

[24] Visik, M. I., "Solvability of the First Boundary-Value Problem for Quasilinear Equations with Rapidly Increasing coefficients in Orlicz Classes," Dokl. Akad. Nauk. SSSR, 151 (1963), 1060-1064.

[25] F. E. Browder, "Nonlinear Elliptic Functional Equations in Nonreflexive Banach Spaces," Bull. Amer. Soc., 72 (1966), 89-95.

[26] J. Leray and J. L. Lions, "Quelques Résultats de Visik sur les Problèmes Elliptiques Nonlinéaires par la Méthode de Minty-Browder," Bull. Soc. Math. de France, 93 (1955), 97-107.

[27] T. K. Donaldson, "Nonlinear Elliptic Boundary-Value Problems in Orlicz-Sobolev Spaces," J. Diff. Equations, 10 (1971), 507-528.

[28] T. K. Donaldson and N. S. Trudinger, "Orlicz-Sobolev Spaces and Embedding Theorems," J. Functional Analysis, 8 (1971), 52-75.

[29] R. A. Adams, "Sobolev Spaces," Academic Press, New York (1975).

[30] J. P. Gossez, "Nonlinear Elliptic Boundary-Value Problems for Equations with Rapidly (or Slowly) Increasing Coefficients," Trans. Amer. Math. Soc., 190 (1974), 163-205.

[31] J. P. Gossez, "A Remark on Strongly Nonlinear Elliptic Boundary-Value Problems," Bol. Soc. Brasil Mat., 8 (1977), 53-63.

[32] J. P. Gossez, "Some Approximation Properties in Orlicz-Sobolev Spaces," Studia Mathematica, T., LXXIV (1982), 17-24.

[33] J. P. Gossez, "Strongly Nonlinear Elliptic Problems in Orlicz-Sobolev Spaces of Order One," Preprint from Université Libre de Bruxelles (1983).

[34] R. Landes and V. Mustonen, "Pseudo-Monotone Mappings in Sobolev-Orlicz Spaces and Nonlinear Boundary-Value Problems on Unbounded Domains," Journ. Math. Anal. Appl., 88, 1 (1982), 25-36.

[35] J. P. Gossez and V. Mustonen, Private Communication by Vesa Mustonen (June 1984).

[36] F. de Thélin, "Régularité Intérieure de la Solution d'un Problème de Dirichlet fortement Nonlinéaire," C. R. Acad. Sc. Paris, 286 (1978), 443-445.

[37] F. de Thélin, "Régularité de la Solution d'une Equation fortement (ou Faiblement) Nonlinéaire dans $\mathbb{R}^n$," Annales Fac. Sc. Toulouse, II (1980), 249-281.

[38] F. de Thélin, "Régularité de la Solution d'un Problème de Dirichlet fortement Nonlinéaire," These d'Etat, Publication Number 269, Université Paul Sabatier, Toulouse, France (1981).

[39] F. de Thélin, "Local Regularity Properties for the Solutions of a Nonlinear Partial Differential Equation," Nonlinear Analysis: Theory, Methods, and Applications, 6, 8 (1982), 839-844.

[40] P. A. Vuillermot, "Solution to Certain Nonlinear Sturm-Liouville Eigenvalue Problems without the Palais-Smale Condition," in preparation (1984).

[41] P. A. Vuillermot, "A Class of Isoperimetric Variational Problem on Certain Orlicz-Sobolev Spaces," North-Holland Math. Studies (R. Lewis and I. Knoles, editors), 92 (1984), 553-559.

DEPARTMENT OF MATHEMATICS
UNIVERSITY OF TEXAS AT ARLINGTON
ARLINGTON, TEXAS 76019

Evolutionary Systems

Lectures in Applied Mathematics
Volume 23, 1986

PATTERN SELECTION AND LOW-DIMENSIONAL CHAOS IN DISSIPATIVE MANY DEGREE-OF-FREEDOM SYSTEMS[o]

A. R. Bishop,[*] J. C. Eilbeck,[†] I. Satija[**] and G. Wysin[††]

ABSTRACT. It is shown numerically that the longtime behavior of a number of driven, dissipative, dispersive, many-degree-of-freedom systems can be characterized by strong mode-locking into a small number of determining (nonlinear) modes. On the basis of the observed profiles, estimates of chaotic attractor dimensions, and projections into nonlinear mode bases, it is argued that the same few modes can (in these extended systems) give a unified picture of spatial pattern selection, low-dimensional chaos, and coexisting coherence and chaos.

In recent years considerable attention has been given to the properties of low-dimensional maps as models for complicated dynamics in higher-dimensional dynamical systems.[1] This attention has been merited by the proof of "universal" properties in classes of one-dimensional maps.[1] However, with few exceptions, the low-dimensionality has been introduced explicitly by restricting consideration to models with a very small number of degrees-of-freedom. On the other hand, equally active research has focused on the subject of spatial pattern selection in non-equilibrium nonlinear systems with many degrees-of-freedom (e.g. convection

[o]One of us (JCE) would like to acknowledge the Royal Society of London for partial financial support during the period when this work was carried out, and the U.S./U.K. Education Commission and the Carnegie Trust (Edinburgh) for further travel grants. Work at Los Alamos was performed under the auspices of the U.S.D.O.E.

0075-8485/86 $1.00 + $.25 per page

cells,[2] reaction-diffusion systems[3]). In these cases mode-locking is very strong and a small number of modes dominate the spatial structure and temporal evolution in a nonlinear partial differential equation (p.d.e.) or large system of coupled ordinary differential equations (o.d.e.'s).

The perspective we wish to emphasize in this report is that the two phenomena of pattern formation and low-dimensional chaos can be intimately connected in perturbed, dissipative dynamical systems with many degrees-of-freedom. More specifically, chaotic dynamics can develop by chaotic motions of the collective coordinates identifying the dominant (determining) patterns in the quiescent regimes. In this way only a small loss of mode-locking is responsible for the temporal chaos coexisting with spatial coherence.

There are many examples of this scenario,[4] which gives the problem of identifying and testing (nonlinear[5]) mode reduction schemes a general mathematical and physical importance. We have chosen to concentrate here on a set of driven and damped nonlinear partial differential equations (and their discrete analogs) in <u>one</u> spatial dimension.[6] In all of these cases, coherent nonlinear modes ("solitons" or near-solitons[7]) are fundamental excitations of the underlying unperturbed Hamiltonian system. Detailed results for a number of examples will be presented elsewhere.[8,9] Here we will abstract <u>typical</u> cases from five examples so as to illustrate a general interconnectedness of pattern selection, low-dimensional chaos, and coexisting coherence and chaos.

Four of our examples are based on perturbed sine-Gordon (SG) equations

$$\phi_{tt} - \phi_{xx} + \sin\phi = F(x,t) - \varepsilon\phi_t \quad . \tag{1}$$

Here ϕ is a scalar field and x, t are space and time, respectively. Subscripts denote derivatives. F(x,t) is a forcing field and ε a damping constant. Note that neither pattern selection nor chaos

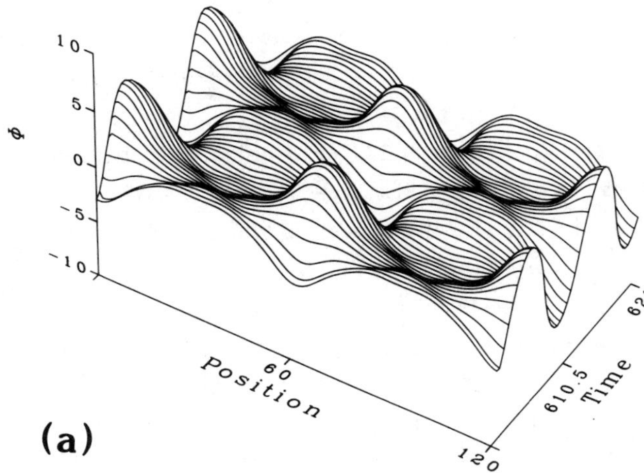

(a)

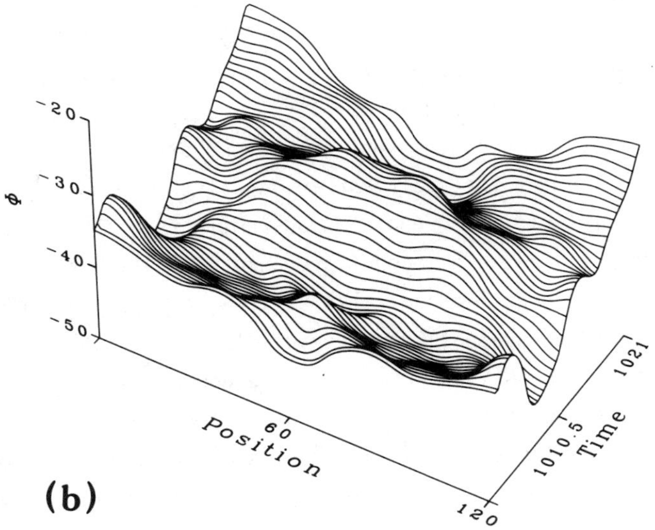

(b)

Figure 1. Space-time evolutions of $\phi(x,t)$ for the SG system (1) through two driving periods for $\varepsilon = 0.2$, $\omega_d = 0.6$, with periodic boundary conditions, and driving strengths (a) $\gamma = 0.8$, which results in periodic time evolution, (b) $\gamma = 1.0$, which results in chaotic kink-antikink motion, nearly repeating every driving period.

occur in (1) without dispersion (i.e. in an overdamped limit.) In the four examples below we have studied eqn. (1) numerically on a finite line (length L) and with a high density spatial mesh (approximating the p.d.e.):

Case A. Here L = 24 with 120 grid points and periodic boundary conditions, $\varepsilon = 0.2$ and $F(x,t) = \gamma \sin \omega_d t$ with $\omega_d = 0.6$. The initial data is a static "pulse" profile (the actual shape of the pulse is not very important -- the number of coexisting attractors is small).[8] As reported elswhere,[8] this system undergoes a spontaneous spatial period doubling for $0.6 \lesssim \gamma \lesssim 0.9$ but remains simply periodic in time. Typical spatial profiles are shown in Fig. 1a, and should be thought of as "breather-soliton" wave-trains.[7,8,10] As $\gamma \tilde{\to} 0.9$, the duration of a "chaotic" initial transient diverges, resulting in temporal chaos for $0.9 \lesssim \gamma \lesssim 1.4$. Accompanying this chaos are large ($>> 2\pi$) variations in the spatial average of $\phi(x,t)$ [denoted by $\bar{\phi}(t)$] and large amplitude ($\simeq 2\pi$) spatial variations in ϕ relative to $\bar{\phi}$. Instantaneous spatial correlation functions suggest[8] strong structural disorder. However, following the evolution in detail through a period of the driving field (Fig. 1b) shows transparently that the basic coherent structures in the quiescent pre-chaotic regime are preserved, but that their mode-locking relative to each other has been (chaotically) broken so that the structure fails to repeat by a small amount after each driver period. Consequently, we can decompose the field at each instant of time into either two "breather solitons" or two "kink-solitons" and two "antikink-solitions" (at instants of kink-antikink collision the field ϕ may appear to be flat -- see Fig. 1b). Furthermore, chaotic evolution of $\bar{\phi}(t)$ through multiples of 2π does not take place via single particle dynamics but rather through the slow diffusion of the kinks (antikinks) -- as with thermally assisted transport in such systems even in the absence of intrinsic chaos.

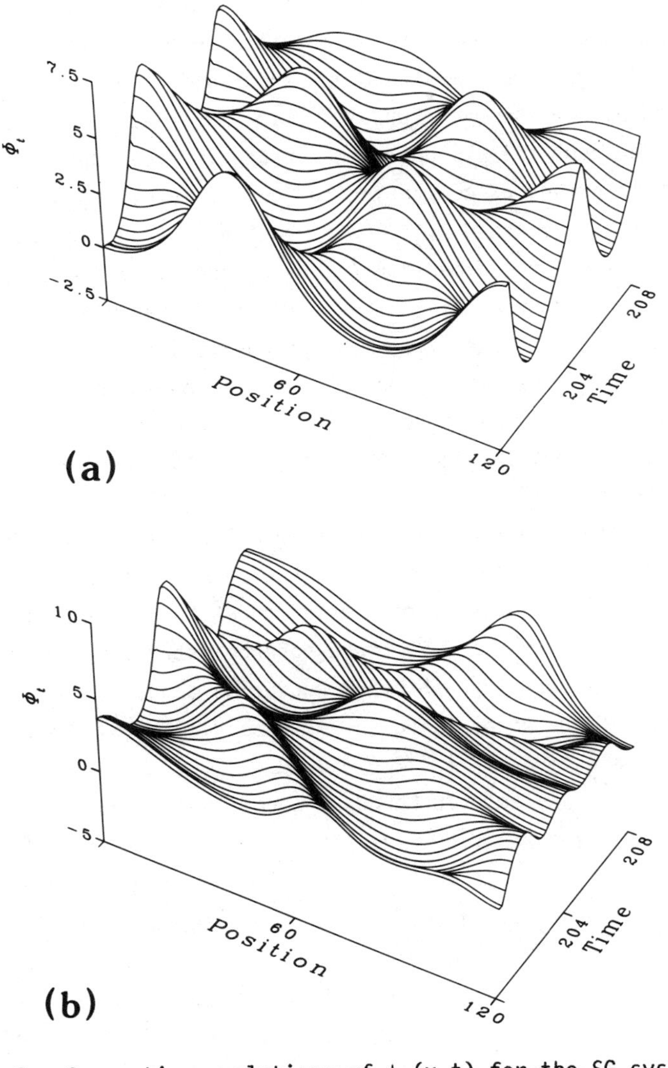

Figure 2. Space-time evolutions of $\phi_t(x,t)$ for the SG system (1) through 1.6 driving periods for $\varepsilon = 0.1$, $\omega_d = 1.25$, with Neumann boundary conditions, dc driving $\gamma_0 = 0.35$, and ac driving strengths (a) $\gamma = 1.2$, which produces a nonchaotic standing wave state on the third ZFS, (b) $\gamma = 1.5$, which produces a chaotic pair of driven pulses.

We have confirmed the identification of the small number of coherent "soliton" modes in the chaotic regime by projecting the field at successive times onto an exact-soliton basis[7] -- the optimal modes for the unperturbed system (1). Results are described elsewhere[11] but fully support the strong implications of Fig. 1. This small number of dominating collective modes suggests that the chaos will be governed by a low-dimensional strange attractor.[12] We have checked the dimension (ν) using the algorithm proposed by Grassberger and Procaccia.[13] We estimate[13] that $\nu(\gamma = 1.0) = 2.5 \pm 0.3$, and in fact the dimension is found to lie within this range throughout the chaotic regime. Thus ν is indeed low.[14] (For $\gamma \lesssim 0.9$, ν is 1.0, as expected. ν also approaches 1.0 for large $\gamma \gtrsim 10$ where the nonlinear potential is a small perturbation on the dynamics.)

Case B. Here conditions are the same as case A, except that the initial data is a single static kink with periodic boundary conditions mod (2π). The attractor is found[8] to be a kink plus a "breather" and for $\gamma \gtrsim 0.9$, these coherent structures move randomly with respect to each other producing temporal chaos. The dimension of the chaos is again correspondingly low: e.g., for $\gamma = 1.0$, we estimate[13] $\nu = 2.6 \pm 0.5$.

Case C. Here[9] we adopted boundary conditions appropriate to a finite Josephson junction oscillator in zero magnetic field,[14] viz. $\phi_x(x = 0,t) = \phi_x(x = L,t) = 0$. The number of grid points = 120, $L = 6$, $\varepsilon = 0.1$, and $F(x,t) = \gamma_0 + \gamma \sin \omega_d t$ with $\gamma_0 = 0.35$. Even with single kink initial conditions i.e. on the first "zero field step" (ZFS)[15] a great variety of dynamical behaviors (including jumping between ZFS's) are observed.[9] For our present purposes we report an example which illustrates a prevalent source of chaos: for $\omega_d = 1.25$ and $\gamma = 1.2$, the kink initial data is attracted to a nonchaotic "symmetric" state[15] on the third ZFS as shown in Fig. 2a. Increasing γ to 1.5, we find a chaotic long-time evolution in which the strongly mode-locked coherent

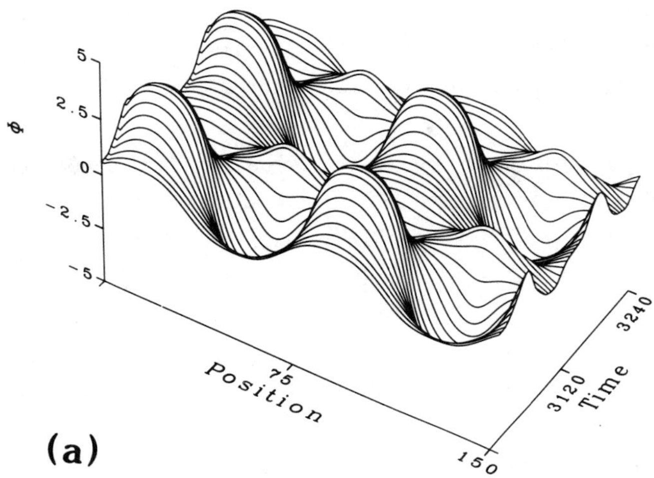

(a)

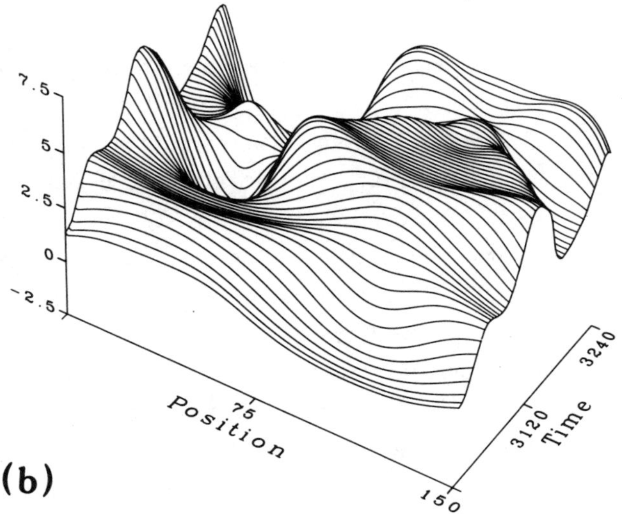

(b)

Figure 3. Space-time evolutions of the in-plane angle $\phi(x,t)$ for the easy-plane ferromagnet (2) through two driving periods for $\varepsilon = 0.1$, $\omega_d = 0.05144$, with periodic boundary conditions, and ac driving strengths (a) $B_0^x = 0.03429$, resulting in a standing wave pattern, (b) $B_0^x = 0.02743$, resulting in intermittency between smooth standing wave structures and chaotic kink-antikink dynamics.

structures of the third ZFS are still dominant but their relative mode-locking has been (chaotically) broken -- see Fig. 2b. Consistently, we estimate[13] the attractor dimension to be low: $\nu = 2.5 \pm 0.4$.

Case D. In this case[8] we used outflow boundary conditions, $L = 40$, 800 grid points, $\varepsilon = 0.05$, and completely flat initial data ($\phi(x,t = 0) = \phi_t(x,t = 0) = 0$). Chaotic evolution is induced by choosing a spatially inhomogeneous driving field $F(x,t) = \gamma$, $x \in [15,25]$, $F(x,t) = 0$ elsewhere. Our estimates of ν for $0 \leq \gamma \leq 5$ show that ν increases from 1.0 at $\gamma = 0$ to 2.8 ± 0.4 for $\gamma = 2$ (with a rapid increase at the chaotic threshold $\gamma \simeq 0.4$) and then rapidly decreases to 1.0 at larger values of γ (as in Case A). Again the low dimension is entirely consistent with an examination of spatial profiles: once more ϕ evolves by the separation of coherent kink-antikink pairs which are periodically nucleated at the center of the line.[8]

Examples A-D are all based on perturbations of the SG system, although with a great variety of initial data, boundary conditions and perturbations. The perturbations are sufficiently strong that this choice is not a limitation on the general phenomena we have described. In particular, the near integrability of the unperturbed Hamiltonian system plays no role.[6] To emphasize this, we describe a non-integrable, two-component field example in:

Case E. Here we studied[9] a strongly perturbed magnetic chain of classical spins $\vec{S}_n = (S_n^x, S_n^y, S_n^z) = (\cos\theta_n \cos\phi_n, \cos\theta_n \sin\phi_n, \sin\theta_n)$, $n = 1, 2, 3 \ldots N$, governed by the Hamiltonian

$$H = -\sum_{n=1}^{N} \vec{S}_n \cdot \vec{S}_{n+1} + \alpha \sum_{n=1}^{N} \left(S_n^z\right)^2 - \vec{B} \cdot \sum_{n=1}^{N} \vec{S}_n \quad , \tag{2}$$

where $\alpha > 0$ is an easy-plane anisotropy (ϕ_n and θ_n are the in-plane and out-of-plane angles) and $\vec{B}$ is an applied magnetic field. A Gilbert-Landau dissipation term[9] of strength ε is added to the equations of motion following from (2). Again a great variety of chaotic and non-chaotic evolutions are observed,[9] depending on magnetic field configurations and parameter values. However, spontaneous pattern formation, consequent low-dimensional chaos, and coexisting coherence and chaos are once again prevalent phenomena. An example is illustrated in Fig. 3. Here we have used periodic boundary conditions, $N = 150$, $\varepsilon = 0.1$, $\alpha = 0.1907$ (for comparisons with the quasi-one-dimensional magnetic material $CsNiF_3$[9]), $\vec{B} = (B^x, B^y, B^z) = (B_0^x \sin(\omega_d t), 0, 0)$, $\omega_d = 0.05144$, and random initial data $(\phi_n(t = 0), \theta_n(t = 0))$. In Fig. 3a, $B_0^x = 0.03429$. The period-½ spatial structure seen in Fig. 3a forms spontaneously as the long time attractor with simply-periodic entrained motion. Decreasing B_0^x to 0.02743 we have entered a chaotic regime characterized (Fig. 3b) by chaotic motions of the coherent nonlinear mode components of the precursor period-½ spatial pattern. The Grassberger-Procaccia estimate[13] of the attractor dimension is correspondingly low -- for $B_0^x = 0.02743$, we find $\nu = 1.7 \pm 0.4$ for S^y and $\nu = 1.9 \pm 0.4$ for S^z.

In conclusion, we have illustrated through a variety of driven, dissipative, dispersive p.d.e.'s that a small number of determining modes may be typically responsible both for spontaneous pattern formation in quiescent regimes and for low-dimensional chaos in subsequent chaotic regimes. Thus coexisting coherence and chaos is a natural corollary. The identification of determining nonlinear modes clearly motivates the choice and study of specific truncated systems of coupled o.d.e.'s to be compared with the dynamics of the full p.d.e.'s. These studies are in progress to analytically substantiate the above unified scenario. General experimental implications (e.g., for spatial correlations, diffusion, fractal coherent structures) are being assessed in specific contexts.[8] Progress with analytic stability and existence proofs is possible in some cases as described in these Proceedings by Ercolani, Forest and McLaughlin.

We are grateful for valuable discussions with D. W. McLaughlin and I. Procaccia.

REFERENCES

1. See articles in Physica 7D (1983).

2. e.g. M. Cross, Phys. Rev. A25, 1065 (1982).

3. e.g. H. Meinhardt, "Models of Biological Pattern Formation" (Academic Press 1982).

4. Examples of coexisting coherence and chaos include clumps and cavitons in turbulent plasmas and large scale structures in turbulent fluids. There are also now laboratory scale observations (e.g. convection cells, surface solitons) and probable biological applications.

5. In some cases rigorous bounds on the number of determining modes have recently been established (e.g. C. Foias, et al., Phys. Rev. Lett. 50, 1031 (1983)). Note also that, for some cases (e.g. for certain reaction-diffusion problems), even a truncated set of linear modes can be accurate (J. C. Eilbeck, J. Math. Biol. 16, 233 (1983); B. Nicolaenko, et. al., Proc. Acad. Sc. Paris 298, 23(1984)). In some cases it has been shown that the maximum number of determining modes bounds the attractor (fractal) dimension. (O. Manley et al., preprint, 1984, B. Nicolaenko and B. Scheurer, preprint, 1984).

6. Our preliminary studies in two dimensions support the same scenario: O. H. Olsen, P. S. Lomdahl, A. R. Bishop, J. C. Eilbeck (Los Alamos preprint LA-UR-84-2852).

7. e.g., R. K. Dodd, et. al. "Solitons and Nonlinear Wave Equations" (Academic Press 1982).

8. A. R. Bishop, et. al., Phys. Rev. Lett. 50, 1095 (1983), and references therein; and in Ref. 1; J. C. Eilbeck, Bull. I.M.A. (in press).

9. A. R. Bishop, et al., APS March Meeting Bulletin (1984). G. Wysin, et. al. APS March Meeting Bulletin (1984);

10. N. Ercolani, et al., preprint (1983).

11. E. A. Overman, D. W. McLaughlin and A. R. Bishop, preprint (1984).

12. e.g. J. D. Farmer, E. Ott and J. Yorke, Physica 7D, 153 (1983).

13. P. Grassberger and I. Procaccia, Phys. Rev. Lett. 50, 346 (1983). The calculation of various attractor "dimensions" remains in an early stage of development (see ref. 12). We emphasize that our error estimates here are realistically conservative. Typically we used an embedding dimension of 8 and 80,000 data points.

14. Assuming only 2 breathers (or 4 kinks) as a truncated modal set, the maximum dimension of the space containing the attractor is 8. Our initial data symmetry reduces this to 4. The presence of dissipation will typically further reduce the "active" dimension. Our estimates of ν are generally in the range 2 - 2.5. This is entirely reasonable in view of our estimates (unpublished) of ν for a chaotic single particle with similar damping and driving strengths: there the maximum dimension is 2 but we generally find ν = 1.1 - 1.3.

15. P. S. Lomdahl, O. H. Soerensen and P. L. Christiansen, Phys. Rev. B25, 5737 (1982).

* CENTER FOR NONLINEAR STUDIES
AND THEORETICAL DIVISION
LOS ALAMOS NATIONAL LABORATORY
LOS ALAMOS, NM 87545, USA

† DEPARTMENT OF MATHEMATICS
HERIOT-WATT UNIVERSITY
RICCARTON, EDINBURGH
EH14 4AS, UK.

** BARTOL RESEARCH INSTITUTE
UNIVERSITY OF DELAWARE
NEWARK, DE 19711, USA.

†† LASSP, CLARK HALL
CORNELL UNIVERSITY
ITHACA, NY 14853, USA.

Lectures in Applied Mathematics
Volume 23, 1986

EXISTENCE, REGULARITY, AND DECAY OF VISCOUS SURFACE WAVES

J. Thomas Beale[1]

We will discuss the basic mathematical properties of the motion of a viscous fluid in a domain like an infinite ocean, bounded above by an atmosphere of constant pressure. Our emphasis is on the regularity and long-time behavior of motions near equilibrium. A great deal of progress has been made in the last several years in understanding the existence and nonexistence of solutions of nonlinear equations near a constant state; cf. the talks at this conference by Klainerman, Shatah, and Giga. The work described here can be thought of in this context. However, the approach used here for existence is somewhat more indirect than in other cases and relies on estimates for the problem transformed in time. The reasons are that the complicated interaction between the free surface and the fluid motion makes it difficult to obtain direct time-dependent estimates, and also that we have little information in general about the decay of the linear problem.

We assume that the fluid is bounded above by an atmosphere whose motion we neglect, and bounded below by a fixed bottom. We write spatial coordiates as (x,y) with $x = (x_1,x_2)$. The vertical direction is of course distinguished by gravity; for the

1980 Mathematics Subject Classification. 35Q10, 35R35, 76B15, 76D05.

[1]Supported by the National Science Foundation.

velocity field $u = (u_1, u_2, u_3)$ the vertical component will be the third. The bottom should have the form $y = -b(x)$ with $b(x) \to b_0 > 0$ as $x \to \infty$. The free upper surface $y = \eta(x,t)$ is an unknown which moves along with the fluid. We choose coordinates so that $\eta = 0$ at equilibrium. The equations of motion are the Navier-Stokes equations of incompressible flow with appropriate boundary conditions. At equilibrium the pressure is $P_0 - gy$, where P_0 is the atmospheric pressure and g is the acceleration of gravity. We write the usual pressure term $\bar{p}$ as

$$\bar{p}(x,y,t) = P_0 - gy + p(x,y,t).$$

The fluid then satisfies

(1) $$u_t + u \cdot \nabla u - \nu \Delta u + \nabla p = 0,$$

(2) $$\nabla \cdot u = 0.$$

On the bottom we have the usual condition

(3) $$u = 0 \qquad (y = -b).$$

There are two kinds of boundary conditions on the free surface $y = \eta$. One is the so-called kinematic boundary condition which states that the surface moves along with the fluid:

(4) $$\eta_t = u_3 - (\partial_1 \eta) u_1 - (\partial_2 \eta) u_2 \qquad (y = \eta).$$

It is obtained by differentiating the equation $y = \eta$ along the path of a fluid particle. The other condition is that the stress on the surface should be zero except for a normal stress due to surface tension. The stress tensor for the Navier-Stokes equation is

$$\sigma_{ij} = \bar{p}\delta_{ij} - \nu(u_{i,j} + u_{j,i}),$$

and the stress on a surface with normal n_j is $\sigma_{ij} n_j$, summed over j. With the pressure rewritten as above, our boundary condition is

$$(5) \quad p n_i - \nu(u_{i,j} + u_{j,i}) n_j = \{g\eta - \beta\nabla \cdot [(1 + |\nabla\eta|^2)^{-1/2}\nabla\eta]\} n_i \qquad (y = \eta).$$

The gravity term, which was removed from the fluid equations, has now reappeared in the surface condition. The normal stress due to surface tension is proportional to the mean curvature of the surface, with β a proportionality constant. (It may be helpful to compare (5) with the linearized version at equilibrium, (18)-(19) below.) Finally, to complete the problem, we suppose that an initial surface $y = \eta_0(x)$ is prescribed and an initial velocity field $u_0(x,y)$ in the domain $\{(x,y): -b(x) < y < \eta_0(x)\}$,

$$(6) \qquad \eta(0) = \eta_0, \qquad u(0) = u_0.$$

Equations (1)-(6) completely determine the motion of the fluid.

Our interest here is to determine to what extent this initial value problem is a good problem in the usual P.D.E. sense. We will summarize the results in general terms: assume that $\eta_0 \in H^{3+\varepsilon}(\mathbb{R}^2)$, some $\varepsilon > 0$, where H^s is the usual Sobolev space of functions with s derivatives in L^2, and $u_0 \in H^{5/2+\varepsilon}$ on $\{-b < y < \eta_0\}$. Assume η_0 and u_0 are sufficiently small in norm, and also impose three compatibility conditions on u_0: $\nabla \cdot u_0 = 0$, $u_0 = 0$ on the bottom, and the tangential stress at $y = \eta_0$ is zero. Then we have the following statements about existence, uniqueness, regularity, and decay:

(I) EXISTENCE. A solution of the initial value problem exists for all time in a moderate regularity class. In particular, $\eta(x,t)$ is C^1, and no kinks or cusps form on the surface.

(II) UNIQUENESS. The solution is unique in the prescribed class, and in fact unique on any finite time interval.

(III) REGULARITY. The solution becomes arbitrarily smooth for $t \geq T_0 > 0$.

(IV) DECAY. In the special case where the bottom is horizontal $(b = b_0)$, and assuming in addition that $\eta_0 \in L^1(\mathbb{R}^2)$, then as $t \to \infty$,

$$|\eta(\cdot,t)|_{L^2} = O(t^{-1/2}) ,$$

$$|\partial_x \eta(\cdot,t)|_{L^2} = O(t^{-1}) ,$$

$$|u(\cdot,t)|_{L^2} = O(t^{-1}) .$$

Parts (I) - (III) are proved in full detail in [1]; see p. 310 for a precise statement of results. Part (IV) is joint work with T. Nishida to appear. The existence statement depends on the inclusion of viscosity, but there are simple nonlinear perturbations of the heat equation for which smooth solutions do not persist even with small data; see below. Although (III) says that the solution becomes regular after an initial time interval, we cannot say that it is very smooth at time zero without imposing further compatibility conditions on the initial data. The decay rates (IV) are best possible in the sense that they cannot be improved for the norms chosen.

Before discussing the surface wave problem further we will illustrate the approach used here in the context of a prototype problem which shows clearly a distinction between existence and nonexistence. We consider the initial value problem

$$\text{(7)} \qquad \begin{aligned} u_t - \Delta u &= u^2 , \\ u(0) &= u_0 , \end{aligned}$$

in $\mathbb{R}^n$ and ask the following question:

> If u is smooth and small in some appropriate norm, will a smooth solution persist for all later time?

If $\mathbb{R}^n$ were replaced by a bounded domain with Dirichlet boundary condition, the answer would be easy. The spectrum of $-\Delta$ is bounded away from zero, so that solutions of the linear heat equation decay exponentially, and it is not hard to argue that solutions of the nonlinear problem remain smooth and decay in the same way. The case of free space is more delicate. The answer to our question is "yes" if $n \geq 3$ but "no" if $n = 1$ or 2. In fact, in the latter case, if $u_0 \in L^p$, $p < \infty$, and $u_0 \geq 0$, the solution must "blow up" in L^p in finite time. For $n = 1$ this was shown long ago by Fujita; $n = 2$ is harder. See [8] for a recent summary. For $n \geq 3$, the existence can be shown in a straightforward manner in the spirit of [2,4,5]. For the linear heat equation we know rates of decay for the solution; e.g.,

$$|u(t)|_{L^2} \leq Ct^{-n/4}|u_0|_{L^1}, \qquad t \geq 1 .$$

We can use this to estimate the growth of the nonlinear solution, treating u^2 as an inhomogeneous term, finally ending up with an inequality which shows that some H^s norm remains bounded if u_0 was small. The dependence on the space dimension comes of course from the fact that the linear solutions decay faster in higher dimensions.

We will outline a deliberately more indirect existence argument for (7) which is closer to the method of [1] and which does not explicitly use a rate of decay. We will obtain the solution from a contraction mapping on the entire time interval $0 < t < \infty$. Assume $u_0 \in H^r \cap L^1$, r large. First, it is easy to show that a solution $u_1(t)$ exists in $H^r \cap L^1$ say for $0 < t < 2$, for u_0 sufficiently small in norm. Let $\rho(t)$ be a cut-off function with $\rho = 1$ for $t \leq 1$, $\rho = 0$ for $t \geq 2$. Write the presumed solution $u(t)$, $0 < t < \infty$, in the form $u = \rho u_1 + v$, so that $v = (1 - \rho)u_1$ for $t \leq 2$. Then v should satisfy

$$\text{(8)} \qquad v_t - \Delta v = v^2 + \rho u_1 v + h, \quad v(0) = 0,$$

where

$$h = -\rho_t u_1 + \rho(1 - \rho)u_1^2 .$$

We have adjusted the problem so that v should vanish near $t = 0$. This will simplify the treatment of the problem by transforms.

It is natural to set up the problem (8) so that the solution is a fixed point of the mapping

$$v \to L^{-1}(v^2 + 2\rho u_1 v + h) \tag{9}$$

where L is the heat operator. We have to decide what choice of norms to use to obtain a contraction mapping. It is in this choice that the difference in space dimension will appear. We concentrate on the linear problem $Lv = f$. Suppose $f \in Y$ for some space Y: we want $v \in X$, where X should be such that

$$L^{-1}: Y \to X. \tag{10}$$

We also have to multiply to form v^2, and multiplication of two elements of X should produce an element of Y:

$$X \cdot X \subseteq Y. \tag{11}$$

These two conditions are the essential features necessary to make the argument work.

To relate v and f we apply a Fourier transform in space-time to the equation $Lv = f$, assuming f vanishes to high order at $t = 0$. Then

$$\hat{v}(\xi,\tau) = \hat{f}(\xi,\tau)/(i\tau + |\xi|^2).$$

For (ξ,τ) away from 0, say $|\xi| + |\tau| \geq \delta$, we can say

$$(1 + |\xi|^2 + |\tau|)|\hat{v}| \leq C_\delta |\hat{f}| \; . \tag{12}$$

This estimate represents a gain of two space derivatives and one time derivative, as we expect for the heat operator. It suggests that we use a Sobolev space K^r of functions $w \in L^2$ of space-time so that $D_x^r w \in L^2$, $D_t^{r/2} w \in L^2$, or the usual generalization if $r/2$ is not an integer. We will write K_0^r for the subspace of functions which vanish for $t \leq 0$. If $\hat{f}$ were zero in the low frequencies, we would have from (12) for $f \in K_0^{r-2}$

$$|v|_{K^r} \leq C|f|_{K^{r-2}} \cdot$$

This determines the gain in smoothness from Y to X, but the low frequencies will determine the long-time behavior. Of course (12) fails as $\xi,\tau \to 0$, but we can at least say that

$$|\hat{v}| \leq |\hat{f}|/(|\tau| + |\xi|^2). \tag{13}$$

Thus $\hat{v}$ will be rougher in the low frequencies than $\hat{f}$. With this in mind, we choose Y to be a subspace of K_0^{r-2} with better behavior in the low frequencies. We take

$$Y = \{f \in K_0^{r-2}: \hat{f} \in L^\infty \text{ on } |\xi| + |\tau| < \delta\}.$$

Next we check that $(|\tau| + |\xi|^2)^{-1} \in L^p$ for $p < (n/2) + 1$. To satisfy both properties (10), (11) it seems critical that we can choose $p = 2$; this is possible for $n \geq 3$ but barely misses for $n = 2$.

We will assume now that $n \geq 3$ and take $p = 2$. From (13) we see that for $f \in Y$, $\hat{v}$ is L^2 in the low frequencies. Combining this with our earlier observation we see that $f \in Y$ implies $v = L^{-1}f \in K^r$. Thus property (10) is satisfied with $X = K_0^r$. (That $v = 0$ for $t \leq 0$ follows from the Paley-Wiener Theorem.) We still need to check (11). For r large enough multiplication preserves K^r as usual for Sobolev spaces. On the other hand, if $v,w \in K^r$, then in particular $v,w \in L^2$, and $v \cdot w \in L^1$. Thus $(v \cdot w)\hat{} \in L^\infty$, so that $vw \in Y$.

It is easily seen that the initial part of the solution ρu_1 is in $K^r \cap L^1$. The solution v of (8) can now be found using the Contraction Mapping Principle in a standard way. The reader may notice that the discussion above could be simplified by taking $Y = K_0^{r-2} \cap L^1$; we have avoided doing this since we have in mind situations where the symbol of the linear operator will behave differently in the high and low frequencies.

We next discuss the behavior of the linear surface wave problem over a horizontal bottom. Without viscosity it is customary to assume the motion is irrotational, since this condition is preserved in time. In this case a surface height of definite wave number determines a traveling wave solution of definite frequency. The speed varies with the wave number, and this is perhaps the most familiar example of dispersive wave motion. The role of viscosity is often described in terms of the effect on these irrotational waves as $\nu \to 0$. Here, however, we need to consider the effect on all wave numbers simultaneously for fixed viscosity. For a domain with a solid boundary above as well as below, viscous flow would decay exponentially in time, as it would for a bounded domain. The reason is that the linear Stokes operator has spectrum away from zero, because of the fact that the domain is finite in one direction. However, the linear surface wave problem has slow decay, as indicated above, as a result of the interaction between free surface and fluid.

To illustrate the decay we consider the analogue for the viscous case of the inviscid traveling waves. If we assume $\eta(x,t) = e^{\lambda t} e^{i\xi x}$ for fixed wave vector ξ and replace $u(x,y,t)$ by $e^{\lambda t} e^{i\xi x} u(y)$ the linear equations reduce to a system of ordinary differential equations in y which constitute an eigenvalue problem with eigenvalue λ. For each ξ there is a sequence of λ's, but as $\xi \to 0$ there is a unique one of slowest decay. This eigenvalue $\lambda(\xi)$ and the corresponding eigenvector can be found by a series expansion about $\xi = 0$ which is not hard to justify. We find that

$$\lambda(\xi) \sim -\frac{gb^3}{3\nu}\xi^2 \quad (14)$$

These slowly decaying modes determine the rate of decay described in (IV) above. In fact, the eigenvectors have velocity component tending to zero as $\xi \to 0$, and this explains why the rate of decay is faster for u than for η. The decay is roughly as though η satisfied a two-dimensional heat equation. For more discussion of the behavior of linear waves, see [3,6,7,9].

The most important part of the problem of existence for the full surface wave problem is to obtain estimates for the linear problem in norms compatible with the nonlinear terms, in analogy with the nonlinear heat equation above. The equations linearized about equilibrium are

(15) $$\eta_t = u_3|_{y=0} \equiv Ru$$

(16) $$u_t - \nu\Delta u + \nabla p = 0$$

(17) $$\nabla \cdot u = 0$$

(18) $$u_{i,3} + u_{3,i} = 0, \quad i = 1,2 \qquad (y = 0)$$

(19) $$p - 2\nu u_{3,3} - (g\eta - \beta\Delta_x\eta) = 0 \qquad (y = 0)$$

(20) $$u = 0 \qquad (y = -b)$$

(21) $$\eta(0) = \eta_0, \quad u(0) = u_0 \ .$$

(Here the bottom may again be arbitrary.) We can put these equations in a more recognizable form by applying a projection onto divergence-free vector fields. For a fluid in a fixed domain this projection eliminates the pressure term. In this case the projection differs from the usual case because of the difference in boundary conditions, and the pressure is not eliminated but reduced to a functional of the velocity and surface. The resulting set of equations has the form

(22) $$\begin{pmatrix} \eta \\ u \end{pmatrix}_t = \begin{pmatrix} 0 & R \\ -R^*S & -A \end{pmatrix} \begin{pmatrix} \eta \\ u \end{pmatrix} .$$

Here R is an operator from vector fields in the fluid domain to functions on the surface, defined by (15) and R^* is its formal adjoint. The operator A is the projection of the Stokes operator; it is self-adjoint and positive definite. The operator S on the top surface is $g - \beta\Delta_x$. The entry $-R^*S$ describes the effect of the surface on the fluid through gravity and surface tension. If the surface were ignored, the equation $u_t + Au = 0$

would behave like a heat equation. On the other hand, if the viscosity were zero, the system would be energy-preserving. In the latter case a single equation can be written for one unknown, second-order in time:

$$u_{tt} + R^*SRu = 0.$$

The operator $B \equiv R^*SR$ is self-adjoint and nonnegative, so that this equation is a "wave" equation. Neither A nor B dominates the other, and the system is mixed in character.

We will now describe the estimates for the resolvent equation

$$\lambda \begin{pmatrix} \hat{\eta} \\ \hat{u} \end{pmatrix} = \begin{pmatrix} 0 & R \\ -R^*S & -A \end{pmatrix} \begin{pmatrix} \hat{\eta} \\ \hat{u} \end{pmatrix} + \begin{pmatrix} 0 \\ \hat{f} \end{pmatrix} . \tag{23}$$

Here $\hat{}$ denotes the Laplace transform in t, so that λ here corresponds to $i\tau$ before. Note that we have not included an inhomogeneous term in the first equation, since it is not necessary for the eventual solution of the nonlinear problem; estimates for the full resolvent would not be as sharp as the ones discussed here. The first equation above is $\lambda\hat{\eta} = R\hat{u}$, and we may substitute this in the second to obtain

$$\lambda\hat{u} + A\hat{u} + \lambda^{-1}B\hat{u} = \hat{f} . \tag{24}$$

Since A and B are both nonnegative, we can estimate for $\text{Re}\,\lambda > 0$ by choosing test functions appropriate for A and then for B in succession. In this way we find for $\hat{u}$

$$|\hat{u}|_r + |\lambda|^{r/2}|\hat{u}|_0 \leq C(|\hat{f}|_{r-2} + |\lambda|^{(r-2)/2}|\hat{f}|_0) \tag{25}$$

for $\text{Re}\,\lambda > 0$. This estimate is a consequence of the positive definiteness of A. It has the expected gain of smoothness of $\hat{u}$ with respect to $\hat{f}$; i.e., two spatial derivatives and one temporal. Having established (25), we can use the positivity of B to estimate $\hat{\eta} = \lambda^{-1}R\hat{u}$. We find for $\text{Re}\,\lambda > 0$ but λ away from zero,

$$|\hat{\eta}|_{r+1/2} + |\lambda|^{(r+1/2)/2}|\hat{\eta}|_0 \leq C\|\hat{f}\| , \tag{26}$$

where $\|\hat{f}\|$ is the right-hand side of (25). The extra smoothness in $\hat{\eta}$ results from the surface tension, which is the highest order part of B. As $\lambda \to 0$, this estimate deteriorates, but we can show that

$$(27) \qquad |\nabla\hat{\eta}|_{r-1/2} + |\lambda|^{1/2}|\hat{\eta}|_0 \leq C\|\hat{f}\| .$$

These three estimates are given in Theorem 1 of [1], p. 318.

It is natural to take $f \in K_0^{r-2}$. In view of (25), we can let λ approach the imaginary axis and find that $u \in K_0^r$. The situation for η is less standard, however, and is analogous to the nonlinear heat equation above. Let $\hat{\hat{\eta}}$ be the space-time transform of η. If, say, $\hat{\hat{\eta}}$ vanished in the low frequencies, we would have $\eta \in K^{r+1/2}$ by (26) and (27). In fact, it is a consequence of (27) that for $\lambda = i\tau$, τ real, and (ξ,τ) small, $\hat{\hat{\eta}}$ has the form, an L^2 function multiplied by the weight $(|\xi|^2 + |\tau|)^{-1/2}$. This weight is L^2, so that $\hat{\hat{\eta}}$ is at least L^1 for small (ξ,τ). It follows that η is in the extended space $\tilde{K}^{r+1/2}$ whose elements are in $K^{r+1/2}$ except that the low frequency part is required only to be L^1. Indeed, such an extension is necessary: the decay estimate (IV) for the case of horizontal bottom shows that η cannot be an L^2 function of space-time.

The choice of spaces used for the time-dependent linear estimates in [1] is motivated by the considerations just discussed. The full problem is rewritten on the equilibrium domain and the solution is obtained through an iteration based on the linearization. The argument follows the outline of the nonlinear heat example above. It is important in estimating the nonlinear terms that the product of a function in $\tilde{K}^s$ with a function in K^s is in K^s for s large, in analogy with (11).

The argument for the rate of decay with a horizontal bottom falls naturally into two parts: we first establish the decay rate for the linearization and then carry it over to the full problem. The first part is done in an essentially classical way. We extend the resolvent estimates slightly into the left half of the plane, write the solution as an inverse Laplace transform, and deform the contour to the left. As we move the contour we pick up residues corresponding to the slow modes of low wave number mentioned above. Their behavior leads to the stated rates. For the nonlinear problem we write the equations in the form

$$v_t + G_v = F(v)$$

where $v = (\eta,u)^T$ and G is the matrix in (22). Estimates are based on the equivalent form

$$v(t) = e^{Gt} v_0 + \int_0^t e^{G(t-s)} F(v(s))ds$$

as has been done in free space problems but here we can take advantage of the fact that regularity has already been established.

BIBLIOGRAPHY

1. J. Thomas Beale, "Large-time regularity of viscous surface waves", Arch. Rational Mech. Anal. 84 (1984), 307-52.

2. S. Klainerman and G. Ponce, "Global, small amplitude solutions to nonlinear evolution equations", Comm. Pure Appl. Math. 36 (1983), 133-41.

3. J. Lighthill, Waves in fluids, Cambridge University Press, Cambridge, 1978.

4. A. Matsumura and T. Nishida,"The initial value problem for the equations of motion of compressible viscous and heat-conductive fluids", Proc. Japan Acad. 55, Ser. A (1979), 337-42.

5. J. L. Shatah, "Global existence of small solutions to nonlinear evolution equations", preprint.

6. J. J. Stoker, Water Waves, Interscience, New York, 1957.

7. J. V. Wehausen and E. V. Laitone, Surface waves, Handbuch der Physik IX, 446-778, Springer-Verlag, Berlin, 1960.

8. F. B. Weissler, "Existence and nonexistence of global solutions for a semilinear heat equation", Israel J. Math. 38 (1981), 29-40.

9. G. B. Whitham, Linear and Nonlinear Waves, Wiley, New York, 1974.

DEPARTMENT OF MATHEMATICS
DUKE UNIVERSITY
DURHAM, N. C. 27706

Lectures in Applied Mathematics
Volume 23, 1986

ATTRACTORS FOR THE KURAMOTO-SIVASHINSKY EQUATIONS

B. Nicolaenko[1], B. Scheurer[2], R. Temam[3]

ABSTRACT. We give sharper upper bounds for the Hausdorff and fractal dimensions of attractors of the Kuramoto-Sivashinsky equation. The latter models various instable cellular patterns. Our main improvement results from extending a remarkable Sobolev type inequality obtained by Lieb and Thirring in the context of Schrödinger operators, using the Birman-Schwinger principle.

0. INTRODUCTION

In this paper, we address the question of constructing an upper bound of the Hausdorff and fractal dimensions $d_H(X)$ and $d_F(X)$ for attractors X of the Kuramoto-Sivashinsky equation. We investigate the large time behavior of the solution $u = u(x,t)$ of:

$$\frac{\partial u}{\partial t} + \nu\Delta^2 u + \Delta u + \frac{1}{2}|\nabla u|^2 = 0 \quad , \quad (x,t) \in R^n \times R_+ \tag{0.1}$$

$$u(x,0) = u_0(x) \qquad\qquad x \in R \tag{0.2}$$

$$u(x+Le_i,t) = u(x,t) \qquad 1 \leq i \leq n \quad , \qquad (0.3)$$

where $\nu>0$ and u_0 is L-periodic. This equation occurs in a large variety of physical situations; it models the formation of cellular patterns whose temporal behavior becomes chaotic, when the typical size L of the cell is large enough. The natural bifurcation parameter of the problem is the adimensional number $\tilde{L} = \frac{L}{2\pi\sqrt{\nu}}$. Here we just refer to the papers by Kuramoto [5-7] and Sivashinsky [1,15-17]; a discussion of these physical situations is given in the introduction of [11]. In the former papers [10-11], we considered solutions of (0.1)-(0.2) with either periodic or Neumann boundary conditions. We obtained general upper bounds for $d_H(X)$, $d_F(X)$ in terms of

$$R \equiv \overline{\lim_{t\to\infty}} \, ||\nabla u(t)||$$

and

$$Y \equiv \overline{\lim_{t\to\infty}} \, \frac{1}{t} \int_0^t ||D^3 u(s)||^2 ds \quad , \quad \text{if } n = 1$$

$$(\text{resp. } Y \equiv \overline{\lim_{t\to\infty}} \, \frac{1}{t} \int_0^t ||D^4 u(s)||^2 ds \text{ if } n= 2,3) \quad .$$

In one dimension, for even solutions of (0.1), (0.2) (including Neumann conditions), we proved the uniform boundedness of orbits. As a consequence:

$$R \leq \text{const } \nu^{-1/4} \, \tilde{L}^{5/2} \qquad (0.4)$$

$$Y \leq \text{const } \nu^{-5/2} \, \tilde{L}^{5} \qquad (0.5)$$

from which we obtained the explicit upper bound

$$d_H(X) \leq d_F(X) \leq 1 + \text{const } \tilde{L}^{13/8} \quad . \tag{0.6}$$

Here, we significantly sharpen the upper bound (0.6) in the one dimensional context. In [11], the main step in the derivation of the (0.6) was a classical lower estimate on the trace of some operator associated to linearized flow of (0.1), (0.2). Our main improvement results from extending a remarkable Sobolev type inequality obtained by Lieb and Thirring (see [9]) in the context of Schrödinger operators. Our extension takes the following form:

$$\sum_{j=1}^{m} \int |\Delta\phi_j|^2 dx > \text{const} \int (\Sigma|\phi_j(x)|^2)^{\alpha_n} dx \tag{0.7}$$

where the $\{\phi_j\}_{j=1}^m$ are any set of m smooth functions, orthonormal in L^2 and L-periodic; $\alpha_n = 1 + \frac{4}{n}$ with $n \leq 3$. The constant depends only upon n. Using (0.7), (0.6) now becomes:

$$d_H(X) \leq d_F(X) \leq \text{const}(1 + \tilde{L}^{\frac{3}{2}}). \tag{0.8}$$

This paper is organized as follows. In the first part, we review the problem and recall some relevant facts in estimating $d_H(X), d_F(X)$. Then, we show how (0.7) is related to bounds for eigenvalues of $\Delta^2 + q(x)$, where $q(x)$ is a smooth negative function. In the second part, using the Birman-Schwinger principle (see [9]), we demonstrate the extended Lieb-Thirring inequality (0.7). The methods derived in [10-11] and sharpened here are in fact generic to a whole class of equations of the type:

$$\frac{\partial u}{\partial t} + P(\frac{\partial}{\partial x})\ u + f\ (\frac{\partial u}{\partial x}) = 0 \tag{0.9}$$

where $P(\frac{\partial}{\partial x})$ is any pseudo-differential operator with even symbol, of order ≥ 2; and

$$f(s) = \frac{s^2}{2} \text{ or } \sqrt{1+s^2} - 1$$

(The latter nonlinearity is a natural one to describe curvature effect in wrinkled flames [15]). Details will appear elsewhere [12].

1. HAUSDORFF DIMENSION OF AN ATTRACTOR

1.1 Review of the Problem

We consider the Kuramoto-Sivashinsky equation, in space dimension $n \leq 3$. That is to find $u = u(x,t)$ solution of:

$$\frac{\partial u}{\partial t} + \nu\ \Delta^2 u + \Delta u + \tfrac{1}{2}|\nabla u|^2 = 0 \qquad (x,t) \in R^n \times R_+ \tag{1.1}$$

$$u(x,0) = u_0(x) \qquad x \in R^n \tag{1.2}$$

$$u(x + Le_i, t) = u(x,t) \qquad 1 \leq i \leq n\ , \tag{1.3}$$

where $\nu > 0$ and u_0 is L-periodic. We address the question of an upper bound of the Hausdorff and fractal dimensions of any attractor of (1.1) -(1.3). First, we recall the general setting of [2,3,4], applied to the Kuramoto-Sivashinsky equation [10,11].

In what follows, we look at any solution of (1.1)-(1.3) such that:

$$||\nabla u(t)|| \leq k < +\infty \quad , \text{ for } t > 0 \quad . \tag{1.4}$$

Such a bound has been proved for even solutions of (1.1)-(1.3), when n = 1 (see [10-11]). We define (see [11]) $\dot{H} = H/R = L^2/R$ the quotient space of $H = L^2$ by R equipped with the natural norm. For any element w of H, $\dot{w}$ will be the corresponding coset in $\dot{H}$. If $S(t):u_0 \to u(.,t)$ is the mapping defined by (1.1)-(1.3), we thus get an associated mapping $\dot{S}(t)$ in $\dot{H}$. The corresponding limit set:

$$X = \bigcap_{t>0} c\ell \, \{\dot{u}(s) \quad , s \geq t\} \tag{1.5}$$

has been shown ([3]) to be a <u>functional</u> <u>invariant</u> <u>set</u> for the dynamical system (1.1), in the terminology of [2,3,4]. In particular:

$$\lim_{t\to\infty} \text{dist } (\dot{u}(t),X) \equiv \lim_{t\to\infty} \inf_{\dot{w}\in X} ||\dot{u}(t) - \dot{w}|| = 0 \quad , \tag{1.6}$$

for any solution satisfying (1.4). The "differential" $D\dot{S}(t,u_0)$ of $\dot{S}(t)$ around a given solution $u = \dot{S}(t)u_0$ of (1.1)-(1.3) is obtained by introducing the linearized evolution equation:

$$\frac{\partial U}{\partial t} + \nu\Delta^2 U + \Delta U + \nabla u.\nabla U = 0 \tag{1.7}$$

$$U(x,0) = \xi(x) \tag{1.8}$$

$$U(x + Le_i,t) = U(x,t) \qquad 1 \leq i \leq n \quad . \tag{1.9}$$

Precisely, $D\dot{S}(t,u_0)$: $\xi \to U(t)$ where $U(t)$ is the value at time t of the solution to (1.7)-(1.8)(1); this is a linear compact operator in $\dot{H}$. Now for $m \in N$ given, to $\xi_1, \ldots, \xi_m$ given in $\dot{H}$ it corresponds $U_1(t), \ldots, U_m(t)$ solutions of (1.7)-(1.8). Then a direct calculation shows that:(2)

$$\frac{d}{dt} \log \|\bigwedge_{j=1}^{m} U_j(t)\| + \text{Trace } (A(u) \circ P_m) = 0 \quad , \tag{1.10}$$

for $u = \dot{S}(t)u_0$ and where by definition:

$$A(u) \equiv \nu\Delta^2 + \Delta + \nabla u.\nabla \tag{1.11}$$

$$P_m \equiv \text{orthogonal projection of } L^2 \text{ onto span } \{U_1(t),\ldots,U_m(t)\}.$$

Note that the projector P_m depends smoothly on $\xi_1, \ldots, \xi_m$, u, t and m. The relation (1.10) is equivalent to:

$$\|\bigwedge_{j=1}^{m} U_j(t)\| = \|\bigwedge_{j=1}^{m} \xi_j\| \exp -\int_0^t \text{Trace } (A(u)_\circ P_m) ds. \tag{1.12}$$

Now, we introduce the norm of $\Lambda^m D\dot{S}(t,u_0)$ in $\Lambda^m \dot{H}$:

$$\omega_m(D\dot{S}(t,u_0)) \equiv \sup \{\|\bigwedge_{j=1}^{m} U_j(t)\| : \xi_j \in \dot{H} \ , \|\xi_j\| = 1 \ , \ 1 \le j \le m\} \ . \tag{1.13}$$

---

(1) The solution U exists for $0 \le t \le t_1(u)$.

(2) Hereafter $\| \ \|$ is the natural norm on $\Lambda^m\dot{H} = \underbrace{\dot{H}\Lambda.\ldots.\Lambda\dot{H}}_{m}$, where Λ denotes the exterior product.

This norm measures the volume variation of a "m-dimensional cube" $\bigwedge_{j=1}^{m} \xi_j$ convected along the orbit $\dot{S}(t)u_0$. From (1.12), (1.13), for u_0 running over X (defined by (1.5)), we then obtain easily (P is any projector of rank m):

$$\bar{\omega}_m(t) \equiv \sup_{u_0 \in X} \omega_m(D\dot{S}(t,u_0))$$

$$\leq \sup_{u_0 \in X} \exp\left(-\int_0^t \operatorname{Inf}_P \operatorname{Trace}(A(u)_\circ P)ds\right) \quad . \qquad (1.14)$$

A general result of [3,4] relates the fractal and Hausdorff dimensions of X, $d_F(X)$ and $d_H(X)$, to the first integer m_o such that $\bar{\omega}_{m_0}(t) < 1$ (for t large) i.e., $\dot{S}(t)$ contracts all the m_0-dimensional cubes at every point u_0 of X. Precisely, if there exists $m_0 \in N$ for which:

$$\overline{\lim_{t\to+\infty}} \frac{1}{t} \log \bar{\omega}_{m_0}(t) < 0 \qquad (1.15)$$

then the following upper bound holds:

$$d_H(X) \leq m_0 + 1 \ , \quad d_F(X) \leq c_1(m_0 + 1) \qquad (1.16)$$

where c_1 is an appropriate constant. We are thus led to find the first $m_0 \in N$ such that:

$$\underline{\lim}_{t\to\infty} \{\operatorname{Inf}_{u_0 \in X} \operatorname{Inf}_P \frac{1}{t}\int_0^t \operatorname{Trace}(A(u)_\circ P)\, ds\} > 0 \quad . \qquad (1.17)$$

As a consequence, to find lower bounds for the expression $\frac{1}{t}\int_0^t \text{Trace } (A(u)_\circ P)\, ds$ will be of crucial importance. This is achieved in the next section.

1.2 The Main Result

We state and prove the main result of this paper; the proof relies on a Sobolev inequality for the operator $\underset{j=1}{\overset{m}{\wedge}} \Delta^2$. As a consequence, we compute $m_0 \in N$ such that (1.15) holds. Recall that X is defined by (1.5), A(u) by (1.11).

Theorem 1.1. Let u be any solution of (1.1)-(1.3) such that (1.4) holds. Denote by P any projector of rank m. Then:

$$\underline{\lim}_{t\to\infty} \{\operatorname{Inf}_{u_0 \in X} \operatorname{Inf}_{P} \frac{1}{t}\int_0^t \text{Trace } (A(u)_\circ P)\, ds\}$$

$$\geq m(L)^{-\frac{n}{\alpha'_n}} \{c_0(n) \frac{\nu}{2} m^{\alpha_n - 1} (L)^{\beta_n}$$

$$- \frac{1}{2} \overline{\lim_{t\to\infty}} \frac{1}{t}\int_0^t ||\Delta u + \frac{1}{\nu}||_{\alpha'_n} ds\} \quad , \qquad (1.18)$$

where $\alpha_n = 1 + \frac{4}{n}$, $\alpha'_n = \frac{\alpha_n}{\alpha_n - 1}$, $\beta = \frac{(\alpha_{n-1})^2}{\alpha_n}$.

Proof. Let us introduce $\{\phi_j\}_{j\in N}$ an orthonormal basis of smooth functions of $\dot{H}$ such that $\{\phi_1, \ldots, \phi_m\}$ is an orthonormal basis of $P\dot{H}$. By definition:

$$\text{Trace } (A(u)_\circ P) = \text{Trace } ((\nu\Delta^2 + \Delta + \nabla u.\nabla)_\circ P)$$

$$= \sum_{j=1}^{m} \nu||\Delta\phi_j||^2 - ||\nabla\phi_j||^2 + \int \nabla u.\nabla\phi_j \; \phi_j \; dx$$

$$\geq \frac{\nu}{2} \sum_{j=1}^{m} ||\Delta\phi_j||^2 - \tfrac{1}{2} \int (\Delta u + \frac{1}{\nu}) \sum_{j=1}^{m} |\phi_j|^2 \; dx,$$

where we used a classical interpolation inequality. We now claim that there exists a constant $c_0(n)$, independent of m and L such that:

$$\sum_{j=1}^{m} ||\Delta\phi_j||^2 \geq c_0(n) \int (\sum_{j=1}^{m} |\phi_j|^2)^{\alpha_n} dx \quad , \; \alpha_n = 1+\frac{4}{n} \; . \tag{1.20}$$

This key inequality will be discussed and proved in the next sections. From (1.19), (1.20), we get, using Hölder inequality:(1)

$$\text{Trace } (A(u)_\circ P) \geq \{c_0(n) \frac{\nu}{2} ||\rho||_{\alpha_n}^{\alpha_n - 1} - \frac{1}{2} ||\Delta u + \frac{1}{\nu}||_{\alpha'_n}\} \; ||\rho||_{\alpha_n} \tag{1.21}$$

where $\alpha'_n = \frac{\alpha_n}{\alpha - 1}$ and $\rho(x) \equiv \sum_{j=1}^{m} |\phi_j(x)|^2$. Since the basis

(1) $|| \; ||_\alpha$ is the norm in L^α.

$\{\phi_1, \ldots, \phi_m\}$ is orthonormal, we have $m \le (L)^{\frac{n}{\alpha'_n}} ||\rho||_{\alpha_n}$; therefore after taking the time average of (1.21) we obtain:

$$\frac{1}{t}\int_0^t \text{Trace } (A(u)_\circ P)\, ds \ge m(L)^{-\frac{n}{\alpha'_n}} \{c_0(n)\, \frac{\nu}{2}\, m^{\alpha_n - 1} (L)^{\frac{-(\alpha_n-1)^2}{\alpha_n}}$$

$$- \frac{1}{2}\frac{1}{t}\int_0^t ||\Delta u + \frac{1}{\nu}||_{\alpha'_n} ds\} \quad . \tag{1.22}$$

The conclusion of the theorem is immediate from (1.22). □

It is now an easy matter to find the smallest $m_0 \in N$ such that the right hand side of (1.21) is strictly positive. As in [10-11], we will restrict ourselves to the case where we proved assumption (1.4) (see Corollary 2.2 of [11]).

Corollary 1.2. Let X (defined by (1.5)) be the functional invariant set associated to (1.1)-(1.3) where n = 1 and u is even. Then, the Hausdorff and fractal dimensions $d_H(X)$ and $d_F(X)$ are finite. Precisely:

$$d_H(X) \le m_0 + 1\,,\quad d_F\,(X) \le c_1(m_0 + 1) \tag{1.23}$$

where c_1 is an absolute constant and m_0 is defined by

$$m_0 = c_0(1)^{-\frac{1}{4}}(2\pi)^{\frac{4}{5}}\,\nu^{\frac{3}{20}}\,\tilde{L}^{\frac{4}{5}}\left((2\pi\,\sqrt{\nu}\,\tilde{L})^{\frac{3}{10}}\,R^{\frac{1}{2}}Y^{\frac{1}{4}} + (2\pi)^{\frac{4}{5}}\,\nu^{-\frac{3}{5}}\tilde{L}^{\frac{4}{5}}\right)^{\frac{1}{4}} \tag{1.24}$$

for $\tilde{L} = \dfrac{L}{2\pi\sqrt{\nu}}$ and:

$$R \equiv \overline{\lim_{t\to\infty}} \, \|Du(t)\| \quad , \; Y \equiv \overline{\lim_{t\to\infty}} \frac{1}{t}\int_0^t \|D^3u(s)\|^2 \, ds \quad . \tag{1.25}$$

Proof. We use (1.18) in the case where $n = 1$, i.e., $\alpha_1 = 5$, $\alpha_1' = 5/4$. The right hand of (1.18) will be strictly positive as soon as:

$$m > c_0(1)^{-\frac{1}{4}} (2\pi)^{\frac{4}{5}} \nu^{\frac{3}{20}} (\tilde{L})^{\frac{4}{5}} \left(\overline{\lim_{t\to\infty}} \frac{1}{t}\int_0^t \|D^2u + \frac{1}{\nu}\|_{\frac{5}{4}} \, ds\right)^{\frac{1}{4}} . \tag{1.26}$$

The following inequalities are easy consequences of Hölder inequality and standard interpolation:

$$\frac{1}{t}\int_0^t \|D^2u\|_{\frac{5}{4}} \, ds \le L^{\frac{3}{10}} \left(\frac{1}{t}\int_0^t \|Du\| \, ds\right)^{\frac{1}{2}} \left(\frac{1}{t}\int_0^t \|D^3u\| ds\right)^{\frac{1}{2}}$$

$$\le L^{\frac{3}{10}} \left(\sup_{0<s<t} \|Du(s)\|\right)^{\frac{1}{2}} \left(\frac{1}{t}\int_0^t \|D^3u\|^2 \, ds\right)^{\frac{1}{4}} \tag{1.27}$$

Therefore, with the definitions (1.25):

$$\overline{\lim_{t\to\infty}} \frac{1}{t}\int_0^t \|D^2u + \frac{1}{\nu}\|_{\frac{5}{4}} \, ds \le L^{\frac{3}{10}} R^{\frac{1}{2}} Y^{\frac{1}{4}} + \nu^{-1} L^{\frac{4}{5}} \tag{1.28}$$

and (1.26) is a fortiori satisfied if $m > m_0$ where m_0 is defined by (1.24). The proof is complete. □

From [11], we know upper bounds, in terms of ν and $\tilde{L}$, for R and Y. (See (2.18), (2.19) in [11].) A direct computation gives

Corollary 1.3. Under the above assumptions, if $\tilde{L} \geq \tilde{L}_0 \simeq 2.44$.

$$m_0 = c_1 \tilde{L}^{\frac{3}{2}} + c_2 \tilde{L} \tag{1.29}$$

where c_1, c_2 are absolute constants.

Remark 1.4. The reason for the condition $\tilde{L} \geq \tilde{L}_0$ has been discussed in [11]. □

1.3 Generalization of a Sobolev inequality of Lieb and Thirring

In this section we prove the inequality (1.20) used during the proof of the Theorem 1.1. The main step in the proof consists of showing, using orthonomality of the ϕ_j's, the equivalence of (1.20) with a bound for the moments of certain eigenvalues. Precisely, these are the eigenvalues of the generalized Schrödinger operator $\Delta^2 + q(x)$, for an appropriate potential q. Here we make precise this equivalence; the bound for moments is discussed in the next paragraph.

Theorem 1.5. Let $\{\phi_1, \ldots, \phi_m\}$ be any set of m (nonconstants) smooth functions, which are orthonomals in L^2 and L-periodic. Then, there exists a constant $c_0(n)$ depending on the space dimension n but not on m such that:

$$\sum_{j=1}^{m} ||\Delta\phi_j||^2 \geq c_0(n) \int \{ \sum_{j=1}^{m} |\phi_j(x)|^2 \}^{\alpha_n} dx \tag{1.30}$$

where $\alpha_n = 1 + \frac{4}{n}$ and $n \leq 3$.

Proof. Let us introduce the operator $A \equiv \Delta^2 + q(x)$ defined on the n-dimensional torus $M \equiv [-\frac{L}{2},\frac{L}{2}]^n$; we will assume $q(x) < 0$ and smooth for simplicity although it is not necessary. The operator A, unbounded in $H = L^2(M)$, admits $H^4_{per}(M)$ for domain and is self adjoint; its spectrum is discrete and denumerable. Let us denote by λ_j, $j \geq 1$, its eigenvalues counted with multiplicities. Finally, we set by definition:

$$N_\alpha(q) = \sum_{\lambda_j<\alpha} 1 \quad . \tag{1.31}$$

The following sequence of inequalities is then straightforward:

$$\begin{aligned}\sum_{j=1}^{m} \|\Delta\phi_j\|^2 + \int q(x) \sum_{j=1}^{m} |\phi_j(x)|^2\, dx &\geq \sum_{j=1}^{m} \lambda j \\ &\geq -\sum_{j=1}^{m} |\lambda_j| \\ &\geq -\sum_{\lambda_j<0} |\lambda_j| \\ &= -\int_0^\infty N_{-\alpha}(q)d\alpha \quad .\end{aligned} \tag{1.32}$$

We thus have to estimate the sum $\sum_{\lambda_j<0} |\lambda_j| = -\int_0^\infty N_{-\alpha}(q)d\alpha$ from above. This will be done in the next section; the result is (recall that we assumed $q(x) < 0$):

$$\int_0^\infty N_{-\alpha}\,(q)\, d\alpha \leq \varepsilon(n) \int |q(x)|^{1+\frac{n}{4}}\, dx \tag{1.33}$$

From (1.32), (1.33) we get:

$$\sum_{j=1}^{m} ||\Delta\phi_j||^2 \geq \int -q(x) \sum_{j=1}^{m} |\phi_j(x)|^2 \, dx - \varepsilon(n) \int |q(x)|^{1+\frac{n}{4}} \, dx. \tag{1.34}$$

Now we choose $q(x) = -c_3(\sum_{j=1}^{m} |\phi_j(x)|^2)^{c_4}$, where the constants c_3, c_4 satisfy $c_4+1 = c_4(1+\frac{n}{4})$, $c_3 = 2\varepsilon(n)\, c_3^{1+n/4}$, i.e., $c_4 = \frac{4}{n}$, $c_3 = (2\varepsilon(n))^{-n/4}$. Therefore, from (1.34) we deduce (1.30), with $\alpha_n \equiv c_4+1$ and $c_0(n) \equiv c_3$. The proof is complete. □

2. BOUNDS FOR MOMENTS OF THE EIGENVALUES

Here we prove the bounds (1.33). The method is general and relies on the Birman-Schwinger principle (as it has been quoted in the introduction) and the computation of the trace for an integral operator. In particular, it could be applied for Δ^2 replaced by any elliptic operator. We recall the result:

Theorem 2.1. Let q be a smooth function defined on the torus M $= [-\frac{L}{2}, \frac{L}{2}]^n$, $n \leq 3$. Then, if $q(x)_- \equiv \text{Min}\,\{q(x),0\}$:

$$\sum_{\lambda_j<0} |\lambda_j| \equiv \int_0^{\infty} N_{-\alpha}(q)\, d\alpha \leq \varepsilon(n) \int |q(x)_-|^{1+\frac{n}{4}} \, dx \quad , \tag{1.33}$$

where the λ_j's are eigenvalues of $\Delta^2 + q(x)$ defined on M and $N_{-\alpha}(q) \equiv \sum_{\lambda_j<-\alpha} 1$. The expression of the constant $\varepsilon(n)$ appears in (2.10) below.

Proof. Due to the min-max principle, we first remark that for $\alpha > 0$

$$N_{-\alpha}(q) \le N_0(q+\alpha) = N_0(t\alpha + q + (1-t)\alpha)$$

$$\le N_0(t\alpha + \text{Min}\ \{q+(1-t)\alpha,0\}) \quad , \quad 0 \le t \le 1 \quad . \qquad (2.1)$$

In what follows, we shall use the shorter notation $(q + (1-t)\alpha)_-$ for Min $\{q + (1-t)\alpha,0\}$. Thus, we are led to estimate the number of negative eigenvalues for $\Delta^2 + t\alpha + (q+(1-t)\alpha)_-$. The Birman-Schwinger principle (see the Appendix) state that the number of negative eigenvalues of $\Delta^2 + t\alpha +(q+(1-t)\alpha)_-$ is equal to the number of eigenvalues greater than one of $|(q+(1-t)\alpha)_-|^{\frac{1}{2}}$ $(\Delta^2 + t\alpha)^{-1}$ $|(q+(1-t)\alpha)_-|^{\frac{1}{2}} \equiv K_{\alpha,t}$. The latter is clearly bounded from above by the number of eigenvalues greater than one of $(K_{\alpha,t})^m$, for $m \in N^*$, and a fortiori by[(1)] Trace $(K_{\alpha,t})^m$. Therefore, from (2.1), we obtain:

$$N_{-\alpha}(q) \le \text{Trace}\ (K_{\alpha,t})^m \quad , \qquad (2.2)$$

where the positive integer m will be chosen below. From the definition of $K_{\alpha,t}$ and classical commutativity properties of the trace, we still get:

$$N_{-\alpha}(q) \le \text{Trace}\ (|(q + (1-t)\alpha)_-|^m\ (\Delta^2 + t\alpha)^{-m}) \quad . \qquad (2.3)$$

By the definition of the trace, we now obtain:

$$N_{-\alpha}(q) \le L^{-n} \sum_{k\in Z^n} \int |(q(x) + (1-t)\alpha)_-|^m \sigma_{\alpha,t}(k)\ dx \quad , \qquad (2.4)$$

(1) One checks (see the Appendix) that $K_{\alpha,t}$ is trace class for $n \le 3$.

where $\sigma_{\alpha,t}(k) = ((\frac{2\pi}{L})^4 |k|^4 + t\alpha)^{-m}$ is the symbol of $(\Delta^2 + t\alpha)^{-m}$.

Finally, (2.4) is equivalent to ($\alpha > 0$, $t \in [0,1]$):

$$N_{-\alpha}(q) \leq \{L^{-n} \sum_{k\in Z^n} \sigma_{\alpha,t}(k)\} \int |(q(x) + (1-t)\alpha)_-|^m \, dx \quad (2.5)$$

In (2.5), the term in brackets is bounded from above by

$$\Gamma_{m,n}(t\alpha) \equiv L^{-n} \int_{R^n} ((\frac{2\pi}{L})^4 \ |\xi|^4 + t\alpha)^m \, d\xi$$

$$= (2\pi)^{-n} \ (t\alpha)^{\frac{n}{4}-m} \int_{S^{n-1}} d\omega_{n-1} \int_0^\infty (\rho^4 + 1)^{-m} \rho^{n-1} \, d\rho \quad (2.6)$$

where $d\omega_{n-1}$ is the measure on the sphere $S^{n-1} = \{\xi \in R^n, |\xi| = 1\}$. The last integral is finite iff $m > \frac{n}{4}$ and (2.5), (2.6) give the desired bound on $N_{-\alpha}(q)$:

$$N_{-\alpha}(q) \leq \Gamma_m(t\alpha) \int |(q(x) + (1-t)\alpha)_-|^m \, dx$$
$$m > \frac{n}{4} \ , \ \alpha > 0 \ , \ t\in[0,1] \quad (2.7)$$

We are now in position to bound from above $\int_0^\infty N_{-\alpha}(q) \, d\alpha$ (and more generally $\int_0^\infty N_{-\alpha}(q) \ \alpha^\ell \, d\alpha$). From (2.7) and Fubini's theorem, we have:

$$\int_0^\infty N_{-\alpha}(q) \, d\alpha \leq \int dx \int_0^\infty |(q(x) + (1-t)\alpha)_-|^m \, \Gamma_{m,n}(t\alpha) \, d\alpha$$

$$= \int dx \int_0^{\frac{1}{1-t}|q(x)_-|} (q(x)_- - (1-t)\alpha)^m \, \Gamma_{m,n}(t\alpha) \, d\alpha$$

$$= \int dx \int_0^1 |q(x)_-|^m (1-\beta)^m \Gamma_{m,n} \left(\frac{t}{1-t}|q(x)_-|\beta\right) \beta \frac{|q(x)_-|}{1-t} d\beta \tag{2.8}$$

Using the value (2.6) of $\Gamma_{m,n}$, we still obtain:

$$\int_0^\infty N_{-\alpha}(q)\, d\alpha \leq \gamma_{m,n} t^{\frac{n}{4}-m} (1-t)^{m-(1+\frac{n}{4})} \int_0^1 (1-\beta)^m \beta^{\frac{n}{4}-m} d\beta$$

$$\cdot \int |q(x)_-|^{1+\frac{n}{4}} dx \quad , \tag{2.9}$$

where

$$\gamma_{m,n} \equiv (2\pi)^{-n} \int_{S^{n-1}} d\omega_{n-1} \int_0^\infty (\rho^4+1)^{-m} \rho^{n-1} d\rho \text{ and } \frac{n}{4} < m < 1+\frac{n}{4}$$

(to have a finite integral in β). It remains to find the best $t \in [0,1]$ and $m \in]\frac{n}{4}, 1+\frac{n}{4}[$ that is t_0, m_0 giving the smallest value for $t^{\frac{n}{4}-m} (1-t)^{m-(1+\frac{n}{4})}$.

The proof of (1.33) is complete; the precise value of $\varepsilon(n)$ in (1.33) is:

$$\varepsilon(n) = \gamma_{m_0,n} t_0^{\frac{n}{4}-m_0} (1-t_0)^{m_0-(1+\frac{n}{4})} \int_0^1 (1-\beta)^{m_0} \beta^{\frac{n}{4}-m_0} d\beta \tag{2.10}$$

with $\gamma_{m_0,n} = (2\pi)^{-n} \int_{S^{n-1}} d\omega_{n-1} \int_0^\infty (\rho^4+1)^{-m_0} \rho^{n-1}\, d\rho.$ □

APPENDIX

Here we recall, for the sake of completeness, some facts concerning the Birman-Schwinger principle (see [8,9]). Let $V(x)$ be a bounded potential, assumed strictly negative and let $\lambda_0 < 0$. On the n-dimensional torus $M = [-\frac{L}{2},\frac{L}{2}]^n$, we consider the following eigenvalue problem:

$$(\Delta^2 - \lambda_0 + V)\phi = \lambda\phi \quad , \tag{A.1}$$

where the biharmonic operator Δ^2 could be replaced by any elliptic operator. Associated to (A.1) is another eigenvalue problem:

$$(\Delta^2 - \lambda_0)\chi = -\mu V\chi \quad . \tag{A.2}$$

Since $\lambda_0 < 0$, $(\Delta^2 - \lambda_0)^{-1}$ is well defined as an operator on M; setting $\psi \equiv |V|^{\frac{1}{2}} \chi$, (A.2) is therefore equivalent ($V < 0$ and smooth), by a direct computation to:

$$S\psi \equiv |V|^{\frac{1}{2}} (\Delta^2 - \lambda_0)^{-1} |V|^{\frac{1}{2}} \psi = \frac{1}{\mu}\psi \quad , \tag{A.3}$$

that is an eigenvalue problem for the operator S.

Theorem A.1: Birman-Schwinger principle.

For $V < 0$, $\lambda_0 < 0$ and if λ_j (resp. ν_j) are the eigenvalues of $\Delta^2 - \lambda_0 + V$ (resp. $|V|^{\frac{1}{2}} (\Delta^2 - \lambda_0)^{-1} |V|^{\frac{1}{2}}$), we have:

$$\sum_{\lambda_j < 0} 1 = \sum_{\nu_j > 1} 1 \tag{A.4}$$

Proof. From (A.1), we deduce:

$$\lambda = ||\phi||^{-2} (||\Delta\phi||^2 - \lambda_0 ||\phi||^2 + \int V(x)|\phi|^2 \, dx)$$

$$= ||\phi||^{-2} (||\Delta\phi||^2 - \lambda_0 ||\phi||^2 - \int |V(x)| \, |\phi|^2 \, dx)$$

$$= ||\phi||^{-2} \int |V(x)| \, |\phi|^2 \, dx \, (\frac{||\Delta\phi||^2 - \lambda_0 ||\phi||^2}{\int |V(x)| \, |\phi|^2 \, dx} - 1)$$

therefore $\lambda < 0$ is equivalent to $\dfrac{||\Delta\phi||^2 - \lambda_0 ||\phi||^2}{\int |V(x)| \, |\phi|^2 \, dx} = \mu < 1$

that is (since $V < 0$) to $\Delta^2\phi + \mu V\phi = \lambda_0\phi$. The equivalence of (A.2) and (A.3) completes the proof. □

Remark A.1. In the proof of Theorem 2.1, we take $\lambda_0 = -t\alpha$, $V = \text{Min}\,\{q + (1-t)\alpha, 0\}$ $t \in [0,1]$. □

Remark A.2. It is worthwhile to notice some properties of the operator $S = |V|^{\frac{1}{2}}(\Delta^2 - \lambda_0)^{-1} |V|^{\frac{1}{2}}$. First of all $S > 0$, by direct computation using Parseval identity. Second S is a compact operator in $H = L^2(M)$, because $(\Delta^2 - \lambda_0)^{-1}$ is compact and $|V|^{\frac{1}{2}}$ is bounded. Finally, for $n \leq 3$, $(\Delta^2 - \lambda_0)^{-1}$ and hence S is trace class, i.e., the serie of its eigenvalues is absolutely convergent. □

ACKNOWLEDGMENTS

This research was supported by the Center for Nonlinear Studies, Los Alamos National Laboratory. Work also performed under the auspices of the U.S. Department of Energy under contract W-7405-ENG-36 and contract KC-04-02-01, Division of Basic and Engineering Sciences.

REFERENCES

1. A. J. Babchin, A. L. Frenkel, B. G. Levich and G. I. Sivashinsky, "Nonlinear Saturation of Rayleigh-Taylor Instability in Thin Films," Phys. Fluids 26, 11 (1983), 3159-3161.

2. P. Constantin and C. Foias, "Global Lyapunov Exponents, Kaplan Yorke Formulas and the Dimension of the Attractors for 2-D Navier-Stokes Equations," Comm. Pure Appl. Math., 38(1985), 1-27.

3. P. Constantin, C. Foias and R. Temam, "Attractors Representing Turbulent Flows," Memoirs of the A.M.S. , vol.53, no.314(1985).

4. C. Foias and R. Temam, "Some Analytic and Geometric Properties of the Solutions of the Navier-Stokes Equations," J. Math. Pures Appl., 58 (1979), 339-368.

5. Y. Kuramoto and T. Tsuzuki, "On the Formation of Dissipative Structures in Reaction-Diffusion Systems," Prog. Theor. Phys. 54 (1975), 687-699.

6. Y. Kuramoto and T. Tsuzuki, "Persistent Propagation of Concentration Waves in Dissipative Media Far From Thermal Equilibrium," Prog. Theor. Phys. 55 (1976), 356-369.

7. Y. Kuramoto, "Diffusion-induced Chaos in Reaction Systems," Suppl. Prog. Theor. Phys. 64 (1978), 346-367.

8. E. Lieb, "On Characteristics Exponents in Turbulence," Comm. Math. Phys., 92 (1984), 473-480.

9. E. Lieb and W. Thirring, "Inequalities for the Moments of the Eigenvalues of the Schroedinger Equations and Their Relation to Sobolev Inequalities," in *Studies in Mathematical Physics: Essays in Honor of Valentine Bergman*, E. Lieb, B. Simon, A. S. Wightman, Eds., Princeton University Press, Princeton, NJ (1976), 269-303.

10. B. Nicolaenko, B. Scheurer and R. Temam, "Quelques propriétés des attracteurs pour l'équation de Kuramoto-Sivashinsky," C. R. Acad. Sc., Paris 298 (1984), 23-25.

11. B. Nicolaenko, B. Scheurer and R. Temam, "Some Global Dynamical Properties of the Kuramoto-Sivashinsky Equations: Nonlinear Stability and Attractors," Physica D 16(1985), 155-183.

12. B. Nicolaenko, B. Scheurer and R. Temam, "Strange Attractors for a Class of Nonlinear Evolution Partial Differential Equations," in preparation.

13. Y. Pomeau, A. Pumir and P. Pelce, "Intrinsic Stochasticity with Many Degrees of Freedom," preprint C.E.A. - S.P.T., Saclay, France.

14. M. Reid and B. Simon, Methods of Modern Mathematical Physics, Academic Press, Vol. IV.

15. G. Sivashinsky, "Nonlinear Analysis of Hydrodynamic Instability in Laminar Flames, Part I. Derivation of Basic Equations," Acta Astronautica 4 (1977), 1177-1206.

16. G. Sivashinsky, "On Flame Propagation under Conditions of Stoichiometry," SIAM J. Appl. Math., 39 (1980), 67-82.

17. G. I. Sivashinsky and D. M. Michelson, "On Irregular Wavy Flow of a Liquid Down a Vertical Plane," Prog. Theor. Phys. 63 (1980), 2112-2114.

1 Theoretical Division, MS-B284
Center for Nonlinear Studies
Los Alamos National Laboratory
Los Alamos, NM 87545

2 Commissariat à l'Energie Atomique
Centre d'Etudes de Limeil-Valenton (Dpt MA) BP 27
94190 Villeneuve St. Georges
France

3 Laboratoire d'Analyse Numérique
CNRS and Université Paris Sud
Bat 425, 91405 Orsay
France

Lectures in Applied Mathematics
Volume 23, 1986

NONLINEAR STABILITY OF THE KELVIN-STUART CAT'S EYES FLOW

Darryl D. Holm, Jerrold E. Marsden[1] and Tudor Ratiu[2]

ABSTRACT. Conditions which ensure the nonlinear stability of the Kelvin-Stuart cat's eyes solution for two dimensional ideal flow are given. The solution is periodic in the x direction and is bounded by two streamlines, which contain the separatrix, in the y-direction. The stability conditions are given explicitly in terms of the solution parameters and the domain size. The method is based on a technique originally developed by Arnold [1969].

1. EQUATIONS OF MOTION AND CONSERVED QUANTITIES.

The Euler equations for an ideal, homogeneous incompressible fluid in a domain D in the plane $\mathbb{R}^2$ are:

$$\left.\begin{aligned} &\frac{\partial \underline{v}}{\partial t} + (\underline{v}\cdot\underline{\nabla})\underline{v} = -\underline{\nabla} p \\ &\operatorname{div} \underline{v} = 0 \\ &\underline{v}\cdot\underline{\hat{n}} = 0 \end{aligned}\right\} \tag{1.1}$$

where $\underline{v} = (v_1, v_2)$ is the velocity field, p is the pressure, and $\underline{\hat{n}}$ is the outward unit normal of the boundary ∂D.

1980 Mathematics Subject Classification 76E30, 58F10.
[1]Partially supported by DOE contract DE-AT03-82ER12097.
[2]Partially supported by an NSF postdoctoral fellowship.

Let $\omega = \hat{\underline{z}} \cdot \text{curl}\, \underline{v} = v_{2,x} - v_{1,y}$ be the scalar vorticity and ψ the stream function, i.e. $\underline{v} = \text{curl}(\psi\, \hat{\underline{z}}) = (\psi_{,y}, -\psi_{,x})$, where $\hat{\underline{z}}$ is the upward unit vector, orthogonal to the xy plane. The existence of ψ is proved in the following manner. Since $\underline{v}$ is tangent to each component $(\partial D)_i$ of ∂D, $i = 0, 1, \ldots, g$, the integral of $v_1\, dy - v_2\, dx$ around each $(\partial D)_i$ is zero. Since $\text{div}\, \underline{v} = 0$, we conclude that its integral around any closed loop is zero. Thus, by elementary calculus, $v_1 dy - v_2\, dx = d\psi$ for some ψ, i.e. $v_1 = \psi_{,y}$ and $v_2 = -\psi_{,x}$. Since $\underline{v}$ is tangent to $(\partial D)_i$, ψ is constant on each $(\partial D)_i$, $i = 0, \ldots, g$, so adding a suitable constant to ψ, we can assume it is zero on $(\partial D)_0$. Since

$$\underline{v} \cdot d\underline{\ell} = \text{curl}(\psi \hat{\underline{z}}) \cdot d\underline{\ell} = (\hat{\underline{z}} \times d\underline{\ell}) \cdot \underline{\nabla}\psi$$

$$= -\underline{\nabla}\psi \cdot \hat{\underline{n}}\, ds = -\frac{\partial \psi}{\partial \underline{n}}\, ds,$$

where $d\underline{\ell}$ and ds are the vectorial and scalar infinitesimal arc elements, we see that the circulations around $(\partial D)_i$, $i = 0, \ldots, g$ have the expressions

$$\Gamma_i := \int_{(\partial D)_i} \underline{v} \cdot d\underline{\ell} = -\int_{(\partial D)_i} \frac{\partial \psi}{\partial \underline{n}}\, ds.$$

In a bounded domain D, given the scalar vorticity ω, the stream function ψ is uniquely determined by the elliptic problem

$$\left.\begin{aligned} &-\nabla^2 \psi = \omega \\ &\psi|(\partial D)_0 = 0 \\ &\psi_i := \psi|(\partial D)_i = \text{constant}, \quad i = 1, \ldots, g \\ &\Gamma_i := -\int_{(\partial D)_i} \frac{\partial \psi}{\partial n}\, ds = \text{constant}. \end{aligned}\right\} \tag{1.2}$$

Applying the operator $\hat{\underline{z}}\cdot$curl to the momentum conservation equation in (1.1) written in the form

$$\partial \underline{v}/\partial t = \underline{v} \times \omega\hat{\underline{z}} - \underline{\nabla}(\tfrac{1}{2}\,|\underline{v}|^2 + p),$$

yields the vorticity equation

$$\partial\omega/\partial t = \{\psi,\omega\}, \tag{1.3}$$

where $\{\psi,\omega\} = \psi_{,x}\,\omega_{,y} - \psi_{,y}\,\omega_{,x}$ is the usual Poisson bracket in $\mathbb{R}^2$.

Fix the vectors $\underline{\psi} = (\psi_1, \ldots, \psi_y)$ and $\underline{\Gamma} = (\Gamma_0, \ldots, \Gamma_g)$, and consider the following space of vorticities (with appropriate smoothness properties):

$$F_{\underline{\psi},\underline{\Gamma}} = \{\omega : D \to \mathbb{R} \mid \text{there exists a function } \psi : D \to \mathbb{R} \text{ satisfying (1.2) with the constants } \psi_1, \ldots, \psi_g, \Gamma_0, \ldots, \Gamma_g\}.$$

For $\omega \in F_{\underline{\psi},\underline{\Gamma}}$, we will write $\psi = -(\nabla^2)^{-1}\omega$ for the unique solution of (1.2). On this space, the total energy takes the form

$$\begin{aligned} H(\omega) &= \frac{1}{2}\int_D |\underline{v}|^2\,dx\,dy = \frac{1}{2}\int_D |\underline{\nabla}\psi|^2\,dx\,dy \\ &= \frac{1}{2}\int_D \psi\,\omega dx\,dy + \frac{1}{2}\int_{\partial D} \psi\,\underline{\nabla}\psi\cdot\hat{\underline{n}}\,ds \qquad (1.4) \\ &= \frac{1}{2}\int_D \omega(-\nabla^2)^{-1}\omega\,dx\,dy - \frac{1}{2}\sum_{i=1}^{g} \psi_i\Gamma_i . \end{aligned}$$

In addition to this conserved energy, from (1.3), the identity

$$\int_D f\{g,h\}\,dx\,dy = \int \{f,g\}h\,dx\,dy - \int_{\partial D} fh\underline{\nabla}g\cdot d\underline{\ell}\ ,$$

and the fact that ω and ψ satisfy (1.2), it follows that the functionals

$$C_\Phi(\omega) = \int_D \Phi(\omega)\,dx\,dy \tag{1.5}$$

are also conserved, for any $\Phi:\mathbb{R} \to \mathbb{R}$.

2. VARIATIONAL PRINCIPLE FOR THE CAT'S EYES. Stationary solutions ω_e, ψ_e of (1.3) are characterized by having $\underline{\nabla}\psi_e$ and $\underline{\nabla}\omega_e$ parallel. A sufficient condition for this to hold is the functional relationship

$$\psi_e = \Psi(\omega_e). \tag{2.1}$$

The stationary solution treated in this paper is the Kelvin [1880]-Stuart [1967] cat's eyes solution given by

$$\psi_e(x,y) = \log[a\cosh y + \sqrt{a^2-1}\cos x], \tag{2.2}$$

in the domain $0 \leq x \leq 2\pi$, $-\infty < y < \infty$, where a is a real parameter satisfying $a \geq 1$ ($a = 1$ gives a shear flow). The streamlines ψ = constant have the form shown in Figure 1. The vorticity is given by

$$\omega_e(x,y) = -\nabla^2\psi_e(x,y) = -e^{-2\psi_e} \tag{2.3}$$

so that Ψ in (2.1) is given by

$$\Psi(\lambda) = -\frac{1}{2}\log(-\lambda),\ \lambda < 0. \tag{2.4}$$

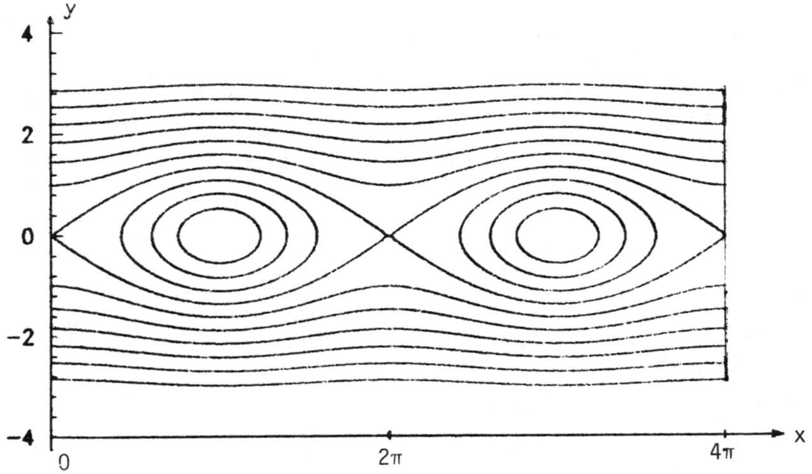

Figure 1. Computer plot of the cat's eyes streamlines for the stream function $\psi(x,y)$ in (2.2) with a = 1.175.

The components of the velocity are

$$\begin{aligned} v_1 &= \psi_{,y} = \frac{a \sinh y}{a \cosh y + \sqrt{a^2-1} \cos x}, \\ v_2 &= -\psi_{,x} = \frac{\sqrt{a^2-1} \sin x}{a \cosh y + \sqrt{a^2-1} \cos x} \end{aligned} \tag{2.5}$$

so that $v_2 \to 0$ as $y \to \pm\infty$, whereas $v_1 \to \pm 1$ as $y \to \pm\infty$, i.e. in the limit, the velocity is a shear flow in each half-plane in the opposite direction. We shall consider in this paper only domains bounded by a pair of streamlines below the upper separatrix and above the lower separatrix, i.e. we shall require that the finite domain D be given by $0 \leq x \leq 2\pi$ and y bounded by the streamlines $\psi = \pm \log[ac + a + \sqrt{a^2-1}]$ for some $c > 0$. The reason for this restriction is that the infinite domain allows arbitrary wave numbers, which prevent the estimates below from being carried out.

In the domain D, we shall seek the stationary solution ω_e as a critical point of the conserved functional

$$H_\Phi(\omega) = \int_D [\tfrac{1}{2}\,\omega(-\nabla^2)^{-1}\omega + \Phi(\omega)]\,dx\,dy$$
$$- \frac{1}{2}\sum_{i=1}^{g} \psi_i \Gamma_i . \tag{2.6}$$

Integrating twice by parts and using the fact that $\delta\psi|(\partial D)_i = 0$, $\int_{(\partial D)_i} \partial(\delta\psi)/\partial \underline{n}\; ds = 0$, $i = 0, \ldots, g$, we get

$$DH_\Phi(\omega_e)\cdot\delta\omega = \int_D (\psi_e + \Phi'(\omega_e)\delta\omega\,dx\,dy$$
$$= \int_D (\Psi(\omega_e) + \Phi'(\omega_e))\delta\omega\,dx\,dy.$$

By (2.4), the function Φ equals (up to a constant)

$$\Phi(x) = -\int_0^\lambda \Psi(s)\,ds$$
$$= \frac{1}{2}\int_0^\lambda \log(-s)\,ds \tag{2.7}$$
$$= \frac{1}{2}\,\lambda(\log(-\lambda) - 1).$$

This function has the graph shown in Figure 2. The function is concave since

$$\Phi''(\lambda) = \frac{1}{2\lambda} < 0 \quad \text{for} \quad \lambda < 0.$$

Bounding the domain in the y direction will keep ω away from the bad point $\lambda = 0$ in Figure 2, where Φ'' is unbounded below.

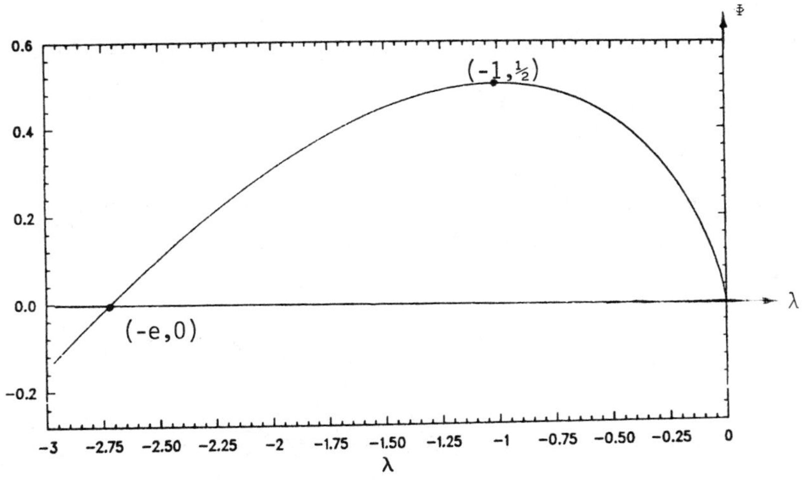

Figure 2. The Casimir function $\Phi(\lambda)$ in equation (2.7) is convex downward.

3. STABILITY ESTIMATES. To study the stability of the Cat's Eyes solution in a finite domain, consider a finite perturbation $\Delta\omega$. The quantity

$$\hat{H}_\Phi(\Delta\omega) := H_\Phi(\omega_e + \Delta\omega) - H_\Phi(\omega_e) - DH_\Phi(\omega_e)\cdot\Delta\omega$$

$$= \int_D [\tfrac{1}{2}\,\Delta\omega(-\nabla^2)^{-1}\Delta\omega + \Phi(\omega_e + \Delta\omega) - \Phi(\omega_e) - \Phi'(\omega_e)\Delta\omega]\,dx\,dy \qquad (3.1)$$

is conserved since $DH_\Phi(\omega_e) = 0$ for Φ given by (2.7)

To establish nonlinear stability, we shall bound the conserved quantity (3.1) above and below, in a way that implies bounds on the L^2 norm of the vorticity perturbation for all time. To get this stability estimate, we modify Φ to a function $\tilde{\Phi}$, in such a manner that Φ'' is bounded above and below and $DH_{\tilde{\Phi}}(\omega_e) = 0$. For these bounds on $\tilde{\Phi}''$, we first compute $\min \omega_e(x,y)$ and $\max \omega_e(x,y)$, where $D = \{(x,y) \in \mathbb{R}^2 \mid 0 \leq x \leq 2\pi,\ |y| \leq \cosh^{-1}\left[c + 1 + \frac{\sqrt{a^2-1}}{a}(1 - \cos x)\right],\ c > 0\}$ is the domain bounded in the y-direction, by the two streamlines $\psi_e(x,y) = \pm\log(ac + a + \sqrt{a^2-1})$ and over a 2π-period in x. Since

$$\omega_e(x,y) = -e^{-2\psi_e(x,y)} = -[a \cosh y + \sqrt{a^2-1} \cos x]^{-2} \tag{3.2}$$

the critical points of the stream function and, thus, of the vorticity are at $x = 0, \pi, 2\pi$ and $y = 0$. The critical values of vorticity are

$$-[a + \sqrt{a^2-1}]^{-2}, \quad \text{for } x = 0, 2\pi$$

and

$$-[a - \sqrt{a^2-1}]^{-2}, \quad \text{for } x = \pi.$$

The value of $\omega_e(x,y)$ on the y-boundary is

$$-[ac + a + \sqrt{a^2-1}]^{-2}$$

and on the vertical boundaries $x = 0, 2\pi$ is

$$-[a + \sqrt{a^2-1}]^{-2} \leq -[a \cosh y + \sqrt{a^2-1}]^{-2} \leq -[ac + a + \sqrt{a^2-1}]^{-2}.$$

Consequently,

$$\begin{aligned} \min_D \omega_e(x,y) &= -[a - \sqrt{a^2-1}]^{-2} \\ \max_D \omega_e(x,y) &= -[ac + a + \sqrt{a^2-1}]^{-2}. \end{aligned} \tag{3.3}$$

Thus, on the interval $[\min_D \omega_e(x,y), \max_D \omega_e(x,y)]$ the function Φ has its second derivative bounded by

$$\frac{1}{2\max_D \omega_e(x,y)} \leq \Phi''(\omega_e) = \frac{1}{2\omega_e} \leq \frac{1}{2\min_D \omega_e(x,y)} < 0.$$

Now define the following C^2-function:

$$\tilde{\Phi}(\lambda) = \begin{cases} -\frac{1}{2}(a - \sqrt{a^2-1})^2\lambda^2 + (\sqrt{a^2-1} - a - 1)\lambda + \alpha, \\ \qquad\qquad \text{for } \lambda \leq -(a - \sqrt{a^2-1})^{-2} \\ \Phi(\lambda) = \frac{1}{2}\lambda(\log(-\lambda) - 1), \\ \qquad\qquad \text{for } -(a - \sqrt{a^2-1})^{-2} \leq \lambda \leq -(ac + a + \sqrt{a^2-1})^{-2} \\ -\frac{1}{2}(ac + a + \sqrt{a^2-1})^2\lambda^2 - (ac + a + \sqrt{a^2-1} - 1)\lambda + \beta, \\ \qquad\qquad \text{for } \lambda \geq -(ac + a + \sqrt{a^2-1})^{-2} \end{cases}$$

where
$$\alpha = (a - \sqrt{a^2-1})^{-2}(\log(a - \sqrt{a^2-1}) + 2) + \frac{1}{2}(a - \sqrt{a^2-1})^{-2}$$
$$+ (\sqrt{a^2-1}) - a - 1)(a - \sqrt{a^2-1})^{-2},$$

and
$$\beta = (ac + a + \sqrt{a^2-1})^{-2}(\log(ac + a + \sqrt{a^2-1}) + 2)$$
$$+ \frac{1}{2}(ac + a + \sqrt{a^2-1})^{-2}$$
$$- (ac + a + \sqrt{a^2-1})(ac + a + \sqrt{a^2-1} - 1)^{-2}.$$

Since $\tilde{\Phi}$ and Φ coincide on the interval $[\min_D \omega_e(x,y),$

$\max_D \omega_e(x,y)]$, it follows that $DH_{\tilde{\Phi}}(\omega_e) = 0$. But unlike Φ, $\tilde{\Phi}$ has its second derivative bounded on the entire axis, namely

$$\frac{1}{2}(a - \sqrt{a^2-1})^2 \leq -\tilde{\Phi}''(\lambda) \leq \frac{1}{2}(ac + a + \sqrt{a^2-1})^2 \tag{3.4}$$

for all $-\infty < \lambda < +\infty$. Consequently, the function $-\tilde{\Phi}$ is convex, i.e.,

$$\begin{aligned}\frac{1}{4}(a - \sqrt{a^2-1})^2(\Delta\omega)^2 &\leq -\tilde{\Phi}(\omega_e + \Delta\omega) + \tilde{\Phi}(\omega_e) + \tilde{\Phi}'(\omega_e)\Delta\omega \\ &\leq \frac{1}{4}(ac + a + \sqrt{a^2-1})^2(\Delta\omega)^2 . \end{aligned}\tag{3.5}$$

Considering the negative of (3.1) with Φ replaced by $\tilde{\Phi}$ we get from (3.5) the estimates

$$-\hat{H}_{\tilde{\Phi}}(\Delta\omega) \leq \frac{1}{2}\int_D [\frac{1}{2}(ac + a + \sqrt{a^2-1})^2(\Delta\omega)^2 + \Delta\omega(\nabla^2)^{-1}\Delta\omega]\,dx\,dy$$

$$-\hat{H}_{\tilde{\Phi}}(\Delta\omega) \geq \frac{1}{2}\int_D [\frac{1}{2}(a - \sqrt{a^2-1})^2(\Delta\omega)^2 + \Delta\omega(\nabla^2)^{-1}\Delta\omega]\,dx\,dy.$$

Let $\Delta\omega_0$ denote the value of the perturbation $\Delta\omega$ at $t = 0$. Then by conservation of $-\hat{H}_{\tilde{\Phi}}$ we get

$$\begin{aligned}\int_D [\frac{1}{2}(a - \sqrt{a^2-1})^2(\Delta\omega)^2 + \Delta\omega\,(\nabla^2)^{-1}\Delta\omega]dx\,dy &\leq -2\,\hat{H}_{\tilde{\Phi}}(\Delta\omega) \\ = -2\hat{H}_{\tilde{\Phi}}(\Delta\omega_0) &\leq \int_D [\frac{1}{2}(ac + a + \sqrt{a^2-1})(\Delta\omega_0)^2 \\ &\quad + (\Delta\omega_0)(\nabla^2)^{-1}(\Delta\omega_0)^2]\,dx\,dy \\ &\leq \int_D [\frac{1}{2}(ac + a + \sqrt{a^2-1})^2(\Delta\omega_0)^2\,dx\,dy,\end{aligned}$$

since $(\nabla^2)^{-1}$ is negative. Thus we have the a priori estimate

$$\frac{1}{2}\,(a - \sqrt{a^2-1})^2 \|\Delta\omega\|^2_{L^2} + \int_D \Delta\omega(\nabla^2)^{-1}\Delta\omega \, dx\, dy \le [\tfrac{1}{2}\,(ac + a + \sqrt{a^2-1})^2] \|\Delta\omega_0\|^2_{L^2}. \tag{3.6}$$

To prove nonlinear stability, we still need an estimate in terms of the L^2-norm of $\Delta\omega$ for the second (negative) integral on the left hand side of (3.6). This will be done by using the following Poincaré type inequality.

LEMMA. *Let* k^2_{min} *be the minimal eigenvalue of* $-\nabla^2$ *in the space* $F_{\psi,\Gamma}$ *on the domain* D. *Then*

$$\int_D \Delta\omega(\nabla^2)^{-1}\Delta\omega \, dx\, dy \ge -k^{-2}_{min} \|\Delta\omega\|^2_{L^2}.$$

PROOF. Let k_i^2 be the eigenvalues of $-\nabla^2$, $i = 0, 1, \ldots,$ with $k_0^2 = k^2_{min}$ and let ϕ_i be an L^2 orthonormal basis of eigenfunctions, i.e.

$$-\nabla^2\phi_i = k_i^2\phi_i, \quad \int_D \phi_i\phi_j \, dx\, dy = \delta_{ij}.$$

Then $-k_i^{-2}$ are the eigenvalues of $(\nabla^2)^{-1}$, i.e.

$$(\nabla^2)^{-1}\phi_i = -k_i^{-2}\phi_i, \quad i = 0, 1, \ldots,$$

setting $\Delta\omega = \sum_{i=0}^{\infty} c_i\phi_i$, we have

$$\int_D \Delta\omega(\nabla^2)^{-1}\Delta\omega \ dx\ dy = \sum_{i,j}^{\infty} c_i c_j \int_D \phi_i (\nabla^2)^{-1}\phi_i \ dx\ dy$$

$$= -\sum_{i,j} c_i c_j k_j^{-2} \int_D \phi_i \phi_j \ dx\ dy$$

$$= -\sum_{i=0}^{\infty} k_j^{-2} c_j^2 \int_D \phi_j^2 \ dx\ dy$$

$$\geq -k_{min}^{-2} \sum_{i=0}^{\infty} c_j^2 \int_D \phi_j^2 \ dx\ dy$$

$$= -k_{min}^{-2} \|\Delta\omega\|_{L^2}^2$$

since $k_j^{-2} \leq k_{min}^{-2}$ for all $j = 0, 1, 2, \ldots$ ■

This lemma, and (3.6) yield the estimate

$$[\tfrac{1}{2}(a - \sqrt{a^2-1})^2 - k_{min}^{-2}]\|\Delta\omega\|_{L^2}^2$$

$$\leq [\tfrac{1}{2}(ac + a + \sqrt{a^2-1})^2]\|\Delta\omega_0\|_{L^2}^2 \tag{3.7}$$

The final requirement for nonlinear stability is to ensure the positivity of the coefficient in the left hand side of (3.7).

According to the characterization of the minimal eigenvalue of the Laplacian on bounded domains by the Rayleigh-Ritz quotient, we see that this minimal eigenvalue is a decreasing function of the size of the domain. Thus, we shall replace k_{min}^2 with the first eigenvalue of the Laplacian on the rectangle $0 \leq x \leq 2\pi$,

$$|y| \leq \ell := \cosh^{-1}\left(c + 1 + \frac{2\sqrt{a^2-1}}{a}\right), \tag{3.8}$$

i.e. the height of the rectangle is the distance between the highest points of the streamlines $\psi = \pm \log(ac + a + \sqrt{a^2-1})$. The minimal eigenvalue of $-\nabla^2$ on the space of functions vanishing on the boundary and having zero circulations on each component of the boundary belongs to the eigenfunction $\cos x \sin \frac{\pi y}{\ell}$ and is $1 + \frac{\pi^2}{\ell^2}$. Thus, for (3.7) to provide a meaningful estimate, we need to satisfy the inequality

$$(a - \sqrt{a^2-1})^2 > 2\Big/\left(1 + \frac{\pi^2}{\ell^2}\right) = \frac{2\ell^2}{\pi^2 + \ell^2} \tag{3.9}$$

Solutions of this inequality exist, since, for example, the pair $a = 1$, $c = 1$ satisfies it, but there is clearly an implicit trade-off between a and ℓ in (3.9), by virtue of (3.8).

We study inequality (3.9) for $c = 0$, i.e. we take the y-boundary of the domain to be the separatrix. The inequality becomes

$$(a - \sqrt{a^2-1})^2 > \frac{2\left[\cosh^{-1}\left(1 + \frac{2\sqrt{a^2-1}}{a}\right)\right]^2}{\pi^2 + \left[\cosh^{-1}\left(1 + \frac{2\sqrt{a^2-1}}{a}\right)\right]^2} \;. \tag{3.10}$$

Numerically one verifies that this inequality holds for $1 \leq a \leq 1.175 \ldots$; see Figure 3.

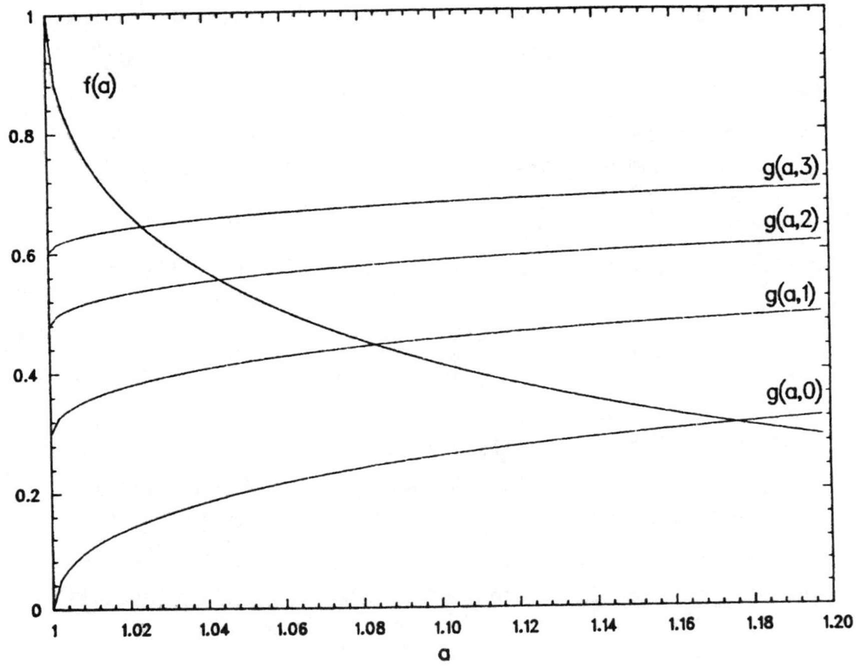

Figure 3. Graphs of $f(a) = (a - \sqrt{a^2 - 1})^2$ and $g(a,c) = 2\ell^2/(\pi^2 + \ell^2)$, where $\ell(a,c) = \cosh^{-1}[1 + c + 2a^{-1}\sqrt{a^2 - 1}]$ for a between 1 and 1.2 and c = 0,1,2,3. The inequality (3.9) is satisfied, so that the cat's eyes flow is stable, for values of (a,c) such that $f(a) > g(a,c)$.

We summarize our results in the following.

THEOREM. *The Kelvin-Stuart cat's eyes solution (2.2) of the Euler equation (1.1) is nonlinearly stable in the* L^2 *norm on vorticities for perturbations of the initial vorticity which preserve the flow rate* (ψ = *constant on the boundaries) and the circulations, in a region bounded by the streamlines* $\psi_e = \pm \log[ac + a + \sqrt{a^2-1}]$, *provided* a *and* c *satisfy (3.8) and (3.9). For* $1 \leq a \leq 1.175, \ldots,$ *this region contains a separatrix in the cat's eyes flow.*

Note that in the special case $a = 1$, the cat's eyes solution reduces to the $v_1 = \tanh(y)$ shear flow, which is stable according to the present analysis, provided the domain is limited in the y direction by $|y| \leq \cosh^{-1}(1 + c)$ where (using 3.9),

$$c < \cosh \pi^2 - 1 = 9{,}665.8 \ldots \tag{3.11}$$

4. FURTHER REMARKS

(i) Variants of the basic flow can be treated by the same method. For example, consider

$$\psi_e = \log[\, a \cosh y + \sqrt{a^2-1} \;\cos x]$$

as before, but on $[0,4\pi]$ rather than $[0,2\pi]$; i.e. include two "eyes" rather than one. The same analysis shows that (3.9) is replaced by

$$(a - \sqrt{a^2-1})^2 > \frac{2}{\frac{1}{4} + \frac{\pi^2}{\ell^2}} \tag{4.1}$$

This restricts the stability region somewhat, but by considering $a = 1$, it holds for $c < \cosh\left(\frac{4\pi^2}{7}\right) - 1 = 139.7 \ldots$ (the analogue of (3.11)), and so (4.1) holds for a nontrivial range of $a > 1$ and $c > 0$ and again we get stability on a region containing the cat's eye separatrix. (These results are consistent with known linearized and nonlinear results; cf. Stuart [1971]).

(ii) Although the computations are more complex, in principle, the method applies to the sinh-Poisson solutions of Ting, Chen and Lee [1984].

(iii) Cat's eye solutions provide interesting equilibria for a plasma confining Grad-Shafranov solution of reduced magnetohydrodynamics (where the current and magnetic potential replace the vorticity and stream function). The present method applies directly to give a nonlinear stability result in that case; see Hazeltine, Holm, Marsden and Morrison [1984]. The known coalescence instability for magnetic islands in that case is avoided by having sufficient transverse constriction for the Poincaré inequality to ensure positivity in (3.7).

ACKNOWLEDGEMENTS. We thank John Gibbon for suggesting this problem and Jerry Kazdan for useful conversations about the Poincaré inequality. We also thank George Nickel for making the figures.

REFERENCES

1. V. Arnold, [1969]. On an a priori estimate in the theory of hydrodynamical stability, Transl. Am. Math. Soc. 79, 267-269.

2. R.D. Hazeltine, D.D. Holm, J.E. Marsden and P.J. Morrison [1984]. Generalized Poisson Brackets and nonlinear Liapunov stability -- Application to reduced MHD. Proc. Lausanne Plasma Phys. Conf., June 1984.

3. D.Holm, J. Marsden, T. Ratiu and A. Weinstein [1984]. Nonlinear stability of fluid and plasma equilibria, Physics Reports (to appear).

4. Lord Kelvin [1880]. On a disturbing infinity in Lord Rayleigh's solution for waves in a plane vortex stratum, Nature 23, 45-46.

5. J.T. Stuart [1971]. Stability problems in fluids, Lectures in Applied Math. AMS 13, 139-155.

6. J.T. Stuart [1967]. On finite amplitude oscillations in laminar mixing layers, J. Fluid Mech. 29, 417-440.

7. A.C. Ting, H.H. Chen and Y.C. Lee [1984]. Exact vortex solutions of two dimensional guiding center plasma (preprint).

CENTER FOR NONLINEAR STUDIES
LOS ALAMOS NATIONAL LABORATORY
MS B 258
LOS ALAMOS, N.M. 87545

DEPARTEMENT OF MATHEMATICS
UNIVERSITY OF CALIFORNIA
BERKELEY, CALIF. 94720

Lectures in Applied Mathematics
Volume 23, 1986

Collapse and Amplification of a Vortex Filament

Eric Siggia

A vortex filament with a variable core size parameter is used to model how a vortex tube breaks down in the Euler equations. Specifically we solve:

$$\vec{v}(r_j(\theta)) = \sum_i \frac{\Gamma_i}{4\pi} \vec{\nabla}\times \int \frac{(d\vec{r}_i/d\theta')\, d\theta'}{((r_j(\theta)-r_i(\theta'))^2+\sigma_i^2+\sigma_j^2)^{1/2}} \quad (1)$$

where Γ is the circulation and $\vec{r}_j(\theta)$ parameterizes the various filaments. The filament nodes move according to (1). The core size σ is a function of time and possibly position. It evolves so as to preserve "volume" globally (2a) or locally (2b) i.e.,

$$\sigma^2_i L_i = \text{cst.} \quad (2a)$$

$$\sigma^2_i \left|dr_i/d\theta\right| = \text{cst.} \quad (2b)$$

where L_i is the total arclength of the ith filament. To make contact with the Euler equations we define

$$\langle\left|\omega\right|^n\rangle = \sum_i \Gamma_i^n \int_0^L \frac{ds_i}{\sigma_i^{2n-2}} \quad (3)$$

where ds is an element of arclength. Equation (3) is reasonable since the circulation is the area integral of the vorticity and is conserved. When the radius of curvature r_c is much greater σ and all pieces of filament are well separated, (1) is an accurate representation of the Euler equations.

The first singularity is a pointwise collapse in which two pieces of filament become antiparallel and propagate down to a

0075-8485/86 $1.00 + $.25 per page

point with r_c and the minimum distance d between them tending to zero approximately as $(t^*-t)^{1/2}$. When d ~ σ the filaments pair antiparallelly along an appreciable fraction of their initial length and the paired object proceeds to rapidly stretch.

The statistical properties of the resulting tangle depend significantly on whether (2a) or (2b) is used to model σ. Under (2a) we have followed a single closed filament out to L ~ $10L_0$ and find the arclength grows faster than exponential eventhough the velocity is always bounded by $\sim\Gamma/\sigma$. This behavior is easily understood if one assumes the curve is a fractal whose length scales as r_c^{1-D}, D>0. If $r_c \sim (\Gamma/\sigma)(t^*-t)$ Then

$$L \sim (t^*-t)^{2(1-D)/(1+D)} \qquad (4)$$

Equation (2b) gave a much more intermittent curve which was difficult to follow much beyond L ~ $3L_0$. It was obvious, however, that $\langle\omega^2\rangle$ grew much faster than $\langle|\omega|\rangle = L$.

We believe the only way to clearly settle whether (1) can generate infinite line or $\langle\omega^2\rangle$ in finite time is to exploit the pairing which makes the velocity a local function of r. A useful preliminary would be to examine models of (1) which use a single space curve to represent the pair. If the velocity is local then:

$$v = \frac{\Gamma}{\sigma}\left(\alpha \hat{n} + \beta \hat{b}\right) \qquad (5)$$

where $\hat{n}$, $\hat{b}$ are the unit normal and binormal and α, β are arbitrary constants or smooth functions of order 1. Investigation of (5) is in progress.

A complete description of our simulations of (1) will appear in Phys. Fluids Winter 1984.

Department of Physics
Cornell University
Ithaca, N.Y. 14853

Lectures in Applied Mathematics
Volume 23, 1986

A PRIORI ESTIMATES AND EXISTENCE RESULTS FOR THE VLASOV AND BOLTZMANN EQUATIONS

C. BARDOS[1]
P. DEGOND[2]
F. GOLSE[3]

ABSTRACT. In this paper we emphasize the similarities between the solutions of the BOLTZMANN or VLASOV equation and the solutions of the non linear hyperbolic systems. In particular we introduce a notion of dispersion which leads to the existence of a smooth, global in time solution in three space variables for small initial data.

INTRODUCTION. The main purpose of this talk is to describe a global (in time) regularity result for the solution of the VLASOV-POISSON, in three space variables, with small initial data (cf. BARDOS-DEGOND [2] for a detailed version). However it is convenient to compare the situation for the VLASOV-POISSON with the situation for other companion equations like VLASOV-MAXWELL, FOKKER PLANCK or, of course, the BOLTZMANN equation. We emphasize the similarity of the situation for these equations, with the situation for the non linear hyperbolic systems or for the non linear wave equation. This idea is in some sense not new, it goes back to BOLTZMANN and HILBERT and the first section is an attempt to present a modern version of their concepts.

On the other hand, the use of the dispersion to prove existence of a smooth solution, with small initial data (cf. PONCE and KLAINERMAN [10] and al.) brings in new tools which also work for our problems.

I. A REMARK ON THE RELATION BETWEEN THE ENTROPY AT THE LEVEL OF FLUID MECHANICS AND THE BOLTZMANN H - THEOREM. In 1971, FRIEDRICHS and LAX made the following remarks concerning a first order system :

$$(1) \qquad \frac{\partial U}{\partial t} + \sum_{i=1}^{n} \frac{\partial}{\partial x_i} F_i (U) = 0 \quad .$$

If (1) has a convex entropy, i.e. if there exists a convex function $U \to \eta (U)$ which satisfies the relation :

$$(2) \qquad \nabla_U \eta (U) \cdot \nabla_U F_i (U) = \nabla_U Q_i (U)$$

(Q_i is the corresponding entropy flux).

Then :

(i) the system (1) admits a symetrizer and is therefore hyperbolic.

(ii) If U_ε is a solution of the equation :

$$\frac{\partial U_\varepsilon}{\partial t} + \sum_{i=1}^{n} \frac{\partial}{\partial x_i} F_i (U_\varepsilon) = \varepsilon \Delta U_\varepsilon \ .$$

which converges in the LEBESGUE sense ($U_\varepsilon \in L^\infty_{loc} (\mathbb{R}^n_x \times \mathbb{R}^+_t)$ and converges almost every where) then the limit is a solution of the equation (1) and satisfies also the entropy inequality.

$$(3) \qquad \frac{\partial}{\partial t} \eta (U) + \sum \frac{\partial}{\partial x_i} Q_i (U) \leqslant 0 \quad .$$

The same remark turns out to be valid for the relation between the BOLTZMANN equation and the compressible EULER Equation.

Expressed in term of conservation of mass ρ, momentum

$m = \rho u$, and internal energy $E = \rho\,(|u|^2/2 + \frac{3}{2}T)$ the compressible EULER equation are :

$$\frac{\partial \rho}{\partial t} + \nabla . m = 0 \tag{4}$$

$$\frac{\partial m_i}{\partial t} + \frac{\partial}{\partial x_j}\left(\frac{m_i\, m_j}{\rho}\right) + \frac{\partial}{\partial x_i}\left(\frac{2}{3}\left(E - \frac{|m|^2}{2\rho}\right)\right) = 0 \quad . \tag{5}$$

$$\frac{\partial}{\partial t} E + \nabla . \left(5\,\frac{m\,E}{3\rho} - \frac{|m|^2}{3\rho^2}\, m\right) = 0 \quad . \tag{6}$$

Now one can prove that the function $S = \rho \operatorname{Log} \frac{\rho^{2/3}}{T}$ is an entropy, which is convex with respect to the variables ρ, m and E.

Therefore a weak solution should satisfy the relation :

$$\frac{\partial}{\partial t} S + \nabla . (u\, S) \leq 0 \quad . \tag{7}$$

On the other hand the collision operator for the BOLTZMANN equation

$$Q\,(f,f) = \int_{\mathbb{R}^3_{v_1} \times S^2} (f' f'_1 - f\, f_1) q(|v - v_1|, \omega) dv_1\, d\omega \tag{8}$$ [1]

satisfies the following relations :

$$\int_{\mathbb{R}^3_v} Q\,(f,f) dv = \int_{\mathbb{R}^3_v} v_i\; Q\,(f,f) dv \tag{9}$$

$$= \int_{\mathbb{R}^3_v} |v|^2\; Q\,(f,f) dv = 0 \quad .$$

and the so called H-theorem :

$$\int_{\mathbb{R}^3_v} Q\,(f,f)\,\frac{d}{df}\,(f \operatorname{Log} f)\; dv \tag{10}$$

[1] The definitions and notations are classical (cf. CHAPMAN-COWLING [4])

$$= \int_{\mathbb{R}^3_v} Q\,(f,f)\,(1 + \text{Log } f)\,dv$$

$$= -4 \iint (f'f'_1 - f f_1)\,(\text{Log } f'f'_1 - \text{Log } f f_1)$$

$$\cdot\, q\,(|v - v_1|, \omega)\,dv_1\,d\omega\,dv \quad .$$

The right hand side of (10) is negative and the only continous solutions of (10) are the so called local maxwellian :

$$(11) \qquad f_M\,(x,v,t) = \frac{\rho(x,t)\; e^{-|v - u(x,t)|^2/2\,T\,(x,t)}}{(2\,\pi\,T\,(x,t))^{\frac{3}{2}}}$$

With these properties one can prove a result which mimics the FRIEDRICHS-LAX Remark of 71.

THEOREM 1 (BARDOS-GOLSE [3]). *Assume that for any* $\varepsilon > 0$, $f_\varepsilon\,(x,v,t)$ *is a smooth function, uniformly bounded in some* L^p *spaces (for details cf.* [3]*) and which converges a.e. to some function* $f(x,v,t)$.

Then if for any $\varepsilon > 0$, $f_\varepsilon\,(x,v,t)$ *is a solution of the BOLTZMANN equation*

$$(12) \qquad \frac{\partial f_\varepsilon}{\partial t} + v.\nabla_x\, f_\varepsilon = \frac{1}{\varepsilon}\, Q\,(f_\varepsilon, f_\varepsilon) \quad ,$$

the limit is a local maxwellian.

$$(13) \qquad f = \frac{\rho}{(2\,\pi\,T)^{\frac{3}{2}}}\, e^{-|v - u|^2/2\,T}$$

and the functions ρ, $m = \rho u$ *and* $E = \rho\,(|u|^2/2 + \frac{3}{2}\,T)$ *satisfy the compressible EULER equation and the entropy inequality.*

PROOF. The basic estimate is obtained by multiplying (12) by $\frac{d}{df}\,(f_\varepsilon \text{ Log } f_\varepsilon)$ and integrating over $\mathbb{R}^3_v \times \mathbb{R}^3_x \times [0,t]$;

this will lead to the formula :

$$(14)\qquad \varepsilon \iint (f_\varepsilon \operatorname{Log} f_\varepsilon)(x,t,v)\, dx\, dt\, dv$$

$$+ 4 \int_0^t \iiint (f'_\varepsilon f'_{\varepsilon 1} - f_\varepsilon f_{\varepsilon 1})(\operatorname{Log} f'_\varepsilon f'_{\varepsilon 1} - \operatorname{Log} f_\varepsilon f_{\varepsilon 1})$$

$$q(|v - v_1|, \omega)\, dv_1\, d\omega\, dv\, dx\, ds = \varepsilon \iint (f_\varepsilon \operatorname{Log} f_\varepsilon) \ .$$

Which implies, for ε going to zero the relation :

$$\iiint (f' f'_1 - f f_1)(\operatorname{Log} f' f'_1 - \operatorname{Log} f f_1)$$

$$q(|v - v_1|, \omega)\, dv_1\, dv\, d\omega \ .$$

Therefore f is almost everywhere equal to a local maxwellian. Now, with the relations (9) and (10) we have :

$$(15)\qquad \frac{\partial}{\partial t} \int f_\varepsilon\, dv + \nabla . \int v f_\varepsilon\, dv = 0$$

$$(16)\qquad \frac{\partial}{\partial t} \int v_i f_\varepsilon\, dv + \sum \frac{\partial}{\partial x_j} \int v_i v_j f_\varepsilon\, dv = 0$$

$$(17)\qquad \frac{\partial}{\partial t} \int |v|^2 f_\varepsilon\, dv + \nabla \int v |v|^2 f_\varepsilon\, dv = 0$$

$$(18)\qquad \frac{\partial}{\partial t} \int f_\varepsilon \operatorname{Log} f_\varepsilon\, dv + \nabla . \int v f_\varepsilon \operatorname{Log} f_\varepsilon\, dv$$

$$= - \frac{1}{\varepsilon} \iint (f'_\varepsilon f'_{1\varepsilon} - f_\varepsilon f_{1\varepsilon})(\operatorname{Log} f'_\varepsilon f'_{1\varepsilon} - \operatorname{Log} f_\varepsilon f_{1\varepsilon})$$

$$. \ q(|v - v_1|, \omega)\, dv_1\, d\omega\, dv \leqslant 0 \ .$$

If we let ε go to zero in these relations we finally obtain expressions involving a local maxwellian, for which the corresponding moments can be computed explicitly in term of ρ, u and T. This gives (4), (5), (6) and (7).

The FRIEDRICHS-LAX remark was first used for a scalar equation and only later turned out to be an important step in the construction of DI PERNA. In the present situation the things are worse, mostly because no existence result in the large, independant of ε is available for the BOLTZMANN Equation. On the other hand, this remark and the eventual appearance of singularities for the Compressible EULER Equation show that one should not expect uniform estimates in regular spaces for solutions of the BOLTZMANN equation, independant of the mean free path.

II. A NOTION OF DISPERSION FOR MICROSCOPIC EQUATIONS. THE EXAMPLE OF THE BOLTZMANN EQUATION. It has been observed by many authors, and in particular by KLAINERMAN and PONCE [10], and SHATAH [11] that one can construct a global (in time) smooth solution for the non linear wave equation provided the following conditions are true.

(i) The solution is defined in the whole space (or in the exterior of a bounded domain) and the dimension of the space is large enough.

(ii) The initial data is small enough and localised.

The proof relies on a priori estimate related to the dispersion rate property of the linearised equation. It turns out that similar ideas can be used for the BOLTZMANN and the VLASOV equations and will lead to analogous results. First, we consider the BOLTZMANN Equation, we assume that the mean free path ε is fixed ($\varepsilon = 1$ for instance) and that the initial data is localised and small enough. In this case we will prove the existence of a global solution which asymptotically behaves like the solution of a transport equation. This result is indeed a special case of a Theorem first given by ILNER and SHINBROT [8] and the improved by HAMDACHE [7].

We will consider the BOLTZMANN equation :

(19) $$\frac{\partial f}{\partial t} + v.\nabla f = Q\,(f,f)$$

as a perturbation of the Transport equation in $\mathbb{R}^d_x \times \mathbb{R}^d_v$.

(20) $$\frac{\partial f}{\partial t} + v.\nabla f = 0 \quad .$$

For (20) the dispersion property is the following : one assumes that the initial data satisfies the following estimate :

(21) $$0 \leqslant f\ (x,v,0) \leqslant h(x)$$

with $h(x) \in L^1(\mathbb{R}^d_x)$; then an explicit calculation, and a change of variable in the integral gives the result :

(22) $$0 \leqslant \int f(x,v,t)\ dv \leqslant \int f(x - vt,v,0)\ dv$$
$$= \int h(x - vt)\ dv$$
$$= \frac{1}{t^d} \int h(X)\ dX$$

with $X = x - vt$. Therefore the total charge (or density) disperses like $1/_{t}d$.

THEOREM 2. Assume that the cross section $q(|v - v_1|,\omega)$ *is uniformly bounded* [1]. *Assume that the initial data is smooth and satisfies the estimate.*

(23) $$0 \leqslant f_o\ (x,v) \leqslant Ae^{-|x|^2}$$

with A *small enough. Then the BOLTZMANN equation :*

(24) $$\frac{\partial f}{\partial t} + v.\nabla_x f = Q\ (f,f)\ ,\quad f(x,v,t) = f_o(x,v)$$

has a unique smooth solution defined in $\mathbb{R}_t \times \mathbb{R}^3_x\ \mathbb{R}^3_v$.

Furthermore, there exist two functions $g^{\pm}\ (x,v,t)$ *solution of the Transport equation :*

(25) $$\frac{\partial g^{\pm}}{\partial t} + v.\nabla g^{\pm} = 0$$

([1]) For more realistic results cf. ILNER and SHINBROT [8] or HAMDACHE [7] .

such that one has :

(26) $$\lim_{t \to \pm \infty} \sup_{x,v} |f(x,v,t) - e^{-|x-vt|^2} g(x,v,t)| = 0 \quad .$$

PROOF. The basic ingredient is the a-priori estimate and it will be our unique concern.

We write :

(27) $$f(x,v,t) = e^{-|x-vt|^2} g(x,v,t) \quad .$$

and we notice that we have :

(28) $$(\frac{\partial}{\partial t} + v.\nabla) (e^{-|x-vt|^2}) = 0 \quad .$$

Therefore we obtain the relation :

(29) $$e^{-|x-vt|^2} (\frac{\partial g}{\partial t} + v.\nabla g) = \iint (e^{-|x-v_1't|^2 - |x-v't|^2} g'g_1' - e^{-|x-vt|^2 - |x-v_1t|^2} g\, g_1)\, q(|v-v_1|,\omega) d\omega\, dv_1 \quad .$$

By construction of the collision operator, we have :

(30) $$|x-vt|^2 - |x-v_1't|^2 - |x-v't|^2 = |x-v_1t|^2 \quad .$$

Therefore we deduce from (29) and (30) the relation :

(31) $$\frac{\partial g}{\partial t} + v.\nabla g = \iint e^{-|x-v_1t|^2} (g'g_1' - g\, g_1)\, q\, dv_1\, d\omega \quad .$$

Finally, with $X(t) = \sup_{(x,v)} g(x,v,t)$ we deduce from (31) the relation :

(32) $$\frac{d}{dt_+} X(t) \leq 2\, C \iint e^{-|x-v_1t|^2} (X(t))^2\, dv_1\, d\omega \quad .$$

The change of variable $X = x - v_1 t$, and a result of local existence (for t small, cf. ILNER and SHINBROT [8]) gives :

$$(33) \qquad X(t) \leqslant X(o)\left[1 - C\left(\int_o^t \frac{ds}{(1+s)^3}\right) X(o)\right]^{-1}$$

Therefore for $X(o)\int_o^\infty \frac{ds}{(1+s)^3} < 1$ we have a uniform (in time) estimate which leads to global existence.

To prove the asymptotic behaviour (for $t \to \pm\infty$) of the solution, we introduce the unitary group e^{tA} spanned by the operator $-v.\nabla_x$, and we will prove the existence of a function $g_+(x,v)$ such that one has :

$$\lim_{t\to\infty} |e^{tA} g_+ - g(x,v,t)|_{L^\infty(\mathbb{R}^3_x \times \mathbb{R}^3_v)} = 0 \quad .$$

Now since e^{tA} is unitary, it is enough to show that :

$$\frac{d}{dt}(e^{-tA} g(x,v,t))$$

is bounded in $L^1(\mathbb{R}_+ ; L^\infty(\mathbb{R}^3_x \times \mathbb{R}^3_v))$

Indeed we have :

$$(34) \qquad \frac{d}{dt}(e^{-tA} g(x,v,t)) = e^{-tA}\left(\frac{\partial g}{\partial t} + v.\nabla_x g\right)$$

and :

$$(35) \qquad \left|\frac{d}{dt}(e^{-tA} g(x,v,t)\right|_\infty = \left|\frac{\partial g}{\partial t} + v.\nabla_x g\right|_\infty \leqslant \frac{C}{t^3} \quad .$$

and this completes the proof.

III. THE DISPERSION EFFECT FOR THE VLASOV EQUATION. In this section, we consider the VLASOV-POISSON equation used to describe the interaction of charged particles in a plasma. N families of particles of charge q_α and mass m_α interact in the plasma. If the collision terms are omitted, we have the following system :

$$(36) \qquad \frac{\partial f_\alpha}{\partial t} + v.\nabla_x f_\alpha + \frac{q_\alpha}{m_\alpha}\nabla_x \Phi.\nabla_v f_\alpha = 0 \qquad 1 \leqslant \alpha \leqslant N \quad .$$

(37) $E = -\nabla\Phi,\ -\Delta\Phi = 4\Pi\rho\ ,\ \rho = \sum_\alpha q_\alpha \int_{\mathbb{R}^d} f_\alpha\ (x,v,t)dv$ ([1])

E is the electric field ; we have assumed that the magnetic field remains very small ; therefore it has been neglected and instead of the MAXWELL equations we have the POISSON equation (37).

Since Φ is independant of v, the only non linear term $\nabla_x \Phi . \nabla_v f_\alpha$ can be written in conservative form :

(38) $\nabla_x \Phi . \nabla_v f_\alpha = \nabla_v (\nabla_x \Phi . f_\alpha)$

The equation (36) means that the function $f_\alpha\ (x,v,t)$ is constant along the trajectories of the vector field :

$$(v,\ -\nabla_x \Phi)\ .$$

This vector field is hamiltonian in the phase space $\mathbb{R}^d_x \times \mathbb{R}^d_v$. Therefore the following quantities are conserved :

$$\sup f_\alpha\ (x,v,t)\ ,\quad \iint |f_\alpha\ (x,v,t)|^p\, dx\, dv \quad 1 \leqslant p \leqslant \infty\ .$$

This implies that the right hand side of the equation :

$$-\Delta\Phi = 4\Pi\rho$$

is uniformly bounded in $L^1\ (\mathbb{R}^d_x \times \mathbb{R}^d_v)$ and with (38) this is the basic ingredient for the proof (ARSENEV [1]) of the existence of a weak solution. On the other hand there is no a priori bound for any quantity of the following type :

$$\left(\int dx\ \left(\int f(x,v,t)dv\right)^p dx\right)^{1/p}\ ,\quad 1 < p \leqslant \infty\ .$$

and this is the main obstacle to the construction of a smooth solution.

Estimates concerning the expression :

$$\sup_x \int f(x,v,t)\ dv$$

([1]) d denotes the dimension of the space, $d = 1,2$ or 3.

can easily be obtained in term of the expression :

$$M(t) = \sup_x |\nabla_x \Phi(x,t)|$$

and this leads to the following GRONWALL type estimate :

$$(39) \qquad \frac{d}{dt} M(t) \leqslant C\left(1 + \left(\int_0^t M(s)\,ds\right)^d\right)^{(d-1)/d}$$

For $d \leqslant 2$ (39) provides an estimate valid for all time while for $d = 3$ we have a non linear inequation and the estimate is valid only for a finite time (depending on the size of the initial data). This leads to a global result of smoothness in dimension less or equal to 2 and to a local (in time) result of smoothness in dimension 3 (cf. UKAI and OKABE [12] or alt. for details).

In dimension three we can obtain the existence of a smooth solution for all time, but in a situation similar to the one considered by KLAINERMANN and PONCE for the wave equation, or to the one considered by ILNER and SHINBROT for the BOLTZMANN equation.

We will consider the VLASOV-POISSON equation as a perturbation of the transport equation :

$$(40) \qquad \frac{\partial f_\alpha}{\partial t} + v.\nabla_x f_\alpha = 0 \qquad (1 \leqslant \alpha \leqslant N) \quad .$$

The charge :

$$\rho(x,t) = \sum q_\alpha \int f_\alpha (x,v,t)\,dv$$

decays in $L^\infty(\mathbb{R}^d_x)$ like t^{-d}.

We will need a series of notations and lemmas.

For a vector field $E(x,t)$, we denote by $(X(s,t,x,v)$, $V(s,t,x,v))$ or, in short (X,V), the solution of the system :

$$(41) \qquad \frac{dX}{ds} = V, \ \frac{dV}{ds} = E(X,s) \ ; \ X(t,t,x,v) = x, V(t,t,x,v) = v.$$

If E is uniformly lipschitz this solution is defined for all (x,v,t).

LEMMA 1. We assume that $E(x,t)$ *satisfies the estimates.*

(41) $\qquad \| \nabla_x E(.,t)\|_\infty \leqslant \eta \,/\, (1+t)^{\frac{5}{2}}$

(42) $\qquad \| \nabla_x^2 E(.,t)\|_\infty \leqslant \eta \,/\, (1+t)^{\frac{5}{2}}$

with $\eta < 1$.

Then for $0 \leqslant s \leqslant t$ *the following estimates are true.*

(43) $\qquad |\frac{\partial X}{\partial v}(s,t,x,v) - (s-t)\ \mathrm{Id}| + |\frac{\partial V}{\partial v}(s,t,x,v)| \leqslant C\,\eta(t-s)$

(44) $\qquad |\frac{\partial X}{\partial x}(s,t,x,v) - \mathrm{Id}| + |\frac{\partial V}{\partial x}(s,t,x,v)| \leqslant C\,\eta$

(45) $\qquad |\frac{\partial^2 X}{\partial x^2}(s,t,x,v)| + |\frac{\partial^2 V}{\partial x^2}(s,t,x,v)| \leqslant C\,\eta\,.$

Where Id *denotes the identity matrix and* C *any arbitrary constant.*

PROOF. Use a simple linear GRONWALL lemma.

LEMMA 2. Under the hypothesis of Lemma 1, for η *small enough one has :*

(46) $\qquad |\det\,(\frac{\partial X}{\partial v}(s,t,x,v))| \geqslant \frac{(t-s)^3}{2}$

and for any fixed $(s,t,x) \in \mathbb{R}_t^+ \times \mathbb{R}_t^+ \times \mathbb{R}_x^3$ *the mapping*

$$\Theta : \ v \longrightarrow X(s,t,x,v)$$

is one to one.

PROOF. From (46) we deduce that it is locally one to one. For s close to t it is one to one ; finally a transversality argument shows that it is one to one for any s ; $0 \leqslant s \leqslant t$.

The Lemma 2 shows that under the hypothesis of the Lemma 1 the projection in the x space of the solution of the system (41) does not generate any caustic.

For the Lemma 1 we need estimates on the L^∞ norm of the second derivative of the solution of the equation $\Delta\,\Phi = \rho$ in term of the L^∞ norm of the first derivatives of ρ. By standard interpolation we prove the following.

LEMMA 3. The following estimate

$$(47) \qquad \| D^2 \Phi \|_\infty \leq C_\Theta \| \Delta \Phi \|_\infty^{3(1-\Theta)/(3+\Theta)}$$

$$. \| \nabla_x (\Delta \Phi) \|_\infty^{3\Theta/(3+\Theta)} . \| \Delta \Phi \|_1^{\Theta/3+\Theta}$$

is valid for any smooth bounded function.

(D^2 denotes any second derivative and C_Θ a constant depending only on $\Theta \in]0,1[$).

Finally we have the

THEOREM 3. We assume that the functions $f_{0,\alpha}(x,v)$ *are twice continuously differentiable, and that they satisfy, the following estimates, for any pair* $(x,v) \in \mathbb{R}^3_x \times \mathbb{R}^3_v$ *and any* α $(1 \leq \alpha \leq N)$:

$$(48) \qquad 0 \leq f_{\alpha,0} (x,v) \leq \varepsilon / (1+|x|)^4 (1+|v|)^4$$

$$(49) \qquad |\nabla_{(x,v)} f_{\alpha,0} (x,v)| + |\nabla^2_{(x,v)} f_{\alpha,0} (x,v)| \leq$$

$$\varepsilon / (1+|x|)^4 (1+|v|)^5 .$$

Then there exists $\varepsilon_0 > 0$, *such that for* $\varepsilon < \varepsilon_0$ *the VLASOV-POISSON equation*

$$(50) \qquad \begin{cases} \dfrac{\partial f_\alpha}{\partial t} + v.\nabla_x f_\alpha + \dfrac{q_\alpha}{m_\alpha} E.\nabla_v f_\alpha = 0 \\ f_\alpha (x,v,0) = f_{\alpha,0} (x,v) \end{cases}$$

$$(51) \qquad \rho(x,t) = \sum_\alpha q_\alpha \int f_\alpha (x,v,t) \, dv$$

$$(52) \qquad E = - \nabla \Phi , \quad - \Delta \Phi = 4 \Pi \rho$$

has a unique smooth solution defined for all time.

PROOF. We use an iterative scheme with

$$E_n = - \nabla \Phi_n \quad , \quad - \Delta \Phi_n = 4 \pi \rho_n \quad \text{and} \quad \rho_n = \sum q_\alpha \int f_\alpha^n(x,v,t)\, dv$$

f_α^{n+1} are the solutions of the linear transport equation.

$$\frac{\partial f_\alpha^{n+1}}{\partial t} + v.\nabla_x f_\alpha^{n+1} + \frac{q_\alpha}{m_\alpha} E_n.\nabla_v f_\alpha^{n+1} = o \tag{53}$$

$$f_\alpha^{n+1}(x,v,o) = f_{\alpha,o}(x,v) \quad . \tag{54}$$

We assume that ρ_n satisfy the estimates :

$$\|\rho_n(.,t)\|_\infty + \|\nabla_x \rho_n(.,t)\|_\infty + \|\nabla_x^2 \rho_n(.,t)\|_\infty \leqslant A / (1+t)^3 \quad . \tag{55}$$

$$\|\rho_n(.,t)\|_1 + \|\nabla_x \rho_n(.,t)\|_1 \leqslant \varepsilon A \quad . \tag{56}$$

and we will prove that the same is true at the order $n+1$.

We apply the Lemma 3 with $\Theta = \frac{3}{5}$ and we obtain :

$$\|\nabla_x E_n(.,t)\|_\infty + \|\nabla_x^2 E_n(.,t)\|_\infty \leqslant C A \varepsilon^{\frac{1}{6}} / (1+t)^{\frac{5}{2}} \quad . \tag{57}$$

Therefore we can apply, with $\varepsilon^{\frac{1}{6}} < 1/C A$ the Lemma 2. For instance we will have, with obvious notations :

$$\begin{aligned} \int f_\alpha^{n+1}(x,v,t)\, dv &= \int f_\alpha(X_\alpha^n, V_\alpha^n, o)\, dv \\ &\leqslant \varepsilon \int \frac{1}{(1+|X_\alpha^n|)^4} \left| \frac{\partial X_\alpha^n}{\partial v} \right|^{-1} dX_\alpha^n \quad . \\ &\leqslant \frac{\varepsilon}{2t^3} \int \frac{1}{(1+|X|)^4}\, dX \quad . \end{aligned} \tag{58}$$

For ε small enough and $t > t_o$ (local existence is classical), we deduce the relation :

$$\|\rho_{n+1}(.,t)\|_\infty \leqslant \frac{A}{3(1+t)^3} \tag{59}$$

Similarly we have

$$(60) \qquad \int \nabla_x f_\alpha^{n+1}(x,v,t)\,dv = \int \nabla_x f_{\alpha,o}(X_\alpha^n, V_\alpha^n, o)\,dv$$
$$\leqslant \int (|\nabla_x f_{\alpha,o}|\, |\frac{\partial X_\alpha^n}{\partial x}| + |\nabla_v f_{\alpha,o}|\, |\frac{\partial V_\alpha^n}{\partial x}|)\,dv$$

therefore we may use (44) and proceed for the rest of the proof as we did above. All the other estimates are similar and it is the end of the essential part of the proof.

It may be possible to extend some of the ideas contained in the present paper to other situations ; this leads to a serie of open problems some difficults other tractables. They are quoted in the following remarks.

REMARK 1. If we do not assume that the magnetic field is small compared to c, the velocity of the light, the VLASOV-POISSON have to be replaced by the so called VLASOV-MAXWELL equation :

$$(61) \qquad \frac{\partial f_\alpha}{\partial t} + v.\nabla_x f_\alpha + \frac{q_\alpha}{m_\alpha}(E + \frac{v\wedge B}{c}).\nabla_v f_\alpha = o \ ; \quad 1 \leqslant \alpha \leqslant N$$

$$(62) \qquad \frac{1}{c}\frac{\partial E}{\partial t} - \nabla\wedge B = -4\pi J \ ; \ \frac{1}{c}\frac{\partial B}{\partial t} + \nabla\wedge E = o$$

$$(63) \qquad J = \sum_\alpha q_\alpha \int_{\mathbb{R}^d} v f_\alpha (x,v,t)\,dv \quad .$$

In this case, in any dimension, only local in time existence in the class of smooth solutions is known. The proof of this result however is not trivial (cf. WOLLMANN [14], or DEGOND [5]).

However it may be possible that with more technical tools one could adapt the ideas of KLAINERMAN [9] to prove the existence of a smooth solution in $\mathbb{R}^3$ with small initial datas ; it may turn out that the method should work more easily for the relativistic version of the problem (61), (62), (63). (Cf. VAN KAMPEN and FELDERHOF [13] for the precise form of these equations).

REMARK 2. One could introduce in the equations (61), (62), (63) a collision operator and obtain for a one family of particles the system.

$$(64) \qquad \frac{\partial f_\varepsilon}{\partial t} + v.\nabla_x f_\varepsilon + \frac{q}{m}(E + \frac{v\wedge B}{c}) \nabla_v f_\varepsilon = \frac{1}{\varepsilon} Q(f_\varepsilon, f_\varepsilon) \quad .$$

$$(65) \qquad \frac{1}{c}\frac{\partial E_\varepsilon}{\partial t} - \nabla\wedge B_\varepsilon = -4\pi J_\varepsilon \ , \ \frac{1}{c}\frac{\partial B_\varepsilon}{\partial t} + \nabla\wedge E_\varepsilon = o \quad .$$

$$(66) \qquad J_\varepsilon = q \int v f_\varepsilon \ (x, v, t) \ dv \quad .$$

Now if the quadratic KERNEL Q is the same as in the BOLTZMANN equation, one could adapt the idea of the section 1 and prove that if f_ε converges in a convenient sense (cf. Theorem 1) to a function f, this function is a local maxwellian $f = \rho/(2\pi T)^{\frac{3}{2}} \exp(-|v-u|^2/2T)$ and ρ, $J = \rho u$, $e = \rho(|u|^2/_2 + \frac{3}{2} T)$ and E, B are solution of the M.H.D. system :

$$(67) \qquad \frac{\partial \rho}{\partial t} + \nabla.J = o$$

$$(68) \qquad \frac{\partial J_i}{\partial t} + \frac{\partial}{\partial x_j}\left(\frac{J_i J_j}{\rho}\right) + \frac{\partial}{\partial x_i}\left(\frac{2}{3}\left(e - \frac{|J|^2}{2\rho}\right)\right) = o$$

$$(69) \qquad \frac{\partial e}{\partial t} + \nabla.\left(\frac{5}{3\rho} J.e - \frac{|J|^2}{3\rho^2} J\right) = o$$

$$(70) \qquad \frac{1}{c}\frac{\partial E}{\partial t} - \nabla\wedge E = -4\pi J \ , \ \frac{1}{c}\frac{\partial B}{\partial t} + \nabla\wedge E = o$$

with the entropy inequality :

$$(71) \qquad \frac{\partial}{\partial t}\left(\rho \operatorname{Log} \frac{\rho^{\frac{2}{3}}}{T}\right) + \nabla.\left(J.\operatorname{Log} \frac{\rho^{\frac{2}{3}}}{T}\right) \leqslant o \quad .$$

However in the case of a gas of charged particles it turns out that the collision operator has to be replaced by a term of the form $c \, \Delta_v f$.

This leads to the so called FOKKER-PLANCK equation. For the FOKKER-PLANCK KERNEL, one has the relation :

(72) $$\int \Delta_v f . \frac{d}{df} (f \operatorname{Log} f)\, dv = - \int \frac{1}{f} |\nabla_v f|^2\, dv \leq o$$

and this implies the decay of the quantity :

$$\int f \operatorname{Log} f\, dv$$

However a limit of a sequence of solutions of (64)-(66) with $Q(f_\varepsilon, f_\varepsilon)$ replaced by $\Delta_v f_\varepsilon$, will satisfy the relation :

(73) $$\int \frac{1}{f} |\nabla_v f|^2\, dv = o$$

and this does not imply that f is a maxwellian, therefore it is impossible to extend to this problem the ideas of section 1.

REMARK 3. Finally one can also consider the FOKKER-PLANCK-POISSON system :

(73) $$\frac{\partial f}{\partial t} + v.\nabla_x f + E.\nabla_v f = \Delta_v f$$

(74) $$E = - \nabla\Phi \ , \ - \Delta\Phi = 4\pi \int f(x,v,t)\, dv \ .$$

and extend for this problem the results existing for the VLASOV-POISSON system. Once again, it may be easy to prove the existence of a weak solution in any space variable. To the best of our knowledge the only regularity result available concerns the two space variables. It is an extension by a probabilistic approach of the result of UKAI and OKABE [12] and it is due to NEUNZERT, PULVIRENTI and TRIOLO [15].

Recently, an other proof of this result has been obtained thanks to standard E.D.P. techniques which do not involve any probabilities (see P. DEGOND "Existence globale de solutions de l'équation de Vlasov-Fokker-Planck, en dimension 1 et 2", C.R.A.S. Paris, to appear).

BIBLIOGRAPHY

[1]. A.A. ARSENEV, "Global existence of a weak solution of VLASOV System of equations". U.S.S.R. Comput. Math. and Math. Phys. 15 (1975), 131-143.

[2]. C. BARDOS and P. DEGOND, "Global existence for the VLASOV-POISSON Equation in 3 space variables with small initial data", Tech. Rep. 101 C.M.A. Ecole Polytechnique 91128 Palaiseau Cedex.

[3]. C. BARDOS and F. GOLSE, "Différents aspects de la notion d'entropie au niveau de l'équation de BOLTZMANN et de NAVIER STOKES", Note C.R.A.S.

[4]. CHAPMAN and COWLING, " The Mathematical Theory of Non-Uniform Gases", Cambridge University Press.

[5]. P. DEGOND, "Existence locale en temps de solutions régulières de l'équation de VLASOV-MAXWELL tridimensionnelle", (to appear).

[6]. FRIEDRICHS and P. LAX, "Proc. Nat. Acad. SC. 68 n° 8", (1971), 1686-1688.

[7]. K. HAMDACHE, " Existence globale et comportement asymptotique pour l'équation de BOLTZMANN à répartition discrète de vitesses", Preprint.

[8]. R. ILLNER and M. SHINBROT, "The BOLTZMANN equation : Global existence for a rare gas in an infinite vacuum", Preprint.

[9]. S. KLAINERMAN, "Long time behaviour of the solution to non linear equations", Arch. Rat. Mech. Anal. Vol. 78, (1982), 73-98.

[10]. S. KLAINERMAN and G. PONCE, "Global small amplitude solutions to non linear evolution equations", Comm. Pure Appli. Math. 36.1 , (1983), 133-141.

[11]. J. SHATAH, "Global existence of small solutions to non linear evolution equations", Preprint.

[12]. S. UKAI and T. OKABE, "On the classical solution in the large in time of the two dimensional VLASOV Equation", Osaka J, of Math. n° 15 245-261.

[13]. VAN KAMPEN and FEDERHOF, "Theoretical Methods in Plasma Physics". North Holland.

[14]. S. WOLLMANN," Existence and uniqueness theory of the VLASOV equation". Internal report of the courant Institute of mathematical sciences.

[15]. H. NEUNZERT, M. PULVIRENTI and L. TRIOLO, "On the VLASOV-FOKKER-PLANCK equation". Preprint n° 77. Universität Kaiserlautern. Jan. 84.

1. C. BARDOS
Centre de Mathématiques Appliquées
E.N.S. Ulm
45 rue d'Ulm
75005 PARIS

2. P. DEGOND
Centre de Mathématiques Appliquées
Ecole Polytechnique
91128 PALAISEAU Cedex

3. F. GOLSE
Centre de Mathématiques Appliquées
E.N.S. Ulm
45 rue d'Ulm
75005 PARIS

Lectures in Applied Mathematics
Volume 23, 1986

INVARIANT MANIFOLDS FOR PERTURBATIONS OF NONLINEAR PARABOLIC SYSTEMS WITH SYMMETRY (PRELIMINARY REPORT)

Peter W. Bates

ABSTRACT. Several physically interesting mathematical models are perturbations of continuous and compact group-invariant parabolic systems. In this note we study one such class of systems and obtain invariant manifolds (tori) as perturbations of group-generated invariant manifolds for the unperturbed system. The approach is to use an infinite dimensional version of the center-manifold theorem.

In Dee and Langer [6] and in Ben-Jacob et al. [4] the equation

(HSPM) $$\frac{\partial w}{\partial t} = [\varepsilon^2 - (\frac{\partial^2}{\partial x^2} + 1)^2]w - w^3$$

is studied as a pattern-forming model (see also Swift and Hohenberg [12] and Pomeau and Manneville [10]). After a change of variables $X = \varepsilon x/2$, $\tau = \varepsilon^2 t$ and $w(x,t) = \varepsilon[(u(X,\tau) + iv(X,\tau))\exp(ix) + (u(X,\tau) - iv(X,\tau))\exp(-ix)]/\sqrt{3}$ the authors reduce equation (HSPM) to "the complex amplitude equations giving the envelope of an oscillating wave form in the near-threshold case, $0 < \varepsilon << 1$, ..."

$(1)_\varepsilon$ $$\begin{aligned} u_\tau &= u_{XX} + (1 - (u^2 + v^2))u + O(\varepsilon) \\ v_\tau &= v_{XX} + (1 - (u^2 + v^2))v + O(\varepsilon) \end{aligned}$$

1980 Math. Subject Classification: 35K55 34G20 34C30 35B32

Similar systems have been used to model physical phenomena such as superconductivity [5] and threshold-excitation phenomena [7], for instance. In the latter category J. M. Lasry [8] proposed the following:

$$(2)_\varepsilon \qquad \begin{aligned} u_t &= u_{xx} + k(1 - (u^2 + v^2)^{1/2})u - \varepsilon\phi(\beta)v \\ v_t &= v_{xx} + k(1 - (u^2 + v^2)^{1/2})v + \varepsilon\phi(\beta)u \end{aligned}$$

where ε, $k > 0$ are parameters and ϕ is a 2π-periodic function of $\beta = \tan^{-1}(v/u)$ which is positive on $(\beta_0, 2\pi)$ and negative on $(0, \beta_0)$ where $0 < \beta_0 << 1$. Notice that systems $(1)_0$ and $(2)_0$ are invariant under rotations about the origin in the uv plane (and under reflections). Furthermore, if we look for spatially L-periodic solutions then these systems are invariant under translations in x modulo L (and under reflections in x). This gives us the torus group T^2, a continuous compact group, under which $(1)_0$ and $(2)_0$ are invariant.

The idea is to find invariant manifolds for $(1)_\varepsilon$ and $(2)_\varepsilon$ with $\varepsilon \neq 0$ as follows:

<u>STEP 1</u> Set $\varepsilon = 0$ and look for steady-state solutions which are spatially L-periodic. This is done by modding out the group action and applying bifurcation theory, using L as the bifurcation parameter.

<u>STEP 2</u> Use Conley's index [11] to find orbits connecting steady state solutions on the bifurcating branch to those on the 'trivial' branch.

<u>STEP 3</u> Apply the group action (in this case T^2) to generate manifolds of equilibria and connecting orbits.

STEP 4 Linearize at each point of a manifold of equilibria and apply the infinite dimensional version of the center manifold theorem (outlined below) to cover the manifold of equilibria for $(1)_0$ (or for $(2)_0$) with patches of invariant manifolds (center manifolds) for $(1)_\varepsilon$ (resp. $(2)_\varepsilon$) with $\varepsilon \neq 0$.

Steps 1 and 3 together with a stability analysis have been done for $(2)_\varepsilon$ by the author and D. Barrow in [1], [2], and [3] and in some unpublished work. The analysis for $(1)_\varepsilon$ is similar and will be summarized below.

Varying the period of steady state solutions to $(1)_0$ is equivalent to varying k in the system

$$(3) \qquad 0 = \mathbf{u}_{xx} + k(1 - |\mathbf{u}|^2)\mathbf{u}, \qquad \mathbf{u} = (u,v),$$

and looking for 2π-periodic solutions (hereafter called solutions). We mod out the T^2 group action by considering (3) in the space $X = \{(u,v)\colon u \text{ is even, } v \text{ is odd}\}$. All 2π-periodic solutions of (3) are obtained from solutions in X by applying the group action. Apart from the trivial solutions $|\mathbf{u}| = 0$ or 1, there are no solutions for $0 < k \leq 1$. Furthermore, 0 has a one dimensional unstable manifold in X and $\mathbf{u} = (1,0)$ is stable. It is easy to see that there are spatially constant orbits $\mathbf{u}(t)$ with $\mathbf{u}(-\infty) = 0$ and $|\mathbf{u}(\infty)| = 1$. At $k = j^2$, $j = 1, 2, \ldots$, there are six solution branches in X bifurcating to the right, one pair of branches of collinear solutions ($v \equiv 0$) and two pairs of branches of circular solutions ($|\mathbf{u}| = $ constant, $u \not\equiv 0$, $v \not\equiv 0$). Using Conley's index (see J. Smoller [11] Chapter 24) one can show that there are orbits from $\mathbf{u} = 0$ to each of the bifurcating branches for fixed $k > j^2$ and sufficiently close to j^2. This is done by computing the indices of the solutions in X:

index (0) = Σ^{2j+1} for $j^2 < k < (j+1)^2$
index (collinear solution) = Σ^{2j-1} for $j^2 < k$
index (circular solution) = Σ^{2j-1} for $j^2 < k < j^2 + 2j - 1/2$
and applying the methods in [11]. Alternatively, one can compute the indices of 0 and the collinear solution in the subspace $\{(u,v) \in X: \ v \equiv 0\}$ obtaining
index (0) = Σ^{j+1} for $j^2 < k < (j+1)^2$ and
index (collinear solution) = Σ^j there. Now apply Theorem 24.14 of [11]. Connections between 0 and the circular solutions are easily constructed of the form $\mathbf{u}(x,t) = a(t)\mathbf{u}_c(x)$ where $\mathbf{u}_c$ is the circular solution and $\dot{a} = k|\mathbf{u}_c|^2(1-a^2)a$ so that $a(t) \to 1$ as $t \to \infty$ and when $|a(0)| < 1$ $a(t) \to 0$ as $t \to -\infty$. Intuitively one expects orbits from the collinear to the circular solutions. This is born out by numerical experiments but we have not yet proved it.

Now as we increase k there is no secondary bifurcation from the branches of collinear solutions, however, from each of the branches of circular solutions there is a sequence of secondary pitchfork bifurcations (into what we call oval solutions) occurring at $k = 3j^2 - n^2/2$, $n = 1, 2, \ldots, 2j-1$. These branches propagate to the right and near $k = 3j^2 - n^2/2$ the index of an oval solution is Σ^n whereas for $3j^2 - n^2/2 < k < 3j^2 - (n-1)^2/2$ the index of the circular solution is Σ^{n-1}, $1 \leq n \leq 2j - 1$. Again, the arguments in [11] may be used to obtain orbits from each oval solution to the circular solution in a neighborhood of the bifurcation point. It is worth noting that the jth circular solution branch and its offshoots consists of solutions with winding number j, that is, in the uv-plane the image of $\mathbf{u}([0,2\pi])$ is a curve which winds around the origin j times. To prove the claim about the secondary bifurcation curves extending to the right we write (3) in polar coordinates $\mathbf{u} = (\rho\cos\beta, \rho\sin\beta)$

and replace k by $(3j^2-n^2/2+\lambda)$, ρ by $1-j^2/(3j^2-n^2/2+\lambda)+\bar{\rho}$, and β by $j+\bar{\beta}$. Simple bifurcation occurs at $(\lambda,\bar{\rho},\bar{\beta}) = (0,0,0)$ and if we parameterize the bifurcating curve we obtain after a lengthy calculation

$(\lambda(s),\ s(\cos nx + r(s)),\ s(-2j/n\sqrt{(6j^2-n^2)/(4j^2-n^2)}\sin nx + \alpha(s)))$

where $(\lambda(0),\ r(0),\ \alpha(0)) = (0,0,0)$, $\lambda'(0) = 0$,

$r'(0) = -A\cos 2nx - B$, $\alpha'(0) = C\sin 2nx$ where A,B,C are positive constants depending only upon j and n, and

$\lambda''(0) = 3(4j^2 + n^2)(6j^2 - n^2)/2n^2$ which is positive for all $n = 1, \ldots, 2j-1$.

The following diagram summarizes the bifurcation in X:

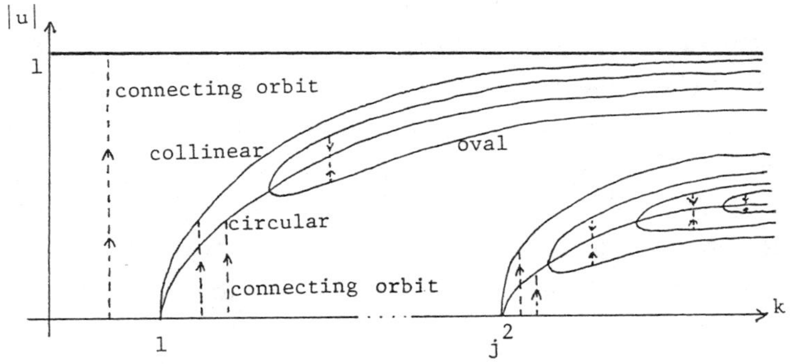

Below we show the images of some of these solutions in the uv-plane:

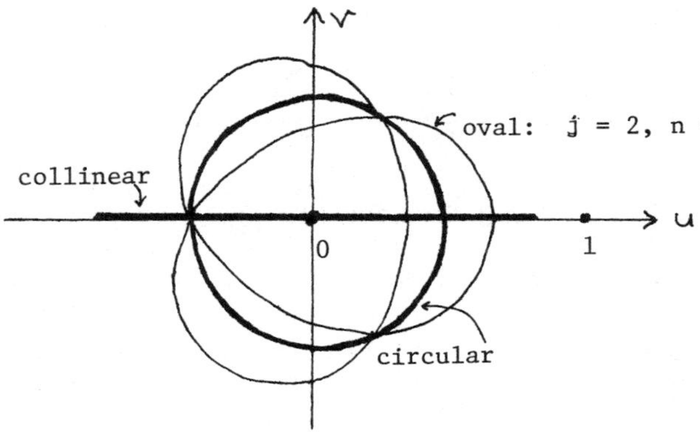

Now, to continue with our plan we employ the T^2 group action to get manifolds of solutions for fixed k. Note that because of the symmetry of some solutions we do not get a torus from each equilibrium. In fact $T^2(0,0)$ is the point (0,0), $T^2(1,0)$ is a circle of constant solutions, and T^2 applied to a circular solution is a circle of circular solutions. The collinear and oval solutions each generate a two-dimensional torus. The connecting orbit between (0,0) and a collinear solution and that between a circular and an oval solution generate three dimensional invariant manifolds.

We now introduce a perturbation and study $(1)_\varepsilon$ for $\varepsilon \neq 0$. It was shown in [1] by using the implicit function theorem that the circle of circular equilibrium solutions becomes a periodic travelling wave when $\varepsilon \neq 0$. Furthermore, for $k > 3j^2 - 1/2$, the circular solution on the jth branch becomes a stable periodic travelling wave with asymptotic phase when $|\varepsilon| \neq 0$ and is sufficiently small. To show the persistence of the tori of collinear and oval equilibrium solutions as invariant tori under the perturbed flow we use the

Center Manifold Theorem.[1] Let X be a Banach space and consider the differential equation

(*) $$du/dt = Au + f(u) \qquad \text{in } X.$$

Assume (i) A generates a C^0 semigroup

(ii) $X = X^+ \oplus X^0 \oplus X^-$ with $\dim (X^+ \oplus X^0) < \infty$, and if $A^* \equiv A|_{X^*}$ then $A^*(X^*) \subset X^*$, for $* = +, 0, -$, $\mathrm{Re}\, \sigma(A^0) = 0$ and for some $\alpha > 0$ $\mathrm{Re}\, \sigma(A^-) < -\alpha$ and $\mathrm{Re}\, \sigma(A^+) > \alpha$

(iii) $f \in C^1(X)$ satisfies $f(0) = 0$, $Df(0) = 0$.

[1] In the above statement of the Center Manifold Theorem it should be assumed that $||S^-(t)|| \leq Me^{-\alpha t}$ for $t > 0$, where $S^-(t)$ is the semigroup generated by A^-. In our application this growth rate is a consequence of the spectral condition (ii).

Then there exists a locally invariant Lipschitz continuous manifold W^c_{loc} tangent to X^0 at 0.

The complete proof of the above formulation of this useful theorem will appear in a forthcoming paper by the author and C. Jones. It was inspired by conversations with C. Conley (see also [9]) and proceeds roughly as follows:

Identify X with $X^- \times X^R$ where $X^R = X^0 \oplus X^+$ and write (*) as a system

$$(**)\quad \begin{aligned} dv/dt &= A^-v + f_1(v,w) \\ dw/dt &= A^Rw + f_2(v,w), \end{aligned}$$

where f_1 and f_2 are the projections of f onto X^- and X^R respectively and modified outside a neighborhood of (0,0). If $K = \{(v,w): |v| \leq |w|\}$ then one can show that the cone, K, is positively invariant and in fact, if $\mathbf{u}_2(0) \in \mathbf{u}_1(0) + K$, then $\mathbf{u}_2(t) \in \mathbf{u}_1(t) + K$ for all $t > 0$. Furthermore, for each fixed v there exists $w(v) \in X^R$ with the solution through $(v,w(v))$ tending to zero exponentially as $t \to \infty$. The argument here uses the Wazewski principle and the moving cone invariance provides the uniqueness of w(v) and the Lipschitz continuity of this manifold. This yields the local stable manifold, W^s_{loc}, for (0,0). Now if we take a Lipschitz continuous function $h: X^R \to X^-$ with graph in K, then the time $T > 0$ map defined by the flow (**) takes the graph of h into a subset of K which can be shown to be the graph of a Lipschitz function. Furthermore, this map is a contraction on the space of Lipschitz functions provided T is sufficiently large. The fixed point of this map is a Lipschitz function whose graph is a local center-unstable manifold W^{cu}_{loc} for (0,0).

Writing (*) in $X^+ \times X^L$ where $X^L = X^0 \oplus X^-$ and performing a similar analysis yields the local unstable manifold, W^u_{loc}, and a local center-stable manifold, W^{cs}_{loc}, for (0,0). Now, $W^c_{loc} = W^{cu}_{loc} \cap W^{cs}_{loc}$ is the desired manifold.

Returning to $(1)_\varepsilon$, we augment this system with the equation $\varepsilon_\tau = 0$. Linearizing this system at an equilibrium on one of our tori and calling the remainder f we obtain (*) with $\mathbf{u} = (u,v,\varepsilon)$. We find through the Center Manifold Theorem a three dimensional local center manifold tangent to the torus $\times\ \mathbb{R}$ at that equilibrium and $\varepsilon = 0$. Now slices through this manifold at constant values of ε are two-dimensional local invariant manifolds for (1). Thus we obtain a family of locally invariant patches, one for each equilibrium, 'above' our original torus. The question arises as to whether or not these patches fit together to form a globally invariant torus 'above' the original. If the center manifold were unique, there would be no problem here since then overlapping manifolds must coincide. However, the center manifold is not unique in general and the one constructed depends upon the modification of the nonlinearity outside a neighborhood of the equilibrium. Fortunately, once the modification is made, the center manifold is unique being the intersection of W^{cs}_{loc} and W^{cu}_{loc} each of which was obtained as the graph of the unique fixed point of a contraction mapping. Hence, as we vary the equilibrium continuously on the torus and have the modification vary continuously, W^{c}_{loc} will vary continuously. This and the compactness of the underlying manifold gives us a positively invariant torus near each torus of equilibria of the unperturbed system. What becomes of the connecting orbits? It is to be expected that they persist but become extremely complicated, spiraling, for instance, from an orbit on the invariant torus of near-oval almost periodic solutions to the orbit of the near-circular periodic solution.

BIBLIOGRAPHY

1. D. L. Barrow and P. W. Bates, Bifurcation and stability of periodic traveling waves for a reaction-diffusion system, J. Differential Equations 50 (1983), 218-233.

2. D. L. Barrow and P. W. Bates, Bifurcation of periodic travelling waves for a reaction-diffusion system, Lecture Notes in Math 964, Springer-Verlag, New York, 1982, 69-76.

3. D. L. Barrow and P. W. Bates, Bifurcation from collinear solutions to a reaction-diffusion system, Contemporary Math 17, AMS, Providence, 1983, 179-187.

4. E. Ben-Jacob, H. Brand, G. Dee, L. Kramer and J. S. Langer, Pattern propagation in nonlinear dissipative systems, Physica D, to appear.

5. K. J. Brown, P. C. Dunne and R. A. Gardner, A semi-linear parabolic system arising in the theory of superconductivity, J. Differential Equations, 40 (1981), 232-252.

6. G. Dee and J. S. Langer, Propagating pattern selection, Phys. Rev. Lett. 50 (1983), 383-386.

7. G. B. Ermentrout and J. Rinzel, Waves in a simple, excitable or oscillatory, reaction diffusion model, J. Math. Biol. 11 (1981), 269-294.

8. J. M. Lasry, Internal working paper of CEREMADE, University of Paris-Dauphine.

9. J. Moser, Stable and random motions in dynamical systems, Princeton Univ. Press, Princeton, 1973.

10. Y. Pomeau and P. Manneville, Stability and fluctuations of spatially periodic convective flow, J. Phys. Lett. 40 (1979), 609-612.

11. J. Smoller, Shock waves and reaction-diffusion equations, Springer-Verlag, New York, 1983.

12. J. Swift and P. C. Hohenberg, Hydrodynamic fluctuations at the convective instability, Phys. Rev. A 15 (1977), 319-328.

Department of Mathematics, Brigham Young University, Provo, Utah 84602.

Lectures in Applied Mathematics
Volume 23, 1986

ANALYSIS OF NONLINEAR PARABOLIC EQUATIONS MODELING PLASMA DIFFUSION ACROSS A MAGNETIC FIELD

James M. Hyman and Philip Rosenau

ABSTRACT. We analyse the evolutionary behavior of the solution of a pair of coupled quasilinear parabolic equations modeling the diffusion of heat and mass of a magnetically confined plasma. The solution's behavior, due to the nonlinear diffusion coefficients, exhibits many new phenomena. In a short time, the solution converges into a highly organized symmetric pattern that is almost completely independent of initial data. The asymptotic dynamics then become very simple and take place in a finite dimensional space. These conclusions are backed by extensive numerical experimentation.

1. INTRODUCTION

We study the asymptotic behavior of a plasma slowly diffusing across a strong magnetic field.[5-8,12,17] In the initial value problem, the plasma has compact support and diffuses into the surrounding vacuum. In the initial boundary value problem, the plasma is confined within a finite domain and convective boundary conditions are imposed. Both models are mathematical idealizations of a more complex physical situation; nevertheless, they provide theoretical insight to the dynamics of a plasma heat and mass diffusion.

In past studies, the decoupled problems for the diffusion of particles in an (essentially) isothermal plasma[3-4] and the diffusion of heat in a stationary plasma[1,13,16,18] have been analyzed. The coupling of these two processes is the source of many new phenomena that are not present in a single diffusion equation.[15]

The equations of motion we will study are

$$\rho_t = (D_1\rho_x)_x \quad \text{and} \tag{1a}$$

$$P_t = (\rho D_2 T_x)_x + (TD_1\rho_x)_x \quad , \tag{1b}$$

where $D_i = d_{oi}\rho^{\alpha_i}T^{\beta_i}$, $i = 1,2$, P is the plasma pressure, ρ is the density, T is ionic temperature, assumed to be equal to that of the electrons, and $P = \rho T$. The initial data are specified over a bounded domain

$$\rho(x,0) = \rho_o(x) \quad , \; P(x,0) = P_o(x) \quad , \; x \in (-x_o,+x_o) \quad . \tag{2}$$

The divergence form of Eqs. (1) guarantee that no additional energy or mass is added (or subtracted) after the process is initialized. An alternative form of (1b) can be obtained for T:

$$\rho T_t = (\rho D_2 T_x)_x + (D_1\rho_x)T_x \quad . \tag{1c}$$

This form reveals the convective nature of the second term on the right-hand side of (1b). The rapid convection of temperature down the density gradients is a dominant force in the asymptotic behavior of the solution.

2. INITIAL VALUE PROBLEM

We first construct a self-similar solution of Eqs. (1) for $x \varepsilon (-\infty,\infty)$ and

$$\rho_o(x) = M_o\delta(x) \quad , \; P_o(x) = E_o\delta(x) \quad . \tag{3}$$

where $\delta(x)$ is the Dirac delta function. The appropriate self-similar solution to (3) satisfies

$$P(x,t) = \rho(x,t)E_o/M_o \quad , \; \rho(x,t) = f(\zeta)/t^{1/(2+\alpha_1)} \quad , \tag{4}$$

where

$$\zeta = x/[(M_o^{\alpha_1}t)^{1/(\alpha_1+2)}]$$

and

$$f(\zeta) = [\alpha_1(\zeta_f^2 - \zeta^2)/(2\alpha_1 + 4)]^{1/\alpha_1} \quad ,$$

if $\zeta \leqq \zeta_f$ and $f(\zeta) = 0$ otherwise.

Note that the position of the diffusing front ζ_f depends only on the total mass M_o of the system and α_1. It follows from (5) that the self-similar solution describes an isothermally diffusing plasma with $T = E_o/M_o$.

Of the many group invariant solutions, the one presented has been selected because of its key role in the late-time evolution of solutions with more complicated initial data. That is, if the self-similar solution shares the same mass M_o and energy E_o as another initial value problem,

$$M_o = \int_{-x_o}^{x_o} \rho_o(x)dx \quad , \quad E_o = \int_{-x_o}^{x_o} P_o(x)dx \quad , \tag{5}$$

then irrespective of the initial distribution of the plasma, this self-similar solution is the leading term in the far-field description of the original problem. This behavior is a natural generalization of the single equation case.

Because we have not yet obtained a rigorous proof of the attractive nature of the self-similar solution, we performed a series of numerical experiments to confirm this property.

The isothermal nature of the asymptotic solution dominates so strongly that the *specific form of the second diffusion coefficient is of no importance in this stage of the problem.* A typical rapid transition to the self-similar regime is shown in Figure 1. After the initial transients, the plasma is isothermal, Eq. (1b) merely duplicates Eq. (1a), and the solution dynamics are almost identical to the single diffusion equation case.

Previously,[16] it was shown that if a finite mass M_o is distributed over the whole space, then the thermal diffusion as given by

$$\rho_o(x)T_t = [A(T)]_{xx}$$

leads to the isothermalization of the medium if A satisfies $A(0) = 0$, $A'(0) \geq 0$ and $A'(T) > 0$ for $T > 0$. That is,

$$T(x,t) \to T_a \equiv \int_{-\infty}^{\infty} \rho_o(x)T(x,0)dx/M_o \quad ,$$

as might be anticipated on the basis of physical considerations. The diffusion of heat in a finite mass medium results in isothermalization of the medium, irrespective of how the mass is distributed.

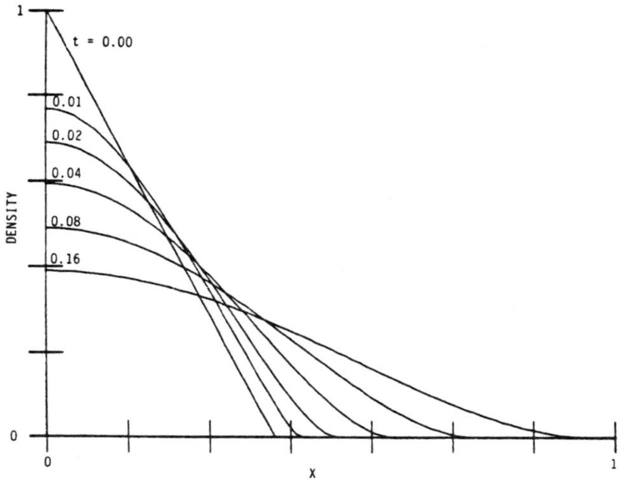

Figure 1a.

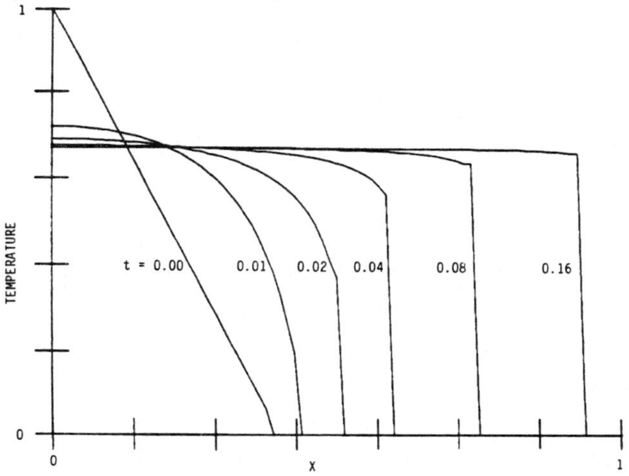

Figure 1b.

Fig 1. The initial transient solution of Eqs. (1) for $\alpha_1 = \frac{1}{2}$, $\beta_1 = \frac{1}{2}$, $\alpha_2 = 1$, $\beta_2 = 1$, $d_{01} = 1$, and $d_{02} = 5$. The initial conditions and solution are symmetric about the origin.

3. INITIAL BOUNDARY VALUE PROBLEMS

3.1 The Reduced Problem.

In order to analyze the behavior of the solution of the initial boundary value problem, we first drop the convective term in (1c). Later this term will be restored and its impact evaluated. Thus for the time being, we consider the evolution of

$$\rho_t = [D_1(\rho,T)\rho_x]_x \quad ; \qquad D_1 = d_{01}\,\rho^{\alpha_1} T^{\beta_1} \quad ; \tag{6a}$$

$$\rho T_t = [\rho D_2(\rho,T)T_x]_x \quad ; \qquad D_2 = d_{02}\,\rho^{\alpha_2} T^{\beta_2} \quad , \tag{6b}$$

and $x \in [-1,1]$. We prescribe initial data for density and temperature together with homogenous convective boundary conditions

$$\rho_x \pm h_1\rho = 0 \quad , \quad T_x \pm h_2 T = 0 \quad \text{at } x = \pm 1 \quad . \tag{7}$$

These convective boundary conditions are physically more relevant and mathematically more tractable than Dirchlet boundary conditions ($h = \infty$).

As in the Cauchy problem, the asymptotic shape of the solution of Eqs. (6) is simple and can be easily classified. The solution evolves very quickly toward a universal diffusion mode which is almost independent of initial data. For this case, however, the highly organized diffusion pattern is mathematically represented by a time-space separable solution. Similar separable solutions have been known to play a key role in the evolution of the solution to a single nonlinear diffusion equation[1,3,4,14].

The analysis of these separable solutions is the central theme of this section. Although such solutions are special cases because they must satisfy special initial data, they attract all initial data and hence play the key role in the asymptotic stage of problems with arbitrary initial data. Although we can prove this proposition rigorously only for a subclass of the considered problem, extensive numerical experimentation has been used to give strong credence to these solutrons being global attractors.

Inserting the separable forms

$$\rho(x,t) = \phi_1(t)N(x) \quad , \; T(x,t) = \phi_2(t)\psi(x) \quad , \tag{8}$$

into Eqs. (6a) and (6b) leads to the following conditions:

$$\dot{\phi}_1 = -\lambda_1 \phi_1^{\alpha_1+1} \phi_2^{\beta_1} \quad ; \qquad \lambda_1 \geq 0 \tag{9a}$$

$$\dot{\phi}_2 = -\lambda_2 \phi_1^{\alpha_2} \phi_2^{\beta_2+1} \quad ; \qquad \lambda_2 \geq 0 \tag{9b}$$

$$d_{01} \frac{d}{dx} N^{\alpha_1} \psi^{\beta_1} \frac{dN}{dx} + \lambda_1 N = 0 \quad ; \tag{10a}$$

$$d_{02} \frac{d}{dx} N^{\alpha_2+1} \psi^{\beta_2} \frac{d\psi}{dx} + \lambda_2 N\psi = 0 \quad ; \tag{10b}$$

and the spatial part of Eq. (7).

The relevant cases of the first integrals of motion for Eqs. (9) are

I. $\alpha_1 \neq \alpha_2 \quad , \; \beta_1 \neq \beta_2 \quad ,$

$$\frac{\lambda_1}{\beta_2 - \beta_1} \phi_2^{\beta_1-\beta_2} + \frac{\lambda_2}{\alpha_2 - \alpha_1} \phi_1^{\alpha_2-\alpha_1} = C_0 \quad ; \tag{11a}$$

II. $\alpha \equiv \alpha_1 = \alpha_2$, $\beta \equiv \beta_1 = \beta_2$,

$$\phi_1 = (\phi_2/T_0)^{\lambda_1/\lambda_2} \quad ; \tag{11b}$$

where C_0, ρ_0, and T_0 are constants.

Even though Case II is degenerate, it is of considerable practical interest in many applications where D_1/D_2 is assumed to be constant (such as for the diffusion of a fully collisional plasma across a magnetic field).

Integration of Eq. (9) yields

$$\phi_1(t) = [T_0^\beta(t_0 + \Omega t)]^{-\lambda_1/\Omega} \quad , \ \Omega = \lambda_1\alpha + \lambda_2\beta \quad , \tag{12}$$

where t_0 is a constant. According to whether Ω is positive, zero, or negative, we refer to the solution ϕ_1 as decaying slowly (algebraic decay), exponentially, or fast (ϕ_1 vanishes in a finite time).

The time dependences of the solutions in Case I is given implicitly as

$$\phi_1 = (\zeta_0 + \lambda_1\alpha_1\zeta)^{-1/\alpha_1} \quad , \ \phi_2 = (\tau_0 + \lambda_2\beta_2\tau)^{-1/\beta_2} \quad , \tag{13a}$$

where ζ_0 and τ_0 are constants of integration and

$$dt = d\zeta/\phi_2^{\beta_1} = d\tau/\phi_1^{\alpha_2} \tag{13b}$$

defines ζ and τ, the stretched time coordinates. Of course, ζ_o and τ_o are not independent, since they are related by Eq. (11a).

Unless either β_1 or α_2 vanishes, τ may be found only after the integration of Eqs. (13b) and (11a). Though the resulting Euler type integrals can be solved only implicitly, ϕ_1 and ϕ_2 can be evaluated asymptotically to determine the large time behavior. The results of this analysis are summarized in Fig. 2.

We can find important features of the solution's temporal part directly from the first integrals of motion. In the $(\tilde{x}, \tilde{y})$ $= (\beta_1-\beta_2, \alpha_1-\alpha_2)$ plane in Fig. 2, the two possible lines where $\Delta \equiv \alpha_1\tilde{x} - \beta_1\tilde{y} = 0$ separate regimes of fast and slow diffusion (the quantifier "fast" means that the process is extinguished within a finite time). The behavior of the temporal part of the solution dramatically changes in each of the four quadrants. In general, only in the second quadrant do both $\phi_1(t)$ and $\phi_2(t)$

$$\tilde{y} = \alpha_1 - \alpha_2$$

$\phi_1(t) \downarrow 0$	$\phi_1(t) \to$ const > 0
$\phi_2(t) \downarrow 0$	$\phi_2(t) \downarrow 0$
$(D_1/D_2) \sim 0(1)$ II	I $C_0 < 0,$ $(D_1/D_2)\downarrow 0$

III	IV $\tilde{x} = \beta_1 - \beta_2$
$\phi_1(t) \downarrow 0$	$C_0 < 0 \Rightarrow \phi_1(t) \downarrow 0, \phi_2(t) \to$ const > 0
$\phi_2(t) \to$ const > 0	$C_0 = 0 \Rightarrow \phi_1(t) \downarrow 0, \phi_2(t) \downarrow 0$
$C_0 > 0, (D_2/D_1)\downarrow 0$	$C_0 > 0 \Rightarrow \phi_1(t) \to$ const $> 0, \phi_2(t) \downarrow 0$

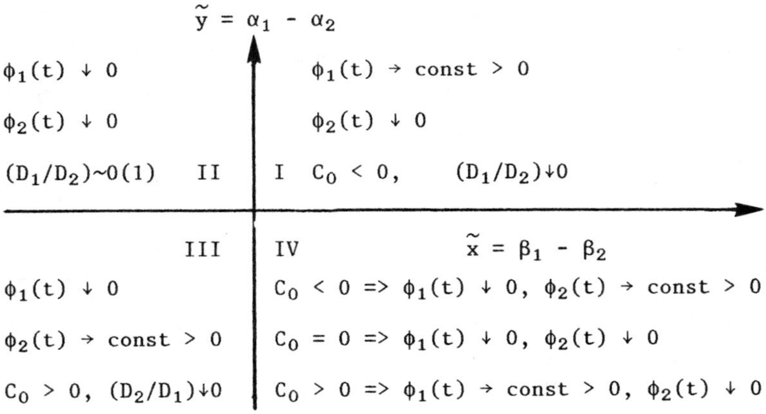

Fig. 2. Solution states of Eqs. (6) in the $(\tilde{x},\tilde{y}) =$ $(\beta_1-\beta_2,\alpha_1-\alpha_2)$ plane. In the first and the third quadrant, the integration constant C_0 must have a definite sign, but its value is irrelevant for solutions in the second quadrant, and crucial in the fourth quadrant. Everywhere, but on the $\Delta \equiv \alpha_2\beta_1 - \alpha_1\beta_2 = 0$ line, the decay is algebraic.

decay to zero, elsewhere one of the ϕ's converges to a positive constant (see 11a).

For large t, the asymptotic form of $\phi_i(t)$ in the second quadrant is given by

$$\phi_i(t) = (t_o+\lambda_i\omega_i t)^{\omega_i}, \; i = 1,2 \quad , \tag{14a}$$

where

$$\omega_1 = (\beta_2-\beta_1)/\Delta \quad , \quad \omega_2 = (\alpha_1-\alpha_2)/\Delta \quad ; \quad \Delta \equiv \alpha_2\beta_1 - \alpha_1\beta_2 \quad . \tag{14b}$$

The decay to zero is algebraic as described by Eq. (14) everywhere but on the lines where Δ is zero, the solution decay is exponential.

The ω_1 and ω_2 which give the temporal decay rates are defined a priori and are independent of the symmetry in which our problem is considered. This is an essential feature of the nonlinear diffusion which has no counterpart in the linear theory.

To obtain the main features of the temporal behavior in the other quadrants, one can use Eqs. (9) along with the fact that the first integral of motion (11a) forces one of the ϕ_i's to approach a nonzero constant everywhere but in the second quadrant. That is, first assume that

$$\phi_1 \cong \phi_{10} = \text{const.} > 0 \quad ; \tag{15a}$$

then from (9b), we have

$$\phi_2 \cong (t_0 + \delta_A t)^{-1/\beta_2} \quad , \; \delta_A \equiv \lambda_2\beta_2\phi_{10}^{\alpha_2} \quad . \tag{15b}$$

Inserting (15b) into (9a), we get a correction to ϕ_{10} and a consistency relation $\beta_1 > \beta_2$ for (15a) to hold.

Proceeding in a similar fashion with ϕ_2, we assume

$$\phi_2 \cong \phi_{20} = \text{const.} > 0 \quad ; \tag{16}$$

then from (9a), we have

$$\phi_1 \cong (t_0 + \delta_B t)^{-1/\alpha_1} \quad , \quad \delta_B \equiv \lambda_1 \alpha_1 \phi_{20}^{\beta_2} \quad , \tag{17a}$$

which in turn, when inserted into (9b), yields

$$\phi_2 \cong [a_1 - a_2 (t_0 + \delta_B t)^{-d_1}]^{-1/\beta_2} \quad , \tag{17b}$$

$$d_1 = (\alpha_2 - \alpha_1)/\alpha_2 \quad , \quad a_i = \text{const.} > 0 \quad ,$$

and a consistency relation, $\alpha_2 > \alpha_1$. In the fourth quadrant, either ϕ_1 or ϕ_2 may tend to a constant.

The rate of the temporal decay is intimely related to the role played by the separation constants λ_1 and λ_2. To clarify this point, consider first the case when Eq. (6) is a linear system whose solution decays as exp $(-\lambda_i t)$, where λ_1 and λ_2 play the role of eigenvalues in Eqs. (10). In a nonlinear diffusive system, the λ_i are nonessential constants in Eqs. (10) whose values depend on the normalization of ψ and N. Indeed, suppose that $\psi(0) = A$ and $N(0) = B$ with $\tilde{\psi}$ and $\tilde{N}$ being the solutions with eigenvalues $\tilde{\lambda}_1$ and $\tilde{\lambda}_2$. For any ψ_0, $N_0 > 0$; we then find that $\psi = \psi_0\tilde{\psi}$ and $N = N_0\tilde{N}$ are also solutions with $\tilde{\lambda}_i \to \tilde{\lambda}_i N_0^{\alpha_i} \psi_0^{\beta_i}$,

$i = 1,2$. Alternatively, let $\Delta = \alpha_2\beta_1 - \alpha_1\beta_2$; then choosing

$$\psi_0^{\Delta} = \tilde{\lambda}_2^{\alpha_1}/\tilde{\lambda}_1^{\alpha_2} \quad , \quad N_0^{\Delta} = \tilde{\lambda}_1^{\beta_2}/\tilde{\lambda}_2^{\beta_1} \quad , \tag{18}$$

normalizes both λ_1 and λ_2 to one, with $\psi(0) = A\tilde{\Psi}_o$ and $N(0) = B\tilde{N}_0$.

Thus the λ's may be reshuffled from the spatial into the temporary part of the solution and are related to the amplitude of the diffusion mode (e.g., see Eqs. (14)). This relationship is fundamentally different from the linear case.

An exception occurs when Δ vanishes. The linear case is a trivial example. In the nonbanal case, where $\alpha_1/\alpha_2 = \beta_1/\beta_2 \neq 0$ (or ∞), only one λ can be eliminated from Eqs. (10); the other λ remains as an essential parameter. In this case, the solutions to Eqs. (6) are invariant with respect to the group of shifts; $T \to AT$, $\rho \to A^{-\beta_1}/\rho^{\alpha_1}$, and $t \to t + t_o$. If $A = \exp(-\lambda t_o)$, this invariance allows solutions of the form

$$T = e^{\lambda t}\psi(x) \quad , \quad \rho = e^{-\beta_1\lambda t/\alpha_1} N(x) \quad , \tag{19}$$

where λ is an eigenvalue that must be determined from the global existence conditions of the separable solution. (A similar situation arises in the problem of imploding shock waves, where the λ is determined uniquely by requiring the global existence of the self-similar solution.[2,18])

A physically interesting case arises when D_1/D_2 is constant and (Case II, Eq. (11b)). Again, (λ_1/λ_2) plays the role of an eigenvalue with the exponential case being a transit solution between fast and slowly diffusing regimes. Here, both the mass and energy decay algebraically at a rate λ_i/Ω, $i = 1, 2$. (See

Eqs. [11b] and [12] that must be found by solving Eqs. [10a] and [10b]).

For given convective boundary condition coefficients h_1 and h_2, the following homologous property,

$$\lambda_2 d_{01} /(\lambda_1 d_{02}) = K_0 \quad , \tag{20}$$

means that λ_1/λ_2 has to be measured only for one pair of d_{01} and d_{02} and then it may be calculated for any other d_{01} and d_{02}. Particularly, if $\alpha\beta < 0$, such as in the fully collisional plasma case wherein $\alpha_1 = \alpha_2 = 1$ and $\beta_1 = \beta_2 = -\frac{1}{2}$, by changing the ratio of d_{01}/d_{02} we may transit from a fast into a slow diffusion regime (or vice versa).

Having delineated the temporal part of the solution, we still need to interpret the fact that in a diffusive process when α and β are not in the second quadrant, one of the solutions (i.e., either ϕ_1 or ϕ_2) does not decay to zero. The time evolution of a particular example is shown in Fig. 3. This behavior is very different from what is expected from a single diffusion equation.

To understand the principle mechanism involved in this somewhat unexpected process, consider the case where β_1 is zero, Eqs. (6a) and (6b) decouple and can be solved separately. The separable solution of Eq. (6a) is a global attractor[1,4,14] and represents a universal mode of diffusion with the temporal behavior

$$\phi_1(t) = (t_0 + \lambda_1\alpha_1 t)^{-1/\alpha_1} \quad , \quad t_0 = \text{const.} \quad ; \tag{21}$$

and $\phi_2(t)$ is given by Eq. (11a). If α_1 is positive, the solution asymptotically converges to the separable form. In

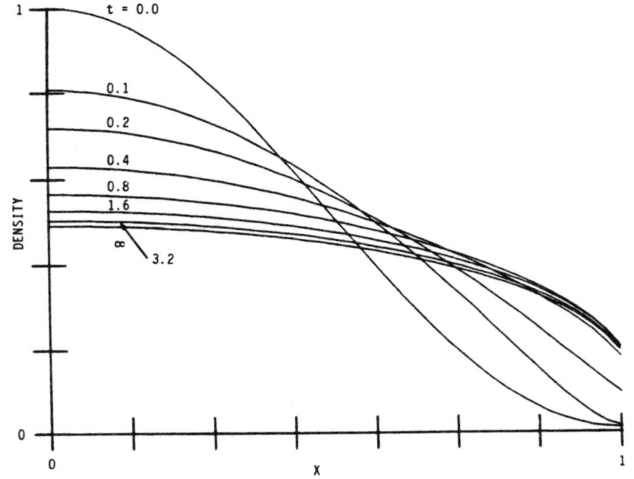

Figure 3a.

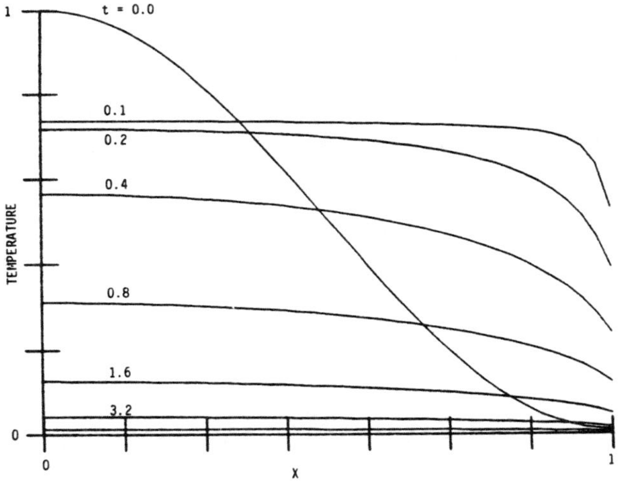

Figure 3b.

Figure 3. For these initial data (symmetric about the origin) and these parameters in the first quadrant, α_1, = 1, $\alpha_2 = \frac{1}{2}$, $\beta_1 = 1$, $\beta_2 = \frac{1}{2}$, $d_{01} = 1$, $d_{02} = 5$, $h = -10$, the decay and diffusion of mass under Eqs. (6) is inhibited by the rapid decay of heat.

numerical tests, the general solution becomes indistinguishable from the separable one after a relatively short time. The constant t_0 depends upon the initial data. For a single equation, t_0 is important only in the case of fast diffusion when $t_0/(\lambda_1|\alpha|_1)$ defines the finite extinction time of the process.

Although $\phi_2(t)$ is known from Eq. (11a), an analysis of the solution of Eq. (6b) directly is instructive. Using the asymptotic form of ρ, known for Eq. (6a), we can treat Eq. (6b) as a separate equation in T with a variable diffusion coefficient. Numerically, we have found that the solution of this equation rapidly converges to this asymptotic separable form. With this expectation, we substitute $\rho = \phi_1(t)N(x)$ and obtain

$$N(x)\phi_1^{-\alpha_2} T_t = N(x)T_\tau = (N^{\alpha_2+1} T^{\beta_2} T_x)_x , \tag{22}$$

where

$$\tau = \int_0^t \phi_1^{\alpha_2}(\eta)d\eta . \tag{23}$$

When $0 < N < \infty$, Eq. (22) is a standard diffusion equation, similar to Eq. (6a) with $\beta_1 = 0$, but measured in τ units.

If $\beta_1 = 0$ and $\alpha_1 \geq \alpha_2$, then $\tau \to \infty$ as $t \to \infty$. For large τ-time, temperature converges to the separable solution $T = \phi_2(\tau)\ \psi\ (x)$ with

$$\phi_2(t) = \tilde{\phi}_2[(\tau(t)] = (\tau_0 + \lambda_2\beta_2\tau)^{-1/\beta_2} , \tag{24}$$

and again τ_0 is an unknown function of the initial conditions.

If $\beta_1 = 0$ and $\alpha_2 > \alpha_1$ the integral in Eq. (23) converges, and

$$\tau = \tau_D[1 - (1 + \lambda_1\alpha_1 t/t_0)^{1-\alpha_2/\alpha_1}] \ , \tag{25a}$$

where

$$\tau_D = t_0^{1-\alpha_2/\alpha_1}/[\lambda_1(\alpha_2 - \alpha_1)] \ . \tag{25b}$$

Thus, $\tau \to \tau_D$ as $t \to \infty$. If τ is bounded, the time needed to attain the separable solution is not available, and $\phi_2(t \to \infty)$ converges to a positive constant. Thus, while $\rho(x,t \to \infty)$ decays to zero, $T(x,t \to \infty) \to T(x,\tau_D)$ is a positive nonzero steady state.

When this is the case, the asymptotic temperature will remember its initial conditions. If, in addition, $\beta_1 = \beta_2 = 0$, then this follows at once by noting $T(x,t) = \sum^{\infty} a_j \cdot \exp[-\delta_j\tau(t)]\psi_j(x)$. Here, δ_j and ψ_j are the jth eigenvalue and eigenfunction, respectively. Using Eqs. (25), we can see from

$$T(x,\ t \to \infty) \to T(x,\ \tau_D) = \sum^{\infty} a_j \exp(-\delta_j\tau_p)\psi_j \tag{26}$$

that none of the harmonics initially present vanish as $t \to \infty$.

For the nonlinear case, we show this property by taking $\psi(x)$, the spatial counterpart of (22), as the initial condition and perturbing it. The perturbed solution of Eq. (22) is

$$T(x,\ t) = \phi(t)\psi(x)\ [1 + u(x,\ t)] \ . \tag{27}$$

If $u = w(t)\ V(x)$, then ψ is the first eigenfunction of V. Again $w(\infty) = w(\tau_D) > 0$ and u cannot return to $\phi_2\psi$.

Thus, in the third quadrant where $\beta_2 > \beta_1$, $\alpha_2 > \alpha_1$, the diffusion of heat is always inhibited by the fast diffusion of density. In the fourth quadrant, depending on the initial data and the values of α_i and β_i, either temperature or density will inhibit the diffusion of the other. Numerical experiments have shown that usually the density decays faster and inhibits the diffusion of heat, as in the third quadrant. If α_1 is negative, the process always terminates on the fast scale. If α_1 is positive, the process is *fast* if the *temperature vanishes* and the plasma becomes cold within a finite time, but it is *slow* if the *density decays to zero*.

When $\beta_1 \neq 0$, the asymptotic analysis of the temporal part is more tedious but confirms the above conclusions. However, for $\beta_1 \neq 0$ we were unable to analytically demonstrate the attractive nature of the separable solution. At this point, we used an extensive numerical experimentation covering all of the four quadrants of the $(\tilde{x}, \tilde{y})$ plane to ensure the attractive nature of the separable solution. This leads us to believe that the lack of rigorous mathematical proof is a technical rather than a fundamental obstacle. Moreover, if $\tau_D < \infty$, unlike the semicoupled case, either both T and ρ come close to their ideal counterparts ψ and N or neither comes close, as $\tau \to \tau_D$. In practice however, for the many cases considered numerically, T and ρ approach their attracting separable solutions very quickly, long before the process "runs out of time." That is, by the time the diffusion coefficient becomes suppressed, the process is extremely close to its universal mode.

In Fig. 4, we show two examples with parameters in the fourth quadrant of how either temperature or density diffusion becomes depressed. The initial conditions and solution are shown for a massive relatively cold plasma where temperature

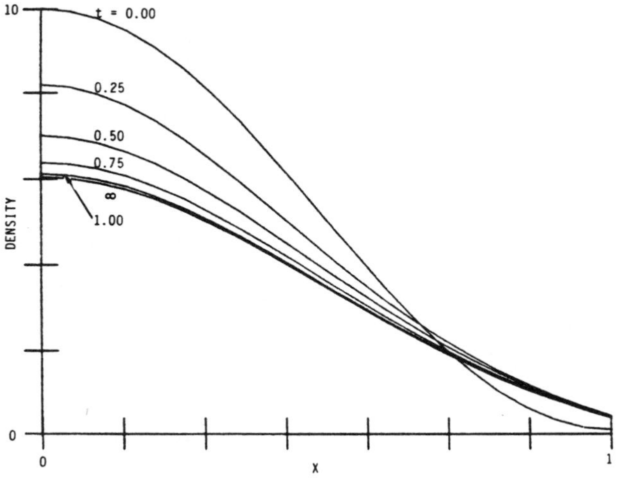

Fig. 4.1a. Density, $\rho(0,0) = 10$.

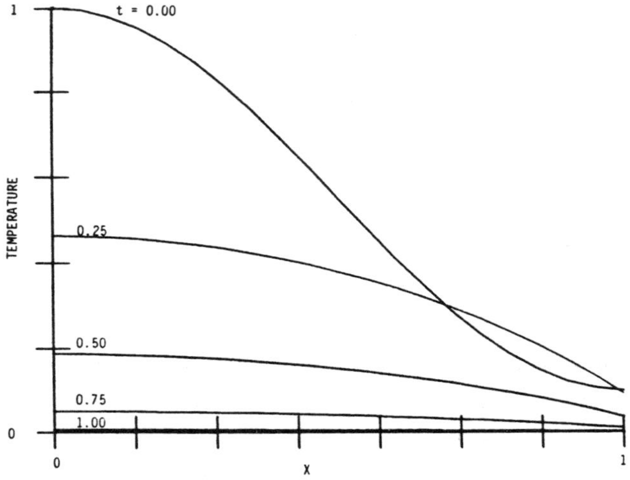

Fig. 4.1b. Temperature, $T(0,0) = 1$.

Fig. 4.1. Symmetric solutions of the diagonal case, Eq. (6), with parameters in the fourth quadrant, $\alpha_1 = -\frac{3}{4}$, $\alpha_2 = -\frac{1}{2}$, $\beta_1 = \frac{1}{2}$, $\beta_2 = -\frac{1}{4}$, $d_{01} = 1$, $d_{02} = 5$, $h = -10$; the temperature decays to zero in a finite time, leaving the density stranded.

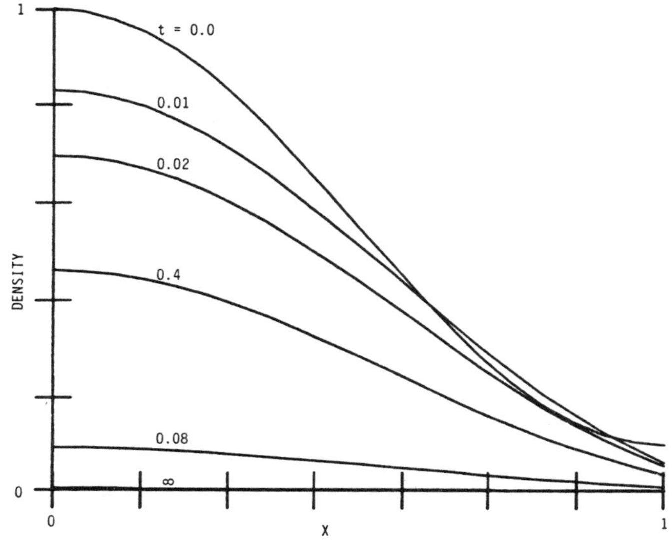

Fig. 4.2a. Density, $\rho(0,0) = 1$.

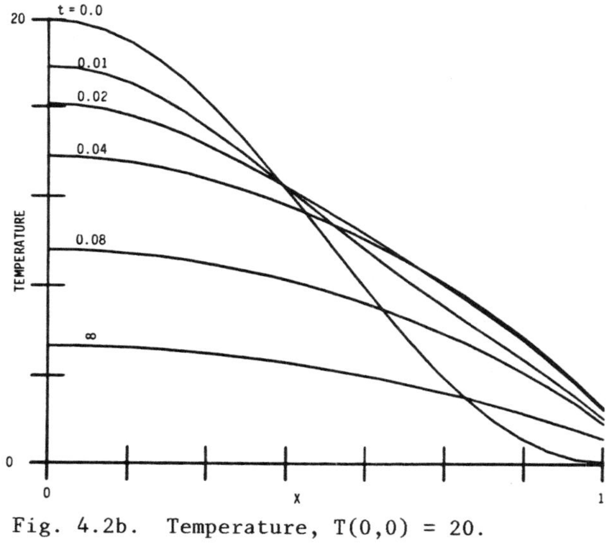

Fig. 4.2b. Temperature, $T(0,0) = 20$.

Fig. 4.2. Symmetric solutions of the diagonal case, Eq. (6), with parameters in the fourth quadrant, $\alpha_1 = -\frac{3}{4}$, $\alpha_2 = -\frac{1}{2}$, $\beta_1 = \frac{1}{2}$, $\beta_2 = -\frac{1}{4}$, $d_{01} = 1$, $d_{02} = 5$, $h = -10$; the density decays to zero in a finite time, leaving the temperature stranded.

vanishes in a finite time (Fig. 4.1), and a hot relatively tenuous plasma, where density decays to zero in a finite time (Fig. 4.2b). In Fig. 4.2b the maximum initial temperature is $T(0,0) = 20$. If $T(0,0) = 10$ then both components decay faster than exponentially and race toward zero between ρ and T ends as it does in Fig. 4.2 but with the final temperature several orders of magnitude smaller.

3.2 <u>The Tensorial Case</u>.

We are now in the position to discuss initial boundary value problems for the tensorial system Eqs. (1). The evolution of the temperature and the effect of the convective term in (1c) is more easily understood by working with this equation rather than (1b).

Substituting the separable form (8) into (1c) yields

$$d_{02} \frac{d}{dx} (N^{1+\alpha_2} \psi^{\beta_2} \frac{d\psi}{dx}) + d_{01} S(t) (N^{\alpha_1} \psi^{\beta_1} \frac{dN}{dx}) \frac{d\psi}{dx} + \lambda_2 N \psi = 0 , \tag{28}$$

where

$$S(t) = \phi_1^{\alpha_1-\alpha_2} \phi_2^{\beta_1-\beta_2} .$$

Compare this equation with (10b). The status of (28) depends critically on the behavior of S(t). In turn, the behavior of S(t) critically depends on the quadrant of the $(\tilde{x},\tilde{y})$ plane where the parameters reside. The possible behaviors are:

1st quadrant: $S(t) \downarrow 0$

2nd quadrant: $S(t) = O(1)$

3rd quadrant: $S(t) \to \infty$

4th quadrant: if $\begin{cases} \phi_1 \to \text{const.}, & S_1(t) \downarrow 0 \\ \phi_2 \to \text{const.}, & S_1(t) \to \infty \end{cases}$.

In the first quadrant, asymptotically the convective term becomes completely suppressed and the shape of both N and ψ remain unaffected by the convective part. In the second quadrant, S(t) is a constant which modifies the shape of the eigenfunctions ψ and N. Otherwise, the characterization of the solution in this quadrant does not change.

S(t) has the most dramatic impact in the third quadrant. Here, S(t) will grow indefinitely unless the temperature becomes isothermal. But, the boundary conditions for T in Eqs. (7) prevent this if $h_2 \neq 0$. Asymptotically, this difficulty is resolved by T converging to a constant everywhere but near the boundary, where an ever thinning boundary layer will be present. A numerical example of such a case is shown in Fig. 5 and should be compared with the diagonal tensor case in Fig. 4. Since the temperature is nearly constant everywhere except for a small boundary layer, the dynamics of the problem are confined primarily to the density Eq. (1a).

In the fourth quadrant the situation is, as in Eqs. (6), either an extension of the first or of the third quadrant.

Finally, note that if $\alpha_1 = \alpha_2$ and $\beta_1 = \beta_2$, $S(t) = O(1)$ and, as in the diagonal case, the decay rate is unknown a priori and the selected pattern depends upon the initial data.

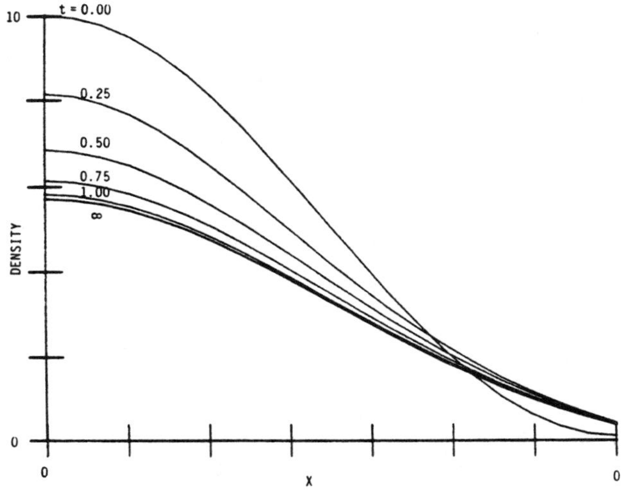

Fig. 5.1a. Density, $\rho(0,0) = 10$.

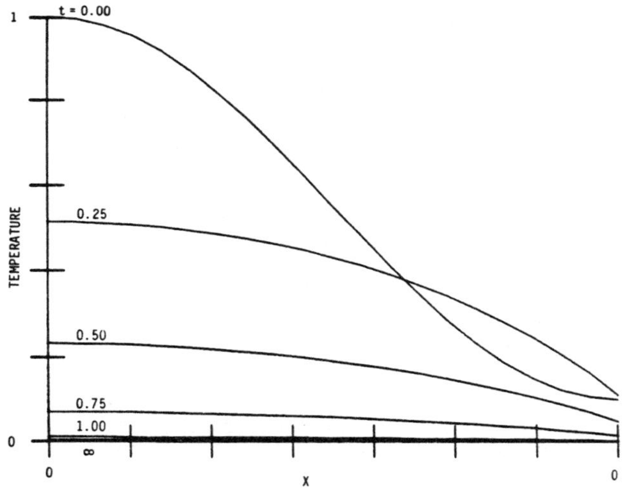

Fig. 5.1b. Temperature, $T(0,0) = 1$.

Fig. 5.1 Symmetric solutions of tensorial case, Eqs. (1), with parameters in the fourth quadrant, $\alpha_1 = -\frac{3}{4}$, $\alpha_2 = -\frac{1}{2}$, $\beta_1 = \frac{1}{2}$, $\beta_2 = -\frac{1}{4}$, $d_{01} = 1$, $d_{02} = 5$, $h = -10$; the temperature decays to zero in a finite time, leaving the density stranded.

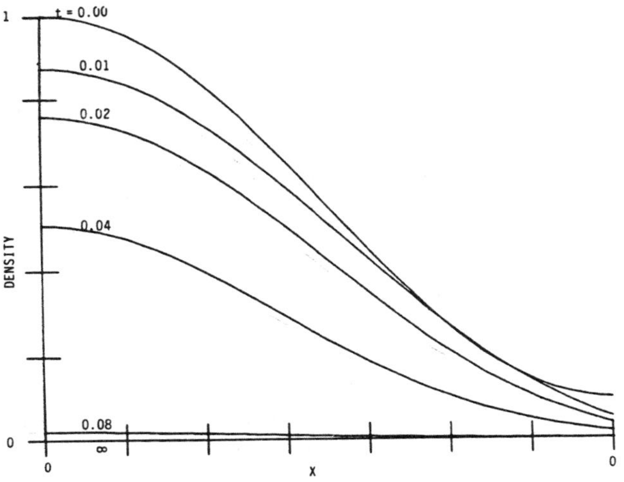

Fig. 5.2a. Density, $\rho(0,0) = 1$.

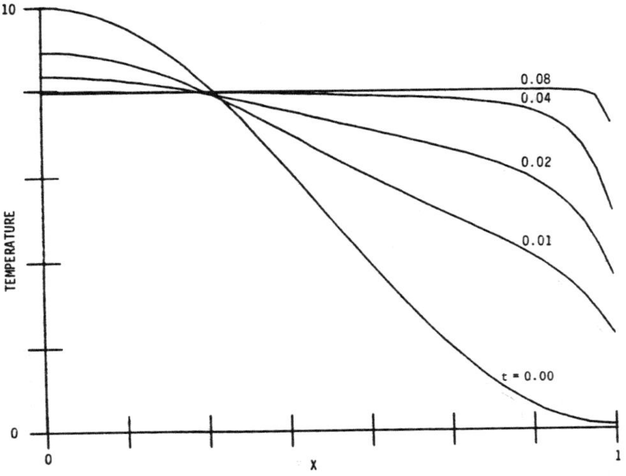

Fig. 5.2b. Temperature, $T(0,0) = 10$.

Fig. 5.2 Symmetric solutions of tensorial case, Eqs. (1), with parameters in the fourth quadrant, $\alpha_1 = -\frac{3}{4}$, $\alpha_2 = -\frac{1}{2}$, $\beta_1 = \frac{1}{2}$, $\beta_2 = -\frac{1}{4}$, $d_{01} = 1$, $d_{02} = 5$, $h = -10$; the density decays to zero in a finite time, leaving the temperature stranded.

4. NUMERICAL CALCULATIONS

Several hundred numerical experiments were performed to support the claims made about the stability and self-similarity of the asymptotic solutions. In the calculations, we used second-, fourth- and sixth-order centered finite difference approximations[11] on grids ranging from 20 to 200 mesh points on a Cray XMP computer. The boundary conditions were incorporated by extrapolating the solution to fictitious points outside the region where the solution was being integrated.[9] The cubic extrapolant satisfied both the boundary conditions and the differential equation at the boundary. The solution was integrated in time using a variable order - variable time step method of lines code, MOL1D,[10] that retained an absolute error tolerance between 10^{-4} to 10^{-6} per unit time step. Many problems were recalculated several times with different order finite difference approximations in space, grid resolution and time truncation error criteria to insure that the numerical solutions were converged within an acceptable accuracy.

5. SUMMARY

The dynamics of the highly coupled quasilinear equations (1) is surprisingly simple. After a relatively short transit time, the dynamics take place in a finite dimensional space and are almost independent of the choice of initial data. In the initial value problem, the medium quickly becomes isothermal and the dynamics are confined to mass diffusion. The initial boundary value problem offers a much wider variety of phenomena, all of which depend on the choice of the nonlinear diffusion coefficients D_1 and D_2. Among the phenomena which do not have a counterpart in the single diffusion equation case are

1) The diffusion is at an unknown a priori rate and the density and temperature are similarity solutions of the second kind.
2) The diffusion rate of one solution component vanishes in favor of the other. The faster decaying solution component is predetermined in quadrants I and III of the $(\beta_1-\beta_2,\alpha_1-\alpha_2)$ plane.
3) In quadrant IV the decay is reminiscent of pattern selection where the winning solution component depends upon the initial data, which in turn decides which diffusion pattern is chosen.

Although we have extensive numerical calculations, the mathematical status of the problem is that we know everything (almost) about the evolution of Eqs. (1) but can prove nothing (almost).

Acknowledgments

The second author wishes to express his gratitude to the Center for Nonlinear Studies and the CTR Division for sponsoring his visit to the Los Alamos National Laboratory.

References

1. D. G. Aronson and L. A. Peletier, J. Diff. Eqs. 39, 378 (1981).

2. G. I. Barenblatt, Similarity, Self-similarity, and Intermediate Asymptotics, New York, Consultants Bureau, 1979.

3. J. Berryman, J. Math. Phys. 18, 2108 (1977).

4. J. Berryman and C. J. Holland, Phys. Rev. Lett. 40, 1720 (1978).

5. D. F. Düchs, D. E. Post, and P. H. Rutherford, Nucl. Fusion 17, 565 (1977).

6. H. Grad, Courant Inst. Report MF-95/1979.

7. J. T. Hogan, "Multifluid Tokamak Transport Models," Methods in Computational Physics (Academic, New York, 1976), Vol. 16, 131.

8. W. A. Houlberg and R. W. Conn, Nucl. Fusion 19, 81 (1979).

9. J. M. Hyman, "Numerical Methods for Nonlinear Differential Equations," Nonlinear Problems: Present and Future, A. R. Bishop, D. K. Campbell, B. Nicolaenko, eds., North-Holland Publishing Co., (1982), 91-107.

10. J. M. Hyman, "The Method of Lines Solution of Partial Differential Equations," Courant Institute of Mathematical Sciences, Vol. COO-3077-139 (October 1976).

11. J. M. Hyman and B. Larrouturou, "The Numerical Differentiation of Discrete Functions Using Polynomial Interpolation Methods," Appl. Math and Comp., Vols. 10-11; also published in Numerical Grid Generation, J. F. Thompson, Ed., Elsevier North-Holland, New York (1982), 487-506.

12. E. F. Jaeger, and C. L. Hedrick, Jr., Nucl. Fusion 19, 443 (1979).

13. S. Kamin and P. Rosenau, "Propagation of Thermal Waves in an Inhomogeneous Medium," Comm. Pure Appl. Math., 34, 831 (1981).

14. S. Kamin and P. Rosenau, "Nonlinear Thermal Evolution in an Inhomogeneous Medium," J. Math. Phys. 23, 1385 (1982).

15. P. Rosenau and J. M. Hyman, "An Analysis of Nonlinear Mass and Energy Diffusion," (1984) submitted to Physical Review A.

16. P. Rosenau and S. Kamin, "Nonlinear Diffusion in a Finite Mass Medium," Comm. Pure Appl. Math., 35, 113 (1982).

17. M. N. Rosenbluth and A. N. Kaufman, Phys. Rev. 109, 1 (1958).

18. Y. A. B. Zeldovich and Yu. P. Raizer, Physics of Shock Waves and High-Temperature Hydrodynamic Phenomena, (Academic, New York, 1966).

James M. Hyman
Center for Nonlinear Studies
Theoretical Division, MS B284
LOS ALAMOS NATIONAL LABORATORY
Los Alamos, NM 87545

Philip Rosenau
Department of Mechanical Engineering
TECHNION
Haifa 3200
ISRAEL

Lectures in Applied Mathematics
Volume 23, 1986

SELF-SIMILAR SOLUTIONS FOR SEMILINEAR PARABOLIC EQUATIONS

Yoshikazu Giga[1]

This note reports some recent examples where self-similar solutions are used to study semilinear parabolic equations. The main purpose is to explain why self-similar solutions are important to describe behavior of solutions. This note is also an informal introduction of my joint work with Kohn [7].

Consider, for example, a semilinear heat equation:

$$u_t - \Delta u - |u|^{p-1}u = 0, \tag{1}$$

where u is real-valued and $p > 1$. This equation has the scaling property, that is, if u solves (1) near $(x,t) = (0,0)$, then so do the rescaled functions

$$u_\lambda(x,t) = \lambda^{2\beta}u(\lambda x,\lambda^2 t),\ \beta = 1/(p-1),$$

for each $\lambda > 0$. If a solution u is invariant under this scaling, i.e.,

$$u(x,t) = u_\lambda(x,t) \quad \text{for } \lambda > 0, \tag{2}$$

u is called <u>a self-similar solution</u>. There are at least two types of self-similar solutions. If $u(x,t)$ solves (1) in $\mathbb{R}^n \times (-\infty,0)$ and satisfies the scaling invariancy (2) for all $x \in \mathbb{R}^n$, $t \in (-\infty,0)$, we say u is <u>a backward self-similar</u>

1980 Mathematics Subject Classification 35B40.
[1]Partially supported by the Sakkokai Foundation.

solution to (1). Similarly, if $u(x,t)$ solves (1) in $\mathbb{R}^n \times (0,\infty)$ and has the property (2), u is called a forward self similar solution to (1).

Backward self-similar solutions are considered to describe the asymptotic behavior of the solution of (1) near blow-up points. To explain the meaning, we recall some aspects of blow up of the solution of (1). Consider the initial value problem for (1) in $\mathbb{R}^n$. If $n(p-1)/2 \leq 1$, the solution blows up in finite time no matter what nonnegative initial data is chosen [5,11]. Of course if the initial data is large, the solution blows up in finite time even if $n(p-1)/2 > 1$; see, for example, the Remarks of Theorem 3 in [6]. If $(0,T)$ is the maximal interval of existence for the classical solution, we call T the blow-up time. Suppose solution v blows up at $x = 0$, $t = T$. Consider rescaling the function around this point.

$$v_\lambda(x,t) = \lambda^{2\beta} v(\lambda x, T+\lambda^2 t), \ t < 0. \tag{3}$$

The asymptotics of blow up are encoded in the behavior of v as $\lambda \to 0$. If the limit exists, the limit should be a backward self-similar solution. Heuristically, the classification of the backward self-similar solution leads to the classification of the asymptotics.

Forward self-similar solutions are useful to describe the asymptotic behavior of the solution of (1) as $t \to 0$, if we consider the initial value problem. Moreover, a forward self-similar solution itself leads to nonuniqueness results to the initial value problem. If v solves (1) on $\mathbb{R}^n \times (0,T)$ and moreover v is in $L^r(0,T;L^q(\mathbb{R}^n))$, $n/q + 2/r = 2/(p-1)$ for large q, such a v is unique provided that the initial value is the same [6]. However if $1 \leq q < n(p-1)/2$, the class $C([0,T),L^q)$ may not guarantee the uniqueness. In [8] Haraux and Weissler proved that if

$$n/2 < (p+1)/(p-1) \quad \text{and} \quad n(p-1)/2 > 1, \tag{4}$$

there is a positive (C^2) forward self-similar solution of (1) on $(0,\infty) \times \mathbb{R}^n$. Because of (2), a forward self-similar solution can be written as

$$u(x,t) = \frac{1}{t^\beta} U(x/\sqrt{t}), \tag{5}$$

where U is a function defined in $\mathbb{R}^n$. If U is a rapidly decreasing function as $|x| \to \infty$, then $u(x,t) \to 0$ in $L^q(\mathbb{R}^n)$ provided that $1 \le q < n(p-1)/2$. The existence of such a forward self-similar solution yields the nonuniqueness of the initial value problem for (1) in $C([0,T), L^q(\mathbb{R}^n))$, $1 \le q < n(p-1)/2$ under (4); this is a main result of [8].

We now classify self-similar solutions. A backward self-similar solution can be written as

$$u(x,t) = \frac{1}{(-t)^\beta} w(x/\sqrt{-t}), \tag{6}$$

where w is a function defined in $\mathbb{R}^n$. Since u solves (1), w solves an elliptic equation

$$\Delta w - (\frac{x}{2} \cdot \nabla w + \beta w) + |w|^{p-1} w = 0 \tag{7}$$

in $\mathbb{R}^n$. The corresponding equation for U in (4) is

$$\Delta U + (\frac{x}{2} \cdot \nabla U + \beta U) + |U|^{p-1} U = 0. \tag{8}$$

Although the only difference between (7) and (8) is the sign in front of the second terms, the nature of the equations are totally different. In fact, there are no bounded solutions of (7) except the constant solutions of $n/2 \le (p+1)/(p-1)$ [7], while (8) has a positive radially rapidly decreasing solution under condition (4) [8]. The elliptic equations (7) and (8) look like

$$\Delta u - u + u^p = 0 \tag{9}$$

which is fully studied; see [9] and references cited there. However, the existence result for (7) is different from (9) because of the first order term.

Since we have classified backward self-similar solutions to (1) to describe asymptotics of the blow up, the next step is to show $\{v_\lambda\}$ have a limit as $\lambda \to 0$. Under a reasonable assumption we prove that this function converges to a backward self-similar solution [7]. As we have seen before, all backward self-similar solutions are $\pm\beta^\beta/(-t)^\beta$ or zero as far as the absolute value of the solution is dominated by $C/(-t)^\beta$ from above. We conclude near the blow up point

$$v(x,t) \sim \frac{\pm\beta^\beta}{(T-t)^\beta}, \quad t \to T$$

if we use a "parabolic microscope." For a precise mathematical statement, see [7]. Although the above conclusion implies that the blow up looks very flat, it does not prevent single point blow up due to Weissler [12]; see also [4] for improvements.

In this note we have just discussed equation (1), but evidently self-similar solutions are useful for other parabolic equations. For instance Brezis, Peletier and Terman [1] construct a forward self-similar solution of

$$u_t - \Delta u + |u|^{p-1}u = 0$$

and construct very singular solutions near $t = 0$, provided that $n(p-1)/2 < 1$. For the Navier-Stokes equation there are no forward self-similar solutions which satisfy the energy inequality [2]. We still do not know whether there is a backward self-similar solution satisfying the energy inequality which is posed by [10]. In [3] Foias and Temam discussed the statistical version of this problem.

BIBLIOGRAPHY

1. H. Brezis, L. A. Peletier and D. Terman, "A very singular solution of the heat equation with absorption," preprint.

2. C. Foias, O. P. Manley and R. Temam, "New representation of Navier-Stokes equations governing self-similar homogeneous turbulence," Phys. Rev. Lett. 51 (1983), 617-620.

3. C. Foias and R. Temam, "Self-similar universal homogeneous statistical solutions of the Navier-Stokes equations," Commun. Math. Phys. 90 (1983) 187-206.

4. A. Friedman and B. McLeod, "Blow-up of positive solutions of semilinear heat equations," preprint.

5. H. Fujita, "On the blowing up of solutions of the Cauchy problem for $u_t = \Delta u + u^{1+\alpha}$," J. Fac. Sci. Univ. Tokyo, Sect. I, 13 (1966), 109-124.

6. Y. Giga, "Solutions for semilinear parabolic equations in L^p and regularity of weak solutions of the Navier-Stokes system," submitted to J. Diff. Eq.

7. Y. Giga and R. Kohn, "Asymptotically self-similar blow up of semilinear heat equations," to appear in Comm. Pure Appl. Math.

8. A. Haraux and F. B. Weissler, "Non-uniqueness for a semilinear initial value problem," Indiana Univ. Math. J. 31 (1982), 167-189.

9. C.K.R.T. Jones, "On the infinitely many standing waves of some nonlinear Schroedinger equations," in this volume.

10. J. Leray, "Sur le mouvement d'un liquid visqueux emplissant l'espace," Acta Math. 63 (1934), 193-248.

11. F. B. Weissler, "Existence and non-existence of global solutions for a semilinear heat equation," Israel J. Math. 38 (1981), 29-40.

12. F. B. Weissler, "Single point blow-up for a semilinear initial value problem," J. Diff. Eq. 55(1984), 204-224.

DEPARTMENT OF MATHEMATICS
UNIVERSITY OF MARYLAND
COLLEGE PARK, MD 20742

Current Address:
Department of Mathematics
Nagoya University
Nagoya 464, Japan

Lectures in Applied Mathematics
Volume 23, 1986

ON BLOW UP OF SOLUTIONS OF EVOLUTION EQUATIONS IN BOUNDED DOMAINS

Philip Korman

This is a report on our work in [3], and so we do not include the proofs here.

Consider a model problem:

(1) $\quad u_{tt} - u_{xx} = u^2, \; 0 < x < \pi, \; t > 0$

(2) $\quad u(x,0) = g(x), \; u_t(x,0) = h(x)$

(3) $\quad u(0,t) = u(\pi,t) = 0.$

<u>Question</u>: for which initial data $(g(x),h(x))$ solution of (1-3) blows up in finite time?

We shall use the technique originally used by S. Kaplan [2] for heat equation, and R. Glassey [1] and H. Levine [4] for wave equation. Namely in contrast to linear problems, one starts by "solving the boundary conditions", i.e. noticing that by (3) solution of (1-3) can be written as

$$u(x,t) = \sum_{n=1}^{\infty} u_n(t) \sin nx,$$

with $u_n(t) = \frac{2}{\pi}\int_0^{\pi} u(x,t) \sin nx \, dx, \; n = 1,2,\ldots$. We shall study the blow up of

$$H(t) \equiv \int_0^{\pi} u(x,t) \sin x \, dx ,$$

i.e. of the first Fourier coefficient of solution.

0075-8485/86 $1.00 + $.25 per page

Integrating by parts twice, compute:

$$(4) \quad H'' = \int_0^{\pi} u_{xx} \sin x \, dx + \int_0^{\pi} u^2 \sin x \, dx = -H + \int_0^{\pi} u^2 \sin x \, dx.$$

By Schwarz's inequality

$$H^2 = \left(\int_0^{\pi} u \sqrt{\sin x} \sqrt{\sin x} \, dx\right)^2 \leq 2 \int_0^{\pi} u^2 \sin x \, dx.$$

Combining this with (4), we get a differential inequality

$$(5) \qquad H'' \geq -H + \frac{H^2}{2}$$

with the initial conditions

$$H(0) = \int_0^{\pi} u(x,0) \sin x \, dx = \int_0^{\pi} g(x) \sin x \, dx \equiv h_1 ,$$

$$H'(0) = \int_0^{\pi} u_t(x,0) \sin x \, dx = \int_0^{\pi} h(x) \sin x \, dx \equiv h_2 .$$

Definition. Consider the differential inquality

$$(6) \qquad \begin{gathered} H'' \geq f(H,H') \\ H(0) = h_1 , \; H'(0) = h_2 . \end{gathered}$$

By the domain of blow up for (6) we mean the set of all pairs (h_1,h_2), such that *any* solution of (6) tends to $+\infty$ in finite time.

Theorem 1. Assume that the function $f(H,H')$ is such as to provide local existence, uniqueness and continuous dependence on data for the equation (7) below (e.g. Lipshitz continuous). The domain of blow up for (6) is described as follows. In (H,H') phase plane dot the region complimentary to the domain of blow up of

$$H'' = f(H,H')$$

(7)

$$H(0) = h_1 \ , \ H'(0) = h_2 \ ,$$

marking the horizontal extreme points (i.e. the ones whose H coordinate is minimal or maximal in the region). Then shade the region which lies below the dotted one (under the left extreme points shade only points with $H' > 0$ and $H' = 0$ if $f(H,0) > 0$, under the right ones, only points with $H' < 0$ and $H' = 0$ if $f(H,0) < 0$). The remainder of the phase plane gives the domain of blow up for (6).

We show next how this theorem applies to our problem (5). The phase portrait for $H'' = -H + \frac{H^2}{2}$ is given by Figure 1. Here the points (0,0) and (0,2) are rest points; the curve Γ_1 is given by $H' = -\sqrt{-H^2 + \frac{H^3}{3} + \frac{4}{3}}$, $H > 2$, and the curves Γ_2, Γ_2 by $H' = \pm \sqrt{-H^2 + \frac{H^3}{3} + \frac{4}{3}}$, $-1 < H < 2$. It is easy to see that

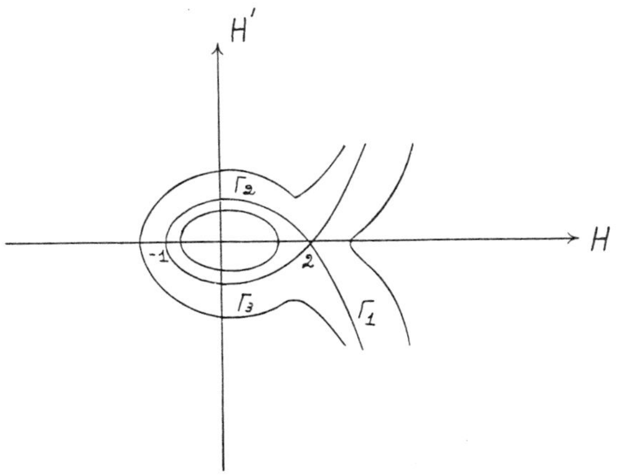

Figure 1

it takes infinite time for points on Γ_1, Γ_2 to enter (2,0), and that Γ_2 and Γ_3 bound a region $\mathcal{D}$ consisting of periodic orbits. Hence the dotted region for (5) is $\mathcal{D} \cup \Gamma_1$ with extreme point (-1,0) being marked. Hence the domain of blow up for (5) is outside of dotted and shaded areas in the Figure 2 (it includes the ray H = -1, H' < 0).

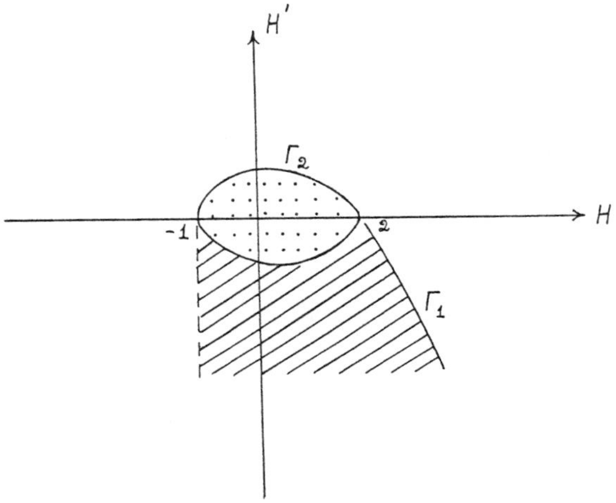

Figure 2

Remark. Our analysis of the problm (1-3) admits considerable generalizations. One can consider operators other than wave, and in higher space dimensions. The only thing we used was nonnegativity of an eigenfunction of the corresponding spatial operator. Also the right hand side of (1) can be changed to any convex f(u) (by using Jensen's inequality).

Next we turn to the "accretive" problem:

$$u_{tt} = \Delta u + u_t^{2n} , \quad n = \text{integer} \geq 1$$

$$(8) \qquad u(x,0) = g(x) , \; u_t(x,0) = h(x)$$

$$u(x,0) = 0 , \; x \in \partial\Omega .$$

Here Ω is a smooth domain in R^n, $g(x), h(x)$ are given data, and again we are interested to know which data will cause blow up for (8). Let $\phi(x)$ and $\lambda > 0$ denote the first eigenfunction and eigenvalue of $\Delta\phi + \lambda\phi = 0$ in Ω, $\phi = 0$ on $\partial\Omega$, with $\phi > 0$ in Ω and $\int_\Omega \phi\, dx = 1$. Define $H = \int_\Omega u\phi\, dx$. Using Holder's inequality and integration by parts, we easily derive

$$\text{(9)} \qquad \begin{aligned} &H'' \geq -\lambda H + H'^{2n} \\ &H(0) = \int_\Omega g(x)\phi(x)dx \equiv h_1, H'(0) = \int_\Omega h(x)\phi(x)dx \equiv h_2 , \end{aligned}$$

and we are interested in the domain of blow up for (9).

The case $n = 1$ is special, since the equation $H'' = -\lambda H + H'^2$ can be integrated explicitly. To see this, we first reduce it by stretching time (which does not affect the blow up of solutions) to

$$\text{(10)} \qquad H'' = -H + H'^2$$

Letting $H' = P(H)$, $H'' = P'P$, we get $\left(\frac{P^2}{2}\right)' = -H + P^2$, which is a linear equation for P^2. Integrating, we get the phase curves for (10)

$$H' = \pm \sqrt{ce^{2H} + H + \frac{1}{2}} \qquad (c = \text{const} \geq -1/2).$$

It is easy to see that the integral curve $\Gamma: H' = \pm\sqrt{H + \frac{1}{2}}$ separates the blow up and global existence regions for (10). Hence by the theorem 1, the blow up region for (9) is the unshaded part of phase plane in the Figure 3.

For $n > 1$ the situation is similar to the one described above, as the following theorem shows.

Theorem 2. There exists an integral curve Γ of $H'' = -\lambda H + H'^{2n}$, separating the blow up and global existence

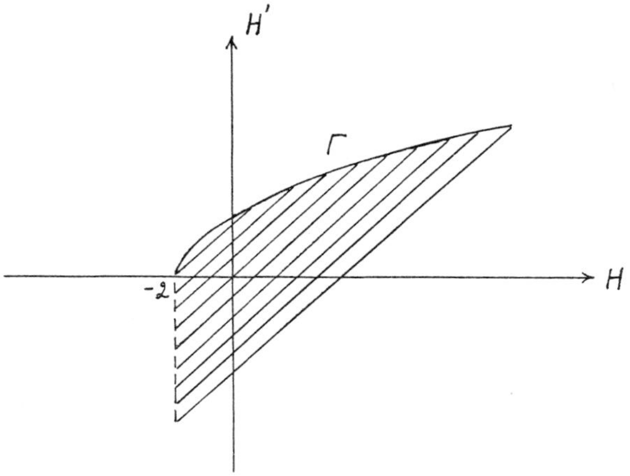

Figure 3

regions of (9). It is symmetric with respect to H axis, intersecting it once at $H_0 < 0$. Along Γ both H,H' tend to ∞ as $t \to \pm\infty$.

Acknowledgements. I wish to thank M. Levi and K. Meyer for important advice, in particular on the theorem 1, to which they contributed.

BIBLIOGRAPHY

1. R. Glassey,"Blow-up theorems for nonlinear wave equations", Math. Z. 132 (1973), 183-203.

2. S. Kaplan, "On growth of solutions of quasilinear parabolic equations", Com. Pure Appl. Math. 16 (1957), 305-330.

3. P. Korman, "On blow up of solutions of nonlinear evolution equations", preprint.

4. H. A. Levine, "Nonexistence of global weak solutions to some properly and improperly posed problems of mathematical physics: the method of unbounded Fourier coefficients", Math. Ann. 214 (1975), 205-220.

5. L. E. Payne, "Improperly Posed Problems in Partial Differential Equations", SIAM, Philadelphia, 1975.

Department of Mathematical Sciences
University of Cincinnati
Cincinnati, OH 45221

Lectures in Applied Mathematics
Volume 23, 1986

NONLINEAR INITIAL-BOUNDARY VALUE PROBLEMS IN $\mathbb{R}^N$

F. A. Howes(*)

1. INTRODUCTION. We wish to consider initial-boundary value problems for the general scalar singularly perturbed equation

$$\varepsilon\nabla^2 u = \underset{\sim}{a}\cdot\underset{\sim}{D}u + b \tag{1.1}$$

as the positive perturbation parameter ε tends to zero. Here $\nabla^2 =: \underset{\sim}{\nabla}\cdot\underset{\sim}{\nabla} = \Sigma_{i=1}^N \partial^2/\partial x_i^2$, and $\underset{\sim}{a} =: (a_1(\underset{\sim}{x},u),\ldots,a_N(\underset{\sim}{x},u))$, $\underset{\sim}{D} = \underset{\sim}{\nabla} =: (\partial/\partial x_1,\ldots,\partial/\partial x_N)$, $b = b(\underset{\sim}{x},u)$ (for the elliptic problem) or $\underset{\sim}{a} =: (a_1(\underset{\sim}{x},t,u),\ldots,a_N(\underset{\sim}{x},t,u),1)$, $\underset{\sim}{D} =: (\underset{\sim}{\nabla},\partial/\partial t)$, $b = b(\underset{\sim}{x},t,u)$ (for the parabolic problem). The space variable $\underset{\sim}{x} = (x_1,\ldots,x_N)$ ranges over a bounded, open set $\Omega \subset \mathbb{R}^N$, while the time variable t is in the finite interval $[0,T]$. Since ε is assumed small we anticipate that certain solutions of the associated hyperbolic (reduced) equation

$$\underset{\sim}{a}\cdot\underset{\sim}{D}U + b = 0 \tag{1.2}$$

are good approximations of solutions of (1.1) (which improve as $\varepsilon \to 0$), except possibly in (N-1)-dimensional subsets of the boundary and/or the interior of the domain. In neighborhoods of such subsets there are thin layers of nonuniform convergence in which solutions of (1.1) and (1.2) differ by an order-one amount, no matter how small ε is taken. These layers are the mathematical counterparts of the familiar boundary and interior layers

---

1980 Mathematics Subject Classification. 35 B 25, 35 Q 20.

(*)Supported, in part, by NSF grant DMS-8319783.

which arise in flows at high Reynolds number and which were first described by Prandtl (cf. [13; Chap. 7], [9; Chap. 4]). Thus it may be helpful in what follows to regard (1.2) as a (very) simplified model of the Euler equations and (1.1) as an equally simplified model of the incompressible Navier-Stokes equations, where ε is the reciprocal of the (large) Reynolds number.

2. AN ILLUSTRATIVE EXAMPLE.

Before considering the various kinds of asymptotic behavior which solutions of the general equation (1.1) display, let us examine a very instructive one-dimensional problem introduced by Lagerstrom and Cole [9; Chap. 2] (cf. also [6])

$$\begin{aligned} \varepsilon u_{xx} &= -uu_x + u, \qquad 0 < x < 1, \\ u(0,\varepsilon) &= A, \qquad u(1,\varepsilon) = B. \end{aligned} \tag{E_1}$$

Formally setting ε equal to zero gives us the reduced equation $UU_x = U$ whose solutions are the straight lines $U(x) = x + \text{const.}$ and the constant solution $U \equiv 0$. It turns out that in view of the smallness of ε solutions of (E_1) (which exist and are unique [6], [12]) can be described by means of these reduced solutions, supplemented by appropriate boundary and/or interior layer corrector terms. There are essentially four cases.

I. Boundary Layer at One Endpoint

Assume that $B > 1$, then the function $U_R = U_R(x) =: x + B - 1$ satisfies the righthand boundary condition and the linearized equation $\varepsilon u_{xx} = (-U_R(x))u_x + u$ has a solution in the form of a decaying exponential function at $x = 0$. Thus, by continuity, if $|A-U_R(0)|$ is sufficiently small then the solution $u = u(x,\varepsilon)$ of (E_1) is close to $U_R(x)$ in $[0,1]$ as $\varepsilon \to 0$ except near $x = 0$. In fact, if $A > 0$ this result obtains without restriction on the size of A, since then the coefficient of u_x, $-u$, is negative along $U_R(x)$ <u>and</u> for all values of u between A and $B - 1 = U_R(0)$.

It is even possible to allow A to be negative and achieve this convergence, provided the integrated effect of $-u$ in the boundary layer is negative, that is, provided

$$\int_A^{B-1} (-s)ds < 0 \quad \text{or} \quad A > -(B-1); \tag{1.3}$$

cf. [3], [6]. This theory also tells us that (if $B > 1$ and $A > -(B-1)$) the solution of (E_1) satisfies as $\epsilon \to 0$

$$u(x,\epsilon) = U_R(x) + w_L(x,\epsilon) \quad \text{in } [0,1],$$

where w_L is a boundary layer corrector at $x = 0$, that is, $w_L(0,\epsilon) = A - U_R(0)$ and $\lim_{\epsilon\to 0} w_L(x,\epsilon) = 0$ for $x > 0$. The existence of w_L is guaranteed by the integral condition (1.3).

By interchanging the roles of $x = 0$ and $x = 1$ one sees easily that if $A < -1$ and $B < -(A+1)$, then the solution of (E_1) has a boundary layer at $x = 1$, that is,

$$u(x,\epsilon) = U_L(x) + w_R(x,\epsilon) \quad \text{in } [0,1],$$

where $U_L(x) = x + A$, $w_R(1,\epsilon) = B - U_L(1)$ and $\lim_{\epsilon\to 0} w_R(x,\epsilon) = 0$ for $x < 1$. For these values of A and B the linearized equation $\epsilon u_{xx} = (-U_L(x))u_x + u$ has a growing exponential solution, and the integral condition $\int_{A+1}^{B} (-s)ds > 0$ obtains.

II. Shock Layer

Assume now that $B > 1$ and $A < -1$ but $-1 < A + B < 1$. Then the solution of (E_1) cannot have boundary layers at either endpoint, since each of the "boundary layer jumps" $|A-U_R(0)|$ and $|B-U_L(1)|$ is too large. In other words, neither U_L nor U_R alone can describe the solution as $\epsilon \to 0$. It turns out that the solution follows U_L up to a point x_0 in $(0,1)$ and then transfers across a shock layer to U_R over the rest of the interval, that is, as $\epsilon \to 0$

$$u(x,\epsilon) = \begin{cases} U_L(x) + v(x,\epsilon), & 0 \le x \le x_0, \\ U_R(x) + v(x,\epsilon), & x_0 \le x \le 1. \end{cases}$$

Here $x_0 = \frac{1}{2}(1-B-A)$ (which is in $(0,1)$ by virtue of the restrictions on A and B) and v is a shock layer corrector at x_0, that is, $v(x_0,\epsilon) = \mathcal{O}(\frac{1}{2}|U_L(x_0)-U_R(x_0)|)$ and $\lim_{\epsilon\to 0} v(x,\epsilon) = 0$ for $x \neq x_0$. The point x_0 is located by means of the integral condition $J[x_0] = 0$, for $J[x] =: \int_{U_L(x)}^{U_R(x)} (-s)ds$, which is nothing more than an "integrated" form of the classical Rankine-Hugoniot shock condition; cf. [10]. As was the case with the boundary layer corrector, the existence of v follows from the integral condition.

III. Corner Layer(s)

Assume next that $-1 < A < 0$, $0 < B < 1$ and $-A < 1 - B$. For this choice of A and B, $U_L = x + A$ is negative in $[0,x_L)$ and positive in $(x_L,1]$ for $x_L =: -A$, while $U_R = x + B - 1$ is positive in $(x_R,1]$ and negative in $[0,x_R)$ for $x_R =: 1 - B$. The points x_L, x_R are in $(0,1)$ and satisfy $x_L < x_R$. Owing to the sign changes of U_L and U_R we anticipate that in the subinterval (x_L,x_R) these functions cannot describe the solution of (E_1), and so we turn to the only other reduced solution, namely $U \equiv 0$. The reduced path

$$U(x) =: \begin{cases} U_L(x), & 0 \le x \le x_L, \\ 0 \quad , & x_L \le x \le x_R, \\ U_R(x), & x_R \le x \le 1, \end{cases}$$

is clearly continuous in $(0,1)$ but it is not differentiable at x_L and x_R. Since the solution of (E_1) is at least twice continuously differentiable in $(0,1)$, we must supplement the function U with interior layer correctors in neighborhoods of these two points so as to smooth out the irregularity there. Indeed, the theory tells us that as $\epsilon \to 0$ the solution of (E_1) satisfies

$$u(x,\epsilon) = U(x) + v_L(x,\epsilon) + v_R(x,\epsilon) \quad \text{in } [0,1],$$

where v_L and v_R are corner layer correctors at x_L and x_R, respectively, that is, $v_{L,x}(x_L,\varepsilon) = \pm \frac{1}{2}$, $v_{R,x}(x_R,\varepsilon) = \pm \frac{1}{2}$ and $\lim_{\varepsilon\to 0} v_L, v_R = 0$ for x in $[0,1]$. We note that if $A = 0$ or $B = 0$ then $x_L = 0$ or $x_R = 1$, and so there is only a single corner layer, that is, as $\varepsilon \to 0$

$$u(x,\varepsilon) \to \max\{0, U_R(x)\} \text{ or } u(x,\varepsilon) \to \min\{0, U_L(x)\}.$$

IV. Boundary Layers at Both Endpoints

Assume finally that $A > 0$ and $B < 0$. We see from the outset that neither U_L nor U_R can participate in describing the solution over any subset of $(0,1)$, since the coefficient of u_x has the incorrect sign along both functions; so we look to the reduced solution $U \equiv 0$. (Note that if we did not know a priori that a unique solution of (E_1) exists for all A and B, we might be tempted to conclude that there is no solution for this choice of boundary conditions. The value of a priori estimates should never be discounted!) The function $U \equiv 0$ is peculiar in that it satisfies neither boundary condition, and so we must supplement it with boundary layer correctors at both endpoints. Fortunately there is a related equation with a solution having precisely this behavior, namely $\varepsilon u_{xx} = u$. Since the coefficient of u_x in (E_1) is identically zero along $U \equiv 0$, the appearance of this linear equation should come as no surprise. The theory then tells us that the solution of (E_1) satisfies as $\varepsilon \to 0$

$$u(x,\varepsilon) = w_L(x,\varepsilon) + w_R(x,\varepsilon) \quad \text{in } [0,1],$$

where $w_L(0,\varepsilon) = A$, $\lim_{\varepsilon\to 0} w_L(x,\varepsilon) = 0$ for $x > 0$, $w_R(1,\varepsilon) = B$ and $\lim_{\varepsilon\to 0} w_R(x,\varepsilon) = 0$ for $x < 1$.

These four types of asymptotic behavior occur also in solutions of the equation (1.1), usually in a more complicated manner owing to the higher dimensionality. This rather rapid review of the Lagerstrom-Cole problem is intended to illustrate these phenomena in a simple yet instructive setting and to provide inspiration for a more general theory.

3. EXTENSIONS TO HIGHER DIMENSIONS. We turn now to a consideration of boundary or initial-boundary value problems for the differential equation

$$\epsilon\nabla^2 u = \underset{\sim}{a}\cdot\underset{\sim}{D}u + b \tag{3.1}$$

in a bounded domain $\Omega \subset \mathbb{R}^N$ or a bounded cylinder $\Pi =: \Omega \times (0,T) \subset \mathbb{R}^{N+1}$, depending on whether the equation is elliptic or parabolic, respectively. For ease of exposition, let us assume that Ω or Π is describable in terms of a smooth, real-valued function F, in the sense that

$$\Omega = \{\underset{\sim}{x}\colon F(\underset{\sim}{x}) < 0\}$$

or

$$\Pi = \{(\underset{\sim}{x},t)\colon F(\underset{\sim}{x},t) < 0\}.$$

Thus the vector $\underset{\sim}{D}F$ is the outer normal to Ω or Π along the boundary $F^{-1}(0)$. (For smooth domains $(-F)$ is simply the distance of a point in the domain from the boundary.) Readers familiar with the asymptotic analysis of partial differential equations will see immediately that the introduction of F obviates the use of rather cumbersome local coordinate systems in boundary layer regions.

The study of (3.1) commences as usual by formally setting ϵ equal to zero and considering solutions of the first-order equation

$$\underset{\sim}{a}\cdot\underset{\sim}{D}U + b = 0 \tag{3.2}$$

as possible approximations of solutions of (3.1). Clearly not all such functions U are suitable for this purpose, and so we ask first (following the linear theory of Levinson [11]) that U also satisfy the given initial and/or boundary data in subsets of $\partial\Omega$ or $\partial\Pi$ (the parabolic boundary) where $\underset{\sim}{a}(U)\cdot\underset{\sim}{D}F < 0$. Along these subsets the characteristic curves of U, which are given as solutions of the system $d\underset{\sim}{x}/ds = \underset{\sim}{a}(\underset{\sim}{x},t,U(\underset{\sim}{x},t))$, $dt/ds = 1$, enter the domain, and so we require that solutions of the hyperbolic equation (3.2) satisfy the given data along such inflow portions of the boundary. The analytical reason for this condition is simply that along inflow portions of the boundary it is not possible to supplement a function U with any type of boundary layer corrector term, that is, no discrepancy between U and a solution of (3.1) is allowed there. We hasten to point out that the inflow boundary may be empty if (3.1) is an elliptic equation; as an example, consider the two-dimensional equation $\epsilon\nabla^2 u = (x,y)\cdot\underset{\sim}{\nabla}u + b$ in the unit disk centered at (0,0). For the parabolic equation, however, the base of Π, $t = 0$, is always a subset of the inflow boundary. Consequently a solution $U = U(\underset{\sim}{x},t)$ of (3.2) satisfying $U(\underset{\sim}{x},0)$ = initial data $(\underset{\sim}{x})$ always exists, at least locally, under the usual smoothness assumptions on $\underset{\sim}{a}$ and b.

With the aid of such reduced solutions we can now proceed to outline in some detail four types of asymptotic behavior displayed by solutions of (3.1) which can be regarded as higher dimensional analogs of the behavior observed in the Lagerstrom-Cole problem.

I. Boundary Layer (Ordinary Layer)

Assume that the recuced problem

$$\begin{gathered}\underset{\sim}{a}\cdot\underset{\sim}{D}U + b = 0 \\ U = \text{given data along } \partial\Omega \text{ or } \partial\Pi,\end{gathered} \tag{R}$$

where $\underset{\sim}{a}(U)\cdot\underset{\sim}{D}F < 0$, has a smooth solution U which is defined in all of Ω or Π. Assume also that along a subset of the boundary where $U \neq$ (given data) the inequality

$$\underset{\sim}{a}(U)\cdot\underset{\sim}{D}F \geq k\|\underset{\sim}{D}F\|^2 \tag{3.3}$$

obtains for a positive constant k. This inequality is the analytical form of the geometric assumption that the characteristic curves of U exit Ω or Π nontangentially along that subset of the boundary (which we call the strict outflow boundary Γ^+). It follows easily from a maximum principle type argument (cf. [7]) that under (3.3) the equation (3.1) has a smooth solution u as $\epsilon \to 0$ satisfying the given data and the estimate

$$u = U + w + \mathcal{O}(\epsilon) \text{ in } \bar{\Omega} \text{ or } \bar{\Pi},$$

where w is a boundary layer corrector along Γ^+, provided $|u-U|_{\Gamma^+}$ is sufficiently small. The function w is such that along Γ^+, $w = \mathcal{O}(|u-U|)$ and $\lim_{\epsilon\to 0} w = 0$ in the remainder of $\bar{\Omega}$ or $\bar{\Pi}$. As was the case with the Lagerstrom-Cole problem, a more precise estimate of the allowable boundary layer jump, $|u-U|_{\Gamma^+}$, is provided by an integral condition, namely the inequality along Γ^+

$$[\text{Data}(\underset{\sim}{x},t)-U(\underset{\sim}{x},t)]\int_{U(\underset{\sim}{x},t)}^{\xi} \gamma(\underset{\sim}{x},t,s)ds > 0, \tag{3.4}$$

which must obtain for all values of ξ between U and the data, $\xi \neq U$. Here γ is a boundary functional defined by $\gamma(\underset{\sim}{x},t,u) = \underset{\sim}{a}(\underset{\sim}{x},t,u)\cdot \underset{\sim}{D}F(\underset{\sim}{x},t)$. Note that (3.4) is satisfied, in particular, if

$$\gamma(\underset{\sim}{x},t,\lambda) > 0 \tag{3.5}$$

for all values of λ between $U(\underset{\sim}{x},t)$ and the data there. In summary, then, we see that a local geometric condition (3.3) supplemented with a global condition like (3.4) or (3.5) in the boundary layer allows us to prove the existence of a solution of (3.1) having a boundary layer along Γ^+. This layer is called an ordinary boundary layer since the assumptions (3.3) and (3.4) enable us to find w as the solution of a certain ordinary differential equation in Γ^+ (cf. [9; Chap. 4] or [5; Chap. 7]). We remark finally that the appropriateness of the inflow boundary conditions in the reduced problem (R) has been considered recently by Bardos et al. [2].

II. Shock Layer (Two Types)

a. Contact Discontinuity

In physical terms a contact discontinuity is a shock layer which sits along a characteristic manifold; cf. [10]. Thus assume that there is a characteristic manifold Σ properly contained in Ω or Π such that $\Sigma =: f^{-1}(0)$, for a smooth, real-valued function f, and that the reduced problem (R) has two smooth solutions U_1 and U_2 with $U_1 \neq U_2$ along Σ. Assume also that Ω or Π is decomposable as

$$\{\Omega,\Pi\} = \{\Omega_1,\Pi_1\} \cup \Sigma \cup \{\Omega_2,\Pi_2\}. \tag{3.5}$$

Here Ω_1 or Π_1 is the subset of Ω or Π where $f < 0$, while Ω_2 or Π_2 is the subset where $f > 0$. Since Σ is characteristic we know that, by definition, $\underset{\sim}{a}(U_i)\cdot\underset{\sim}{D}f \equiv 0$ along Σ, for $i=1,2$. We ask, in addition, that for all values of λ between U_1 and U_2

$$\underset{\sim}{a}(\underset{\sim}{x},t,\lambda)\cdot\underset{\sim}{D}f(\underset{\sim}{x},t) \geq 0 \tag{3.6}$$

for all $(\underset{\sim}{x},t)$ in Ω_1 or Π_1 whose distance from Σ is less than some small number $\delta > 0$, and that

$$\underset{\sim}{a}(\underset{\sim}{x},t,\lambda)\cdot\underset{\sim}{D}f(\underset{\sim}{x},t) \leq 0 \tag{3.7}$$

for all $(\underset{\sim}{x},t)$ in Ω_2 or Π_2 whose distance from Σ is less than δ. These two inequalities are nothing more than analytical formulations of the geometric requirement that the characteristics of U_1 and U_2 do not diverge from Σ within a (2δ)-neighborhood of Σ. Then if the problem for equation (3.1) admits a maximum principle (for example, if $\partial b/\partial u > 0$ in the region of interest), it is possible to prove that (3.1) has a solution u as $\epsilon \to 0$ satisfying the given data and the estimates

$$u = U_1 + v + \mathcal{O}(\epsilon) \quad \text{in } \bar{\Omega}_1 \text{ or } \bar{\Pi}_1$$

and (3.8)

$$u = U_2 + v + \mathcal{O}(\epsilon) \quad \text{in } \bar{\Omega}_2 \text{ or } \bar{\Pi}_2;$$

cf. [7]. Here v is a shock layer corrector along Σ, that is,

$v = \mathcal{O}(\frac{1}{2}|U_1-U_2|)$ on Σ and $\lim_{\varepsilon\to 0} v = 0$ in $\{\Omega,\Pi\}\backslash\Sigma$. We note that the inequalities (3.6), (3.7) may be replaced by the single condition that for all λ between U_1 and U_2

$$\underset{\sim}{a}(\underset{\sim}{x},t,\lambda)\cdot \underset{\sim}{D}f = \mathcal{O}(|f|) \tag{3.9}$$

in a (2δ)-neighborhood of Σ. Condition (3.9) thus allows characteristics near Σ to diverge from Σ provided they do so "slowly" enough. The restrictions (3.6), (3.7) or (3.9) serve to provide an estimate of the allowable shock strength, $|U_1-U_2|_\Sigma$, under which the convergence described in (3.8) obtains.

b. Compressive Shock

This case corresponds to the classical compressive discontinuity (shock layer) which propagates at a certain speed (shock speed) determined by the Rankine-Hugoniot condition (cf. [14; Chap. 2]). The shock speed and hence, the location of the discontinuity manifold Σ, are not known a priori, since Σ is not a characteristic manifold. Let us proceed as in the case of the contact discontinuity and assume that Ω or Π is expressible in the form (3.5). The only difference is that now Σ is regarded as unknown initially; it must be determined as part of the solution. To this end, assume that the reduced problem (R) has two solutions U_1, U_2 ($U_1 \neq U_2$), and consider the functional

$$J[\underset{\sim}{x},t] =: \int_{U_1(\underset{\sim}{x},t)}^{U_2(\underset{\sim}{x},t)} \sigma(\underset{\sim}{x},t,s)ds,$$

for $\sigma(\underset{\sim}{x},t,u) =: \underset{\sim}{a}(\underset{\sim}{x},t,u)\cdot\underset{\sim}{D}f(\underset{\sim}{x},t)$. Then we determine Σ by assuming that the equation

$$J[\underset{\sim}{x},t] = 0 \tag{3.10}$$

has a solution given by $f(\underset{\sim}{x},t) = 0$ with the property that $\underset{\sim}{D}J[\underset{\sim}{x},t] \neq (\underset{\sim}{0},0)$ on $\Sigma =: f^{-1}(0)$. Equation (3.10) may be regarded as an "integrated" form of the classical Rankine-Hugoniot condition; cf. Example (E_2) below. Finally if the (weak) entropy condition that

$$\underset{\sim}{a}(\underset{\sim}{x},t,U_1(\underset{\sim}{x},t))\cdot\underset{\sim}{D}f(\underset{\sim}{x},t) \geq k > 0$$

and (3.11)

$$\underset{\sim}{a}(\underset{\sim}{x},t,U_2(\underset{\sim}{x},t))\cdot\underset{\sim}{D}f(\underset{\sim}{x},t) \leq -k < 0$$

on Σ holds for a positive constant k, then it is possible to prove the existence of a solution of (3.1) satisfying the estimates (3.8); cf. [8]. The conditions (3.10), (3.11) again provide an estimate of the allowable shock strength $|U_1-U_2|_\Sigma$.

III. Corner Layer

This type of asymptotic behavior is described by making the same assumptions on the reduced solutions U_1, U_2, the characteristic manifold Σ and Ω or Π as in the case (II.a) of the contact discontinuity, except that now $U_1 = U_2$ on Σ but the normal derivatives $\partial U_1/\partial n$ and $\partial U_2/\partial n$ are unequal there. Thus this case corresponds to the physical situation of the propagation of a rarefaction wave, across which the flow variables are continuous but not differentiable. By arguing as in II.a, one can show the existence of a solution of (3.1) as $\varepsilon \to 0$ which satisfies the estimates (3.8); cf. [7]. The only difference is that the shock layer corrector v is replaced by an interior layer corrector $\tilde{v}$ with the properties that $\partial\tilde{v}/\partial n = \mathcal{O}(\frac{1}{2}|\partial U_1/\partial n - \partial U_2/\partial n|)$ on Σ and $\lim_{\varepsilon\to 0} \tilde{v} = 0$ in $\overline{\Omega}$ or $\overline{\Pi}$. We note that such a corner layer always sits along a characteristic manifold, since it is only across such manifolds that solutions of the reduced equation (3.2) can be continuous but not differentiable; cf. [4; Chap. 2].

IV. Boundary Layer (Characteristic Layer)

The final asymptotic phenomenon we consider involves a boundary layer which is thicker ($\mathcal{O}(\varepsilon^{\frac{1}{2}})$) than the ordinary layer of thickness $\mathcal{O}(\varepsilon)$ discussed in I. Assume that the reduced equation (3.2) has a smooth solution U such that $\underset{\sim}{a}(U)\cdot\underset{\sim}{DF} \equiv 0$ along a portion Γ^o of the boundary, that is, Γ^o is an (N-1)-dimensional characteristic manifold. In general, it is not possible to ask that U satisfy the given data along Γ^o (cf. [4; Chap. 2]), and so we anticipate the appearance of a boundary layer along Γ^o. On the other hand, since the characteristic curves of U are tangent to Γ^o, we do not expect that this layer is an ordinary one. In order to proceed let us assume that

$$\underset{\sim}{a}(\underset{\sim}{x},t,\lambda)\cdot\underset{\sim}{D}F(\underset{\sim}{x},t) \geq 0 \tag{3.12}$$

for all values of $(\underset{\sim}{x},t)$ in a δ-neighborhood of Γ^o and for all values of λ between U and the data. Then if the problem for (3.1) admits a maximum principle one can show (cf. [7]) that there is a smooth solution u as $\epsilon \to 0$ satisfying in a neighborhood of Γ^o

$$u = U + \tilde{w} + \mathcal{O}(\epsilon).$$

Here the boundary layer corrector $\tilde{w}$ has the defining properties that $\tilde{w} = \mathcal{O}(|u-U|)$ on Γ^o and $\lim_{\epsilon\to 0} \tilde{w} = 0$ in $\{\Omega,\Pi\}\backslash\Gamma^o$; however, $\tilde{w}$ is usually found as a solution of a semilinear equation like $\epsilon\nabla^2\tilde{w} = (\partial B/\partial u)\tilde{w}$ (if we assume that $\partial B/\partial u \geq m^2 > 0$, for $B(\underset{\sim}{x},t,u) =: \underset{\sim}{a}(\underset{\sim}{x},t,u)\cdot\underset{\sim}{D}U + b(\underset{\sim}{x},t,u))$. Thus we can often find $\tilde{w}$ in explicit form, from which it follows that

$$\tilde{w} = \mathcal{O}(|u-U|_{\Gamma^o} \exp[mF(\underset{\sim}{x},t)/\epsilon^{\frac{1}{2}}])$$

near Γ^o, that is, the boundary layer along Γ^o has thickness of order $\epsilon^{\frac{1}{2}}$.

Having isolated these four basic types of asymptotic behavior let us now illustrate the occurrence of some of them in the solution of an actual problem.

4. AN INITIAL-BOUNDARY VALUE PROBLEM FOR BURGERS' EQUATION.

Consider the following problem for Burgers' equation

$$\begin{aligned} &\epsilon u_{xx} = uu_x + u_t, \quad 0 < x < 1,\ 0 < t < T \\ &u(0,t) = B_0(t),\ u(x,0) = A(x),\ u(1,t) = B_1(t), \end{aligned} \tag{E_2}$$

where A, B_0 and B_1 are smooth functions; cf. [9; Chap. 4], [14; Chap. 4]. Let us assume first that $A \equiv B_1$ are negative constants. Then the sides $t = 0$ and $x = 1$ are the inflow boundaries, and so the correct reduced solution is $U(x,t) \equiv A$. Along the outflow boundary, $x = 0$, we anticipate the occurrence of a boundary layer if B_0 $(\neq A)$ is suitably restricted, since the characteristics of $U = A$ leave the side $x = 0$ nontangentially, that is, (3.3) obtains. Now the boundary functional $\gamma =: (u,1) \times (-1,0) = -u$, and so the restriction on B_0 is determined by the condition that (cf. (3.4))

$$[B_0-A] \int_A^{\xi} (-s)ds = -(B_0-A)(\xi^2-A^2)/2 > 0$$

for all values of ξ between A and B_0, $\xi \neq A$, that is, $(A-B_0)^2(A+B_0) < 0$. Thus if $B_0(t) < -A$ (and $A \equiv B_1 < 0$) the problem (E_2) has a solution $u = u(x,t,\varepsilon)$ as $\varepsilon \to 0$ satisfying

$$\lim_{\varepsilon \to 0} u(x,t,\varepsilon) = A \quad \text{in } [\delta,1] \times [0,T].$$

In particular, if $B_0(t) \leq \beta < 0$ in $[0,T]$ then it follows easily that in $[0,1] \times [0,T]$

$$u(x,t,\varepsilon) = A + \mathcal{O}(|B_0(t)-A| \exp[-kx/\varepsilon]),$$

where $k =: \min\{-A,-\beta\}$, since $\gamma(0,t,\lambda) \geq k > 0$ for all values of λ between $B_0(t)$ and A.

Assume next that $A(x)$ and $B_0(t)$ are identically constant, say $A(x) \equiv A$ and $B_0(t) \equiv B_0$, with $0 < A < B_0$. Then the sides $x = 0$ and $t = 0$ are the inflow boundaries for the reduced solutions $U_1 \equiv B_0$ and $U_2 \equiv A$, respectively. Since $0 < A < B_0$ the characteristics of U_1 and U_2 converge toward each other, and we anticipate the occurrence of a (compressive) shock layer along a curve $x = \varphi(t)$. The location of this shock curve Σ is determined by the condition that (cf. (3.10))

$$J[x,t] = 0,$$

for $J[x,t] =: \int_{B_0}^{A} (s,1)\cdot(1,-\varphi'(t))ds$. Since $J[x,t] = (A-B_0)\{\frac{1}{2}(A+B_0)-\varphi'(t)\}$ we find that $J = 0$ iff $\varphi'(t) = \frac{1}{2}(A+B_0)$, that is, Σ is given by the equation $x = \frac{1}{2}(A+B_0)t$. Thus the problem (E_2) has a solution with a shock layer along Σ, as well as a boundary layer along the exit boundary, $x = 1$, if $B_1(t) > -A$ in $[0,\frac{1}{2}]$ and $B_1(t) > -B_0$ in $[\frac{1}{2},T]$. Note that $\underset{\sim}{D}J = \underset{\sim}{0}$ along Σ since the states U_1 and U_2 are identically constant; however, a suitable shock layer corrector can still be constructed (cf. [9; Chap. 4]).

Assume finally that $A(x) = 1 + x$, $B_0(t) \equiv 1$ and $B_1(t) \equiv 2$. Then the initial value problem $U_t + UU_x = 0$, $U(x,0) = 1 + x$, has the simple wave solution $U_2(x,t) = (1+x)/(1+t)$ which is defined for $t \leq x$. Moreover, since $B_0(t) > 0$ the side $x = 0$ is also an inflow boundary, and so the function $U_1 \equiv 1$ is another reduced solution in the region $t \geq x$. Thus the continuous composite function

$$U(x,t) =: \begin{cases} 1, & t \geq x, \\ (1+x)/(1+t), & t \leq x, \end{cases}$$

satisfies all of the inflow data, and it is a solution of reduced equation away from the characteristic line $t = x$. Along $t = x$, however, U is not differentiable, and so we must supplement it there with an interior layer term. Letting $f(x,t) =: (x-t)/\sqrt{2}$ and $\sigma(x,t) =: (U(x,t),1)\cdot\underset{\sim}{D}f = (U(x,t)-1)/\sqrt{2}$, we see that $\sigma \equiv 0$ for $t \geq x$ and $\sigma(x,t) = \delta/[\sqrt{2}\,(x+1)] + \mathcal{O}(\delta^2)$ for $0 \leq x - t \leq \delta$; cf. (3.9). Consequently, for such data the solution of (E_2) has a corner layer along $t = x$ and an ordinary boundary layer along the outflow boundary $x = 1$, since $(\lambda,1)\cdot(1,0) = \lambda > 0$ for all values of λ between B_1 $(\equiv 2)$ and $U(1,t)$ (cf. (3.5)). We note that for these values of A, B_0 and B_1 the problem (E_2) does indeed satisfy a maximum principle for ϵ sufficiently small. This follows directly by making the change of variable $v =: e^{-\alpha x}u$ in

(E_2) and choosing the positive constant α sufficiently large.

The complete asymptotic solution of (E_2) for all values of the data would be an ambitious project. We do know, however, that there is a solution u for all $\varepsilon > 0$ which satisfies the estimate $\min\{A,B_0,B_1\} \le u \le \max\{A,B_0,B_1\}$ in $[0,1] \times [0,T]$; this follows from a theorem of Amann [1]. There is also the question of the stability of solutions as $t \to \infty$. We leave such considerations to future papers.

ACKNOWLEDGMENTS. It is a pleasure to thank the NSF for its financial support and the typist, Mrs. Ida Zalac, for her secretarial support.

BIBLIOGRAPHY

1. H. Amann, Periodic Solutions of Semilinear Parabolic Equations, in Nonlinear Analysis, ed. by L. Cesari et al., Academic Press, New York, 1978, pp. 1-29.

2. C. Bardos, A. Y. LeRoux and J. C. Nedelec, First Order Quasilinear Equations with Boundary Conditions, Comm. Partial Diff. Eqns. 4(1979), 1017-1034.

3. E. A. Coddington and N. Levinson, A Boundary Value Problem for a Nonlinear Differential Equation with a Small Parameter, Proc. Amer. Math. Soc. 3(1952), 73-81.

4. R. Courant and D. Hilbert, Methods of Mathematical Physics, vol. II, Interscience, New York, 1962.

5. W. Eckhaus, Asymptotic Analysis of Singular Perturbations, North-Holland, Amsterdam, 1979.

6. F. A. Howes, Boundary-Interior Layer Interactions in Nonlinear Singular Perturbation Theory, Memoirs Amer. Math. Soc., no. 203, 1978.

7. F. A. Howes, Perturbed Boundary Value Problems Whose Reduced Solutions are Nonsmooth, Indiana U. Math. J. 30(1981), 267-280.

8. F. A. Howes, Perturbed Elliptic Problems with Essential Nonlinearities, Comm. Partial Diff. Eqns. 8(1983), 847-874.

9. J. Kevorkian and J. D. Cole, Perturbation Methods in Applied Mathematics, Springer-Verlag, New York, 1981.

10. P. D. Lax, Hyperbolic Systems of Conservation Laws and the Mathematical Theory of Shock Waves, CBMS Series in Appl. Math., vol. 11, SIAM, Philadelphia, 1973.

11. N. Levinson, The First Boundary Value Problem for $\epsilon\Delta u + A(x,y)u_x + B(x,y)u_y + C(x,y)u = D(x,y)$ for Small ϵ, Ann. Math. 51(1950), 428-445.

12. J. Lorenz, Nonlinear Boundary Value Problems with Turning Points and Properties of Difference Schemes, in Springer Lecture Notes in Math., vol. 942, Springer-Verlag, New York, 1982, pp. 150-169.

13. M. D. Van Dyke, Perturbation Methods in Fluid Mechanics, Parabolic Press, Stanford, 1975.

14. G. B. Whitham, Linear and Nonlinear Waves, Interscience, New York, 1974.

DEPARTMENT OF MATHEMATICS
UNIVERSITY OF CALIFORNIA AT DAVIS
DAVIS, CA 95616

Lectures in Applied Mathematics
Volume 23, 1986

A NON-LINEAR, SINGULAR PERTURBATION PROBLEM WITH SOME UNUSUAL FEATURES

C.M. Brauner, W. Eckhaus, M. Garbey, A. van Harten

ABSTRACT. We consider the singular perturbation problem in $\Omega \subset \mathbb{R}^2$

$$-\varepsilon\Delta u_\varepsilon + \frac{u_\varepsilon}{1+u_\varepsilon} = f, \qquad u_\varepsilon\big|_{\partial\Omega} = 0.$$

There is a region where u_ε blows up as $\varepsilon \to 0$, whose boundary S is solution of a Free Boundary Problem. There is a sharp transition layer phenomenon across S. A uniformly valid asymptotic development of u_ε is constructed, and its validity is proved using barrier-function techniques. The structure of the asymptotic expansions near the free surface S turns out to be rather unusual involving fractional orders of ε multiplied with powers of $\ln \varepsilon$. Earlier convergence statements are improved.

1980 Mathematical Subject Classification: 35B25, 35J60, 35R35, 35B40, 35B50.

1. INTRODUCTION.

Let Ω be a regular domain in $\mathbb{R}^2$, Δ the Laplace operator, and Φ a mapping from $\mathbb{R}$ into $[-1,1]$, $\lim_{t\to\pm\infty} \Phi(t) = \pm 1$, $\Phi(0) = 0$, Φ smooth, increasing, in the interval where $\Phi(t) \in (-1,1)$. f is a (smooth) given function.

We consider the nonlinear singular perturbation problem:

$$(1.1) \qquad -\varepsilon\Delta u_\varepsilon + \Phi(u_\varepsilon) = f, \quad u_\varepsilon\big|_{\partial\Omega} = 0.$$

The reduced problem, obtained by putting ε equal to 0, is

$$(1.2) \qquad \Phi(u_o) = f.$$

It admits a smooth solution if f(x) belongs to the range of Φ whenever $x \in \Omega$.

This singular perturbation problem is even more singular when f is, in some sense "incompatible" with the nonlinearity Φ; it has been studied by Brauner and Nicolaenko [1-3]. Two cases were considered:

<u>Case A</u>: $f(x) \notin \operatorname{ran}\Phi \;\forall x \in \Omega$ (global incompatibility). Then u_ε blows up as $\varepsilon \to 0$.

<u>Case B</u>: $f(x) \notin \operatorname{ran}\Phi \;\forall x \in E^c$, subdomain of Ω (<u>partial incompatibility</u>). Then the remarkable feature of the problem is that $u_\varepsilon(x)$ converges only in a subdomain Ω_0 of Ω, strictly included in E. Ω_0 in determined as the "coincidence set" of some Free Boundary Problem.

Let us describe briefly some of the main results in case B. As simplification we restrict ourselves to the situation $f \geq 0$, namely

$$(1.3) \qquad f \in C^\infty(\overline{\Omega}),\; f \geq 0,\; \max_{x\in\overline{\Omega}} f(x) > 1,\; f < 1 \text{ on } \partial\Omega.$$

Let $w_\varepsilon = \varepsilon u_\varepsilon > 0$. Then w_ε satisfies the following "penalized problem" (see [4] for a general theory):

$$(1.4) \qquad -\Delta w_\varepsilon + \Phi\left(\frac{w_\varepsilon}{\varepsilon}\right) = f, \; w_{\varepsilon/\partial\Omega} = 0.$$

It can be shown that w_ε converges for $\varepsilon \downarrow 0$ to a function $w_0 \in C^{1,1}(\overline{\Omega})$, where the convergence takes place strongly in $H_0^1(\Omega)$ and weakly in $W^{2,p}(\Omega)$. Note that $w_0 \geqq 0$.

One introduces:

$$(1.5) \qquad \Omega_+ \underset{\text{def}}{=} \{x \in \Omega | w_0(x) > 0\}, \qquad S = \partial\Omega_+$$
$$\Omega_0 = \{x \in \Omega | w_0(x) = 0\}.$$

In this paper we consider the situation where the open set Ω_+ has no holes, its boundary is a Jordan curve. S is called the free boundary and we suppose that S has a positive distance to $\partial\Omega$. Regularity results on S in two dimensions have been established by Lewy and Stampacchia [15], Caffarelli and Rivière [7] (see also [11, p. 138]), which infer that S has a smooth parametrization.

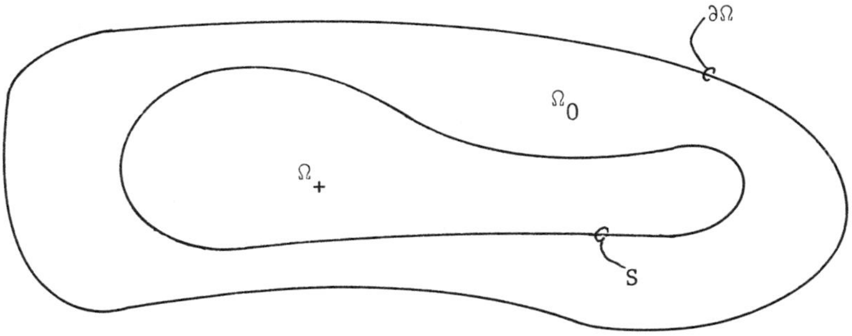

In this situation w_0 restricted to Ω_+ is $\in C^\infty(\overline{\Omega}_+)$ and w_0 solves the obstacle problem:

$$(1.6)\qquad \begin{aligned} -\Delta w_o &= f-1 \text{ in } \Omega_+ \\ w_o &= 0 \text{ in } \Omega_o \\ w_o\big|_{\partial\Omega} &= 0 \\ w_o\big|_S &= \frac{\partial w_o}{\partial n}\Big|_S = 0. \end{aligned}$$

In the sequel we shall require

$$(1.7)\qquad f < 1 \text{ on } \Omega_0$$

hence $\max_{x\in S} f(x) < 1$. This result may be demonstrated under further assumptions on the set $\{x \in \Omega,\ f(x) = 1\}$, see [2], [12].
Now it holds that:

$$(1.8)\qquad \begin{aligned} u_\varepsilon(x) &\to U_0(x) = \Phi^{-1}(f(x)) \text{ a.e. } x \text{ in } \Omega_0 \\ u_\varepsilon(x) &\to +\infty\ \forall x \in \Omega_+. \end{aligned}$$

From now on, *as a prototype equation*, we shall consider

$$(1.9)\qquad -\varepsilon\Delta u_\varepsilon + \frac{u_\varepsilon}{1+u_\varepsilon} = f,\ u_\varepsilon\big|_{\partial\Omega} = 0.$$

This "homographic" function $\Phi(t) = t/(1+|t|)$ is stemming from the Michaëlis-Menten law in enzyme kinetics. Other choices may be considered such that $\Phi(t) = 1$ if $t \geq 1$, t if $-1 < t < 1$, -1 if $t \leq -1$, or $\Phi(t) \to \pm 1$ exponentially. Although the general feature of the asymptotic development does not depend on Φ, the "homographic" case appears to contain more technical difficulties than others. Also this case is definitely a good example of applications of matching principles.

Let us now sketch the contents of the next sections:

Section 2: the regular expansion in Ω_0 and the layer at $\partial\Omega$.

Section 3: the principal terms in the internal layer at S and in the expansion in Ω_+.

Section 4: higher order terms near S and in Ω_+

Section 5: matching relations of the internal layer at S and the expansions in Ω_0, Ω_+.

Section 6: composition of global approximation Z_N and estimation of the error $u - Z_N$.

From a point of view of asymptotic expansions section 2 will be rather straightforward. However, in the next sections the construction process contains some surprising elements. For example, the structure of the asymptotic expansions near the free surface S and in Ω_+ turns out to be unusual: not only fractional orders of ε appear as orderfunctions, but also fractional orders of ε multiplied with powers of $\ln \varepsilon$. Partial results in this sense were also announced by Frank, cf. [10].

One of the non-trivial aspects of the construction will be to determine which orderfunctions are needed in the local expansions near S and in Ω_+. For precise information on these orderfunctions, see section 3 and 4. Another interesting phenomenon in the construction is the unavoidable occurrence of singularities in the higher order terms of the expansion in Ω_+ when the variables approach the free surface. This requires a rather delicate analysis of these terms dealing with the singularities separate from the regular part of the term. Nevertheless, these singularities match with the layer at S and the layer changes the singularities coming from Ω_+ into regular terms, see section 4 and 5. To demonstrate the validity of the constructed global approximation we use an estimation result based on barrier-function techniques, see section 6. Using the results in that section we can improve the convergence statements made earlier in this introduction. For example, if K is an ε-independent compact subset of Ω_0, then

(1.10) $$\max_{x\in K} |u - U_0| = O(\varepsilon) \quad \text{for} \quad \varepsilon \downarrow 0.$$

Instead of an estimate $\|w - w_0\|_{H_0^1(\Omega)} = O(\sqrt{\varepsilon})$ and weak convergence in $W^{2,p}(\Omega)$ we obtain here

(1.11) $$\|w - w_0\|_{C^1(\overline{\Omega})} = O(\sqrt{\varepsilon}) \quad \text{for} \quad \varepsilon \downarrow 0$$

(1.12) $$\|w - w_0\|_{W^{2,p}(\Omega)} = O(\varepsilon^q) \quad \text{for} \quad \varepsilon \downarrow 0, \text{ with } q = 1/(2p).$$

In the case of general nonlinearities Φ, i.e. for the full problem (1.1), the structure of the approximation near S in Ω_+ will heavily depend on how $\Phi(x,t)$ tend to its limit for $t \to \infty$. In cases, where Φ has the following asymptotics:

(1.13) $$\Phi(x,t) \simeq 1 - \lambda_1(x)t^{-1} + \sum_{k\geq 2} \lambda_k(x)t^{-k} \quad \text{for} \quad t \to \infty,$$

with coefficients $\lambda_k \in C^\infty(\overline{\Omega})$, $\lambda_1 > 0$, the approximation will still have a structure as in this paper and its construction is analogous. In cases where Φ has completely different asymptotics for $t \to \infty$, the structure of the approximation will also be different. For example, for $\Phi = 1 - e^{-t}$ the structure of the approximation will be easier, since only orderfunctions of the type $\varepsilon^{p/2}$, $p = -2,0,1,..$ will then appear, see [12].
The method to prove validity given in section 6 extends without difficulties to these more general non-linearities.

N.B. In this paper, many proofs are only sketched, or avoided. For further details we refer the reader to [6] (see also [12], [5]).

2. THE REGULAR EXPANSION IN Ω_0 AND THE LAYER AT $\partial\Omega$.

In Ω_0 we construct a local formal approximation of the solution u as

$$(2.1) \qquad u \simeq U^M \underset{\text{def}}{=} \sum_{k=0}^{M} \varepsilon^k U_k(x).$$

Substitution in (1.9) provides us with a recursive system of equations

$$(2.2) \qquad U_0/(1+U_0) = f, \quad \text{i.e. } U_0 = f/(1-f)$$

and

$$(2.3) \qquad U_{n+1} = (1+U_0)^2.\{\Delta U_n + F_{n+1}(U_0,\dots,U_n)\}$$

$$\text{with} \qquad F_{n+1} = \left[\frac{\partial^{n+1}}{\partial\varepsilon^{n+1}}\left(1 + U_0 + \sum_{k=1}^{n} \varepsilon^k U_k\right)^{-1}\right]_{\varepsilon=0} =$$

$$= \sum_{m=2}^{n} \frac{(-1)^m}{(1+U_0)^{m+1}} \sum_{\substack{\vec{k}\in\mathbb{N} \\ |\vec{k}|=n}} \prod_{i=1}^{m} U_{k_i}.$$

The system for the U_n's can be solved uniquely and each U_n is $\in C^\infty(\bar{\Omega}_0)$, because of (1.3), (1.7).

In general U_0 and also the other U_n's will not satisfy the Dirichlet boundary condition on $\partial\Omega$. As a consequence we need a layer along $\partial\Omega$ to correct this.

As usual (cf. [8], [14]) we put

$$(2.4) \qquad u \simeq Z_0^M \underset{\text{def}}{=} U^M + G^M\, H\!\left(\frac{d(x)-d_0}{d_0}\right) \text{ with } G^M = \sum_{k=0}^{2M+1} (\sqrt{\varepsilon})^k G_k(\zeta,\omega).$$

Here ζ is the layer variable:

$$(2.5) \qquad \zeta = \frac{d(x)}{\sqrt{\varepsilon}}$$

where d(x) denotes the distance of a point x to $\partial\Omega$. H is a C^∞ cut-off function:

(2.6) $H(s) = 1$ for $s \leqq \frac{1}{2}$, 0 for $s \geqq 1$, and $H'(s) \leqq 0$.

One has for the G_n's:

$$\frac{\partial^2 G_0}{\partial \zeta^2} = \frac{1}{1+\gamma} - \frac{1}{1+\gamma+G_0} \tag{2.7}$$

with $\gamma(\omega) = U_0\big|_{\partial\Omega}(\omega) \geqq 0$,

$$G_0\big|_{\zeta=0} = -\gamma, \quad G_0 \to 0 \quad \text{for } \zeta \to \infty, \tag{2.8}$$

$$\frac{\partial^2 G_{n+1}}{\partial \zeta^2} - \frac{G_{n+1}}{(1+\gamma+G_0)^2} = F_{n+1}(G_0,\ldots,G_n)$$

(2.9)

$$G_{n+1}\big|_{\zeta=0} = 0 \text{ if } n+1 \text{ odd}, = -U_{\underline{n}}\big|_{\partial\Omega} \text{ if } n+1 \text{ is even}, \; \underline{n} = \tfrac{1}{2}(n+1).$$

To conclude this section we remark that the approximation constructed upto now, i.e. $Z_M^{(0)}$ as given in the righthandside of (2.4), is such that

$$-\varepsilon\Delta Z_M^{(0)} + \frac{Z^{(0)}}{1+Z_M^{(0)}} = f - r \tag{2.10}$$

where the error term r is $O(\varepsilon^{M+1})$ uniformly on $\overline{\Omega}_0$.

3. THE PRINCIPAL TERMS IN THE INTERNAL LAYER AT S AND IN THE EXPANSION IN Ω_+.

In order to describe the local approximation of the solution in the layer at the free surface S we use the following coordinates: (ρ,θ) as introduced in figure 2.

In these coordinates the operator Δ is given by

$$\Delta u = \frac{1}{\sqrt{g}} \frac{\partial}{\partial\rho}\left(\sqrt{g}\,\frac{\partial u}{\partial\rho}\right) + \frac{\partial}{\partial\theta}\left(\frac{1}{\sqrt{g}}\,\frac{\partial u}{\partial\theta}\right) \tag{3.1}$$

where $g(\rho,\theta) = 1 + 2\rho a(\theta) + \rho^2 b(\theta)$, with $a = \langle n',x'\rangle$, $b = \langle n',n'\rangle = a^2$ (' denotes differentiation w.r.t. θ).

Fig. 2

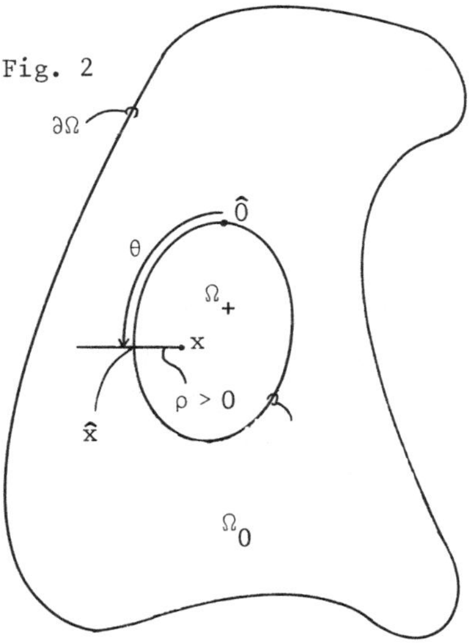

$x = \hat{x}(\theta) + \rho\, n(\theta)$ with n the normal on S in the direction of Ω_+;
$\langle n,\hat{x}'\rangle = 0$, $\langle \hat{x}',\hat{x}'\rangle = \langle n,n\rangle = 1$.
Note that:
$\rho > 0$ in Ω_+, $\rho < 0$ in Ω_0;
for points sufficiently close to S:
$|\rho|$ = distance to S,
$0 \leq \theta < \theta_0$.

The ρ coordinate is stretched in a significant way:

$$\xi = \rho/\sqrt{\varepsilon}.$$

Expanding the coefficients in (3.1) in a Taylor series, the operator $\varepsilon\Delta$ can be expressed as follows

$$\varepsilon\Delta = \frac{\partial^2}{\partial\xi^2} + \sqrt{\varepsilon}M_1 + \varepsilon M_2 + \ldots \tag{3.2}$$

where

$$\sqrt{\varepsilon}M_1 + \varepsilon M_2 + \ldots = \varepsilon g^{-1}\frac{\partial^2}{\partial\theta^2} + \sqrt{\varepsilon}(\sqrt{\varepsilon}\xi b + a)g^{-1}\frac{\partial}{\partial\xi} -$$

$$- (\sqrt{\varepsilon}\xi a' + \tfrac{1}{2}\varepsilon\xi^2 b')g^{-2}\frac{\partial}{\partial\theta}$$

with $g(\xi,\theta) = 1 + 2\sqrt{\varepsilon}\xi a + \varepsilon\xi^2 b$.
For the local approximation of the solution near S we put

$$u \simeq \psi_0(\xi,\theta) + \delta_1\psi_1(\xi,\theta) + \delta_2\psi_2(\xi,\theta) + \ldots \,. \tag{3.3}$$

In this expansion the magnitudes of the higher order terms $\delta_1(\varepsilon)$, $\delta_2(\varepsilon)$, etc. are unknown at the start of the construction. They

will be determined during the construction process.
Recalling (1.6) it is clear that the local approximation in Ω_+ will be of the following type

$$u \simeq \varepsilon^{-1} w_0 + \overline{\delta}_1 \overline{w}_1 + \dots \tag{3.4}$$

Again the magnitude $\overline{\delta}_1(\varepsilon)$ has to be found during the construction.
Let us now first deduce the principal term in the layer.
The equation for ψ_0 is obtained from the O(1) terms in (1.1) substituting (3.2) for $\varepsilon\Delta$. It is the following O.D.E.

$$\frac{\partial^2 \psi_0}{\partial \xi^2} = 1 - \hat{f} - \frac{1}{1 + \psi_0} . \tag{3.5}$$

The variable θ acts only as a parameter. Here $\hat{f}$ denotes $f|_S$. This function depends only on θ, it is smooth and periodic and $0 < \hat{f} < 1$.
In addition to (3.5) matching provides us with "boundary conditions" for $\xi \to -\infty$ and $\xi \to +\infty$.
Expanding the regular expansion in Ω_0 in the ξ,θ variables we obtain:

$$\lim_{\xi \to -\infty} \psi_0 = \hat{\gamma} \tag{3.6}$$

with $\hat{\gamma} \underset{\text{def}}{=} U_0|_S = \hat{f}/(1-\hat{f})$.
Using (1.6) it follows that $w_0 = \frac{1}{2}(1-\hat{f})\rho^2 + O(\rho^3)$ for $\rho \downarrow 0$.
As a consequence expansion of (3.4) in ξ,θ coordinates leads us to:

$$\psi_0 = \tfrac{1}{2}(1-\hat{f})\xi^2 + o(\xi^2) \text{ for } \xi \to \infty. \tag{3.7}$$

The solutions of (3.5)-(3.7) can easily be found. Namely, (3.6), (3.5) imply that $\frac{\partial^2 \psi_0}{\partial \xi^2}$ and also $\frac{\partial \psi_0}{\partial \xi}$ vanish for $\xi \to -\infty$.
Multiplication of (3.5) with $\frac{\partial \psi_0}{\partial \xi}$ and integration yields

(3.8) $$\psi_0 = \hat{\gamma} + (1 + \hat{\gamma})v\,\frac{\xi}{1+\hat{\gamma}}$$

$$\int_{v_0}^{v} \frac{dz}{\sqrt{2[z-\ln(1+z)]}} = \eta \text{ with } v_0 > 0.$$

Here the lower endpoint $v_0(\theta)$ plays the role of a free constant, possibly depending on θ. Eventually $v_0(\theta)$ will be determined by a matching argument. The function $v(\eta,\theta)$ is strictly increasing on $(-\infty,\infty)$ from 0 to $+\infty$.

Let us have a closer look at the asymptotic behaviour of ψ_0 for both $\xi \to -\infty$ and $\xi \to +\infty$.

We find that

(3.9) $$v = c\, e^{\eta}(1 + O(e^{\eta})) \text{ for } \eta \to -\infty$$

with $c = v_0 \exp\left(\int_0^{v_0} W(g)dg\right) > 0$. Since

(3.10) $$\frac{\partial\psi_0}{\partial\xi}(\xi,\theta) = \sqrt{2[v-\ln(1+v)]}\left(\frac{\xi}{1+\hat{\gamma}}\right)$$

it is also clear that

(3.11) $$\frac{\partial\psi_0}{\partial\xi} = c\, e^{\xi/(1+\hat{\gamma})}\,(1 + O(e^{\xi/(1+\hat{\gamma})})) \text{ for } \xi \to -\infty.$$

For the analysis of the behaviour of the behaviour for $\xi \to +\infty$ we introduce

(3.12) $$I(z) = (2z)^{-\frac{1}{2}}.\{(1-z^{-1}\ln(1+z))^{-\frac{1}{2}} - 1\}.$$

Note that $I(z) = O(z^{-3/2}\ln z)$ for $z \to +\infty$, hence $I(z)$ is integrable at $+\infty$.

Substituting

(3.13) $$\frac{1}{\sqrt{2[z-\ln(1+z)]}} = \frac{1}{\sqrt{2z}} + I(z).$$

The implicit formula for v in (3.8) can be written as:

(3.14) $$v = \tfrac{1}{2}\{\eta + (\sqrt{2v_0} - \int_{v_0}^{\infty} I(z)dz) + \int_{v}^{\infty} I(z)dz\}^2$$

i.e.

(3.15) $$v(\eta,\theta) = \tfrac{1}{2}\eta^2.\{1+2\eta^{-1}(\sqrt{2v_0} - \int_{v_0}^{\infty} I(z)dz) + o(\eta^{-1})\}$$
for $\eta \to +\infty$.

For ψ_0 this means

(3.16) $$\psi_0(\xi,\theta) = \tfrac{1}{2}(1-\hat{f})\xi^2 + \{\sqrt{2v_0} - \int_{v_0}^{\infty} I(z)dz\}\xi + o(\xi)$$
for $\xi \to +\infty$.

Next we observe that the linear term in ξ in ψ_0 gives rise to an $O(\varepsilon^{-\frac{1}{2}})$ term when reexpanded in the ρ,θ variables in Ω_+.
Now let us take in (3.4)

(3.17) $$\overline{\delta}_1 = \varepsilon^{-\frac{1}{2}}.$$

Then $\overline{w}_1$ has to satisfy the equation

(3.18) $$\Delta\overline{w}_1 = 0.$$

Matching with the layer at the free surface, which has no $O(\varepsilon^{-\frac{1}{2}})$ term, makes it necessary, that

(3.19) $$\overline{w}_1 = 0 \text{ on } S.$$

The conclusion is:

(3.20) $$\overline{w}_1 \equiv 0 \text{ in } \overline{\Omega}_+.$$

Another consequence of the matching is that the linear term in in (3.16) equals $\left.\frac{\partial \overline{w}_1}{\partial n}\right|_S$ and hence vanishes, i.e.

(3.21) $$\sqrt{2v_0} = \int_{v_0}^{\infty} I(z)dz.$$

Thus the value of v_0 follows indeed by a matching argument. Somewhat surprisingly $v_0 > 0$ does not depend on θ.
The leading term ψ_0 in the layer is now completely known. It will be important to have a precise description of the asymptotics of ψ_0 for $\xi \to +\infty$.

Lemma 1.

Let us denote

$$(3.22) \qquad \eta_1 = \frac{\ln \xi}{\xi^2}, \quad \eta_2 = \frac{1}{\xi^2}.$$

Then there exists a $\xi_0 > 0$ *such that for* ξ *sufficiently large,* $\xi \geqq \xi_0$:

$$(3.23) \qquad \psi_0 = \tfrac{1}{2}(1-\hat{f})\xi^2 + \frac{2}{1-\hat{f}} \ln \xi + T(\theta,\eta_1,\eta_2)$$

where for η_1, η_2 *sufficiently small, say* $|\eta_1| < \nu$, $|\eta_2| < \nu$, *and* $\theta \in [0,\theta_0)$

(i) T *is smooth, and* T *is periodic in* θ;
(ii) T *has a convergent powerseries in* η_1 *and* η_2*:*

$$T = \sum_{k=0}^{\infty} \sum_{l=0}^{\infty} T_{kl}\, \eta_1^k \eta_2^l.$$

The proof of this lemma is based on a somewhat tricky, implicit function argument, see [6].

As a consequence of the behaviour of ψ_0 for $\xi \to \infty$ we need in the expansion in Ω_+ a term of magnitude $\ln(\varepsilon)$.

$$(3.24) \qquad u = \varepsilon^{-1} w_0 + \ln \varepsilon . w_1 + \ldots$$

where w_1 is the solution of

$$(3.25)\qquad \begin{aligned} &\Delta w_1 = 0 \\ &w_1 = 2(1-\hat{f})^{-1} \text{ on } S. \end{aligned}$$

But this reflects back on the layer at S. It is easy to see that matching requires, that the next order term in the layer expansion corresponds to

$$(3.26)\qquad \delta_1 = \sqrt{\varepsilon}.\ln \varepsilon$$

where ψ_1 is a solution of the homogeneous, linearized layer equation

$$(3.27)\qquad \frac{\partial^2 \psi_1}{\partial \xi^2} - \frac{1}{(1+\psi_0)^2}\psi_1 = 0$$

i.e. ψ_1 is a (θ-dependent) multiple of $\frac{\partial \psi_0}{\partial \xi}$.

Now an obvious question is how the other order functions in the asymptotic sequence in the layer are generated. "Minimal" requirements for this sequence of orderfunctions $S' = \{\delta_n | n \in \mathbb{N}\}$ are:

1° 1, $\sqrt{\varepsilon}\ln\varepsilon$ and $(\sqrt{\varepsilon})^k$, $k \in \mathbb{N}$ are in S'.
The first magnitudes are those corresponding to ψ_0, ψ_1, the others are generated by the Taylor series of f in the ξ,θ coordinates.

2° stability under multiplication: $\delta_n, \delta_m \in S' \Rightarrow \delta_n . \delta_m \in S'$, especially $\delta_n \in S' \Rightarrow \delta_n\sqrt{\varepsilon} \in S'$.
The reason is the structure of $\varepsilon\Delta$ in (3.3) and the fact that such products appear automatically in the Taylor series of the non-linearity $u/(1+u)$.

These conditions 1° and 2° naturally lead us (at least) to

$$(3.28)\qquad S' = \{\delta_1^k \delta_2^l | k \in \mathbb{N} \text{ and } l \in \mathbb{N}\}$$

with: $\delta_1 = \sqrt{\varepsilon}\ln\varepsilon$, $\delta_2 = \sqrt{\varepsilon}$.

In the sequel we shall demonstrate that this sequence S' is indeed sufficient.
On the other hand, if the layer term $\delta_1^k\delta_2^l\psi_{k,l}(\xi,\theta)$ contains a constant term $\sim \delta_1^k\delta_2^l$ for $\xi \to \infty$ we expect that all these order-functions $\delta_1^k\delta_2^l$ will also be present in the expansion in Ω_+.
In combination with (3.31) this leads us to the following sequence S_+ of order functions needed in Ω_+.

$$S_+ = \{\varepsilon^{-1}, \ln \varepsilon\} \cup S'. \tag{3.29}$$

This sequence contains all the orderfunctions generated by $\psi_0 + \delta_1\psi_1$ in (ρ,θ) coordinates. It is also closed under Taylor series expansion of the non-linearity. Below we shall show, that the sequence S_+ is sufficient to construct the expansion in Ω_+. One could have the impression that S_+ contains a lot of super-fluous orderfunctions. It can not be denied, that S_+ contains a few redundant orderfunctions, for example δ_1. However, the redundancy in S_+ is not large, as we shall see.

4. HIGHER ORDER TERMS NEAR S AND IN Ω_+.

Let us denote the expansion in the free surface layer by

$$u \simeq \psi^N \underset{\text{def}}{=} \sum_{k=0}^{k+l\leq N} \sum_{l=0} \delta_1^k\delta_2^l\, \psi_{k,l}(\xi,\theta) \tag{4.1}$$

with $\delta_1 = \sqrt{\varepsilon}.\ln \varepsilon$, $\delta_2 = \sqrt{\varepsilon}$ and $\xi = \rho/\sqrt{\varepsilon}$, as before.
The notation for the expansion in Ω_+ will be

$$u \cong \Phi^N \underset{\text{def}}{=} \varepsilon^{-1}w_0(x) + \ln \varepsilon.w_1(x) + \sum_{k=0}^{k+l\leq N} \sum_{l=0} \delta_1^k\delta_2^l\phi_{k,l}(x). \tag{4.2}$$

In this section we shall present an iterative scheme by which the $\psi_{k,l}$'s and $\phi_{k,l}$'s can be determined in a unique way. To start with we discuss in §4.1 the construction of the $\psi_{k,l}$'s, while in

each $\psi_{k,1}$ an additive term $A_{k1} \frac{\partial\psi_o}{\partial\xi}$, solution of the homogeneous equation, is still free. Furthermore we analyze some of the properties of the $\psi_{k,1}$'s and we show that some of the A_{k1} have to be zero because of a simple matching argument. In §4.2 the $\phi_{k,1}$'s are constructed, while in each $\phi_{k,1}$ a linear degree of freedom corresponding to a choice of a boundary value function $g_{k,1}$ on S, is built in. The $\phi_{k,1}$'s will contain a singular part for $\rho \downarrow 0$. The freedom $g_{k,1}$ will arise in the boundary values on S of the regular part of $\phi_{k,1}$.
Next, in §4.3 we describe the scheme by which all free functions $A_{k,1}$ and $g_{k,1}$ can be uniquely determined. As a matter of fact this scheme will be based on a partial matching relation. The full matching will be considered in the next section.

4.1. *The $\psi_{k,1}$'s with free $A_{k,1}$'s.*

Collecting the $O(\delta_1^k\delta_2^1)$ terms in (1.1) with $\varepsilon\Delta$ as in (3.2) we find the following inhomogeneous, linear O.D.E. for $\psi_{k,1}$, $k + 1 > 0$.

$$\frac{\partial^2\psi_{k,1}}{\partial\xi^2} - \frac{1}{(1+\psi_0)^2}\psi_{k,1} = \hat{f}_{k,1}. \tag{4.1.1}$$

with $\hat{f}_{k,1} = (k!1!)^{-1}\left[\frac{\partial^{k+1}}{\partial\delta_1^k\partial\delta_2^1}\check{f}_{k,1}\right]_{\delta_1=0,\delta_2=0}$

where $\check{f}_{k,1} = -f(\delta_2\xi,\theta) - \delta_2 B\check{\Sigma}\,\delta_1^r\delta_2^s\psi_{r,s} - (1+\check{\Sigma}\,\delta_1^r\delta_2^s\psi_{r,s})^{-1}$,

$$\delta_2 B = \delta_2^2 g^{-1}\frac{\partial^2}{\partial\theta^2} + \delta_2 g^{-1}.(\delta_2\xi b + a)\frac{\partial}{\partial\xi}$$

$$- g^{-2}.(\delta_2\xi a' + \tfrac{1}{2}\delta^2\xi^2 b')\frac{\partial}{\partial\theta}.$$

$\check{\Sigma}$ denotes summation over all indices r,s with

$0 \leq r \leq k,\ 0 \leq s \leq 1,\ r + s < k + 1.$

In addition matching with the expansion in Ω_0 enhances the following conditions for $\xi \to -\infty$:

(4.1.2) $\psi_{0,1}$ is bounded by a polynomial for $\xi \to -\infty$

$\psi_{k,1} \to 0$ for $\xi \to -\infty$ for $k > 0$.

If $\hat{f}_{k,1}$ is polynomially bounded for $\xi \to -\infty$, then the solutions of (4.1.1), which don't grow exponentially for $\xi \to -\infty$ are given by:

$$(4.1.3)\quad \psi_{k,1}(\xi,\theta) = \frac{\partial\psi_0}{\partial\xi}(\xi,\theta)\cdot\int_{\xi_0}^{\xi}\left[\frac{\partial\psi_0}{\partial\xi}(\eta,\theta)\right]^{-2}\cdot$$

$$\int_{-\infty}^{\eta}\frac{\partial\psi_0}{\partial\xi}(\zeta,\theta)\hat{f}_{k,1}(\zeta,\theta)\,d\zeta\,d\eta + A_{k,1}(\theta)\cdot\frac{\partial\psi_0}{\partial\xi}(\xi,\theta).$$

Here we shall take ξ_0 a fixed, sufficiently large number. The function $A_{k,1}(\theta)$ is not determined by (4.1.2) because of the exponential decay of $\frac{\partial\psi_0}{\partial\xi}$ for $\xi \to -\infty$. This simply means that at the moment (4.1.3) contains an amount of freedom.

Note, that, except for this freedom, (4.1.3) allows us to calculate the $\psi_{k,1}$'s recursively. This can be done in several ways, for example:

a: calculate the $\psi_{k,1}$'s with $k + 1 = 1$, next those with $k + 1 = 2$, etc.

or

b: calculate all $\psi_{0,k}$ by increasing k, next all $\psi_{1,k}$ with $1 = 1$, etc.

Though the $\psi_{k,1}$'s are not yet uniquely determined we can already derive some of their properties.

When the $A_{k,1}$'s are chosen as smooth periodic functions in θ, all $\psi_{k,1}$'s are smooth functions of ξ and θ in $\mathbb{R}\times[0,\theta_0)$, periodic in θ.

As for the behaviour of $\psi_{k,1}$ for $\xi \to -\infty$ we can derive the following results:

Lemma 2.

(4.1.4) $\psi_{0,1} = P_1$ + exponentially small terms for $\xi \to -\infty$

Here P_1 *is a polynomial in* ξ *of degree* ≤ 1 *with, coefficients depending smoothly and periodic on* θ. *Moreover* P_1 *has the same parity as* 1.

Further, for $k > 0$ *there is an* $m \in \mathbb{N}$, *such that:*

(4.1.5) $\psi_{k,1} = O(\xi^m \exp(\frac{\xi}{1+\hat{\gamma}}))$ for $\xi \to -\infty$.

Moreover, the derivatives $\dfrac{\partial^{r+s}\psi_{k,1}}{\partial\xi^r \partial\theta^s}$ *behave in an analogous way.*

The derivation of (4.1.1) and (4.1.5) is an interesting, rather easy exercise in induction using recursion as in b and using (3.10-11), further details are left to the reader.
Note that the conditions in (4.1.2) are indeed fulfilled, regardless of the choice of the $A_{k,1}$'s.

Let us now consider the asymptotic behaviour of $\psi_{k,1}$ for $\xi \to \infty$. It will be convenient to introduce the following concept:

a function χ *of* $(\xi,\theta) \in \mathbb{R}\times[0,\theta_0)$ *is called of type p with* $p \in \mathbb{Z}$ *if:*

(i) χ *is smooth in* ξ *and* θ *and periodic in* θ.

(ii) for ξ *sufficiently large,* $\xi \geq \xi_0$, χ *can be represented in the following way*

(4.1.6) $\chi = \xi^p X(\eta,\theta) + \eta_2 Y(\eta_1,\eta_2,\theta)$

with: $\eta_1 = \dfrac{\ln \xi}{\xi^2}$, $\eta_2 = \dfrac{1}{\xi^2}$.

- X *is polynomial of degree* $\leq \frac{1}{2}p$ *in* η_1 *with coefficients depending smoothly and periodic on* θ. *For* $p < 0$ *this means that* $X \equiv 0$.
- Y *and all its derivatives with respect to* θ *are analytic in*

η_1, η_2 *for* $|\eta\ | < \nu$, $|\eta\ | < \nu$ *with coefficients of the powerseries in* η_1, η_2 *depending smoothly and periodicly on* θ.

In this definition ξ_0 and ν are numbers > 0, which are fixed in accordance with there values in lemma 1. It is not difficult to verify the following extension of lemma 1:

(4.1.7) ψ_0 is of type 2,

$\frac{\partial\psi_0}{\partial\xi}$ is of type 1.

This demonstrates already the relevance of the type p concept. Further, the other solution of the homogeneous version of (4.1.1) is

(4.1.8) $\Phi \underset{\text{def}}{=} \frac{\partial\psi_0}{\partial\xi} \int_\xi^\infty [\frac{\partial\psi_0}{\partial\xi}(\eta,\theta)]^{-2} d\eta.$

Now a little calculation shows, that

(4.1.9) $\xi^2 [\frac{\partial\psi_0}{\partial\xi}]^{-2}$ is of type 0

and

(4.1.10) Φ is of type 0.

The type p concept is nicely compatible with algebraic and analytic operations and notions. We mention a few useful rules, which can be checked in an elementary way.

(4.1.11) χ of type p $\Rightarrow \xi^r\chi$ is of type p+r for $r \in \mathbb{N} \cup \{0\}$,

$\frac{\partial\chi}{\partial\theta}$ is of type p,

$\frac{\partial\chi}{\partial\xi}$ is of type p-1

$\frac{\partial^{r+s}\chi}{\partial\xi^r\partial\theta^s}$ is of type p-r.

(4.1.12) χ_1 of type p_1 and χ_2 of type $p_2 \Rightarrow \chi_1\chi_2$ is of type p_1+p_2

and $p_2 \leqq p_1$, $p_1 - p_2$ even $\Rightarrow \chi_1+\chi_2$ is op type p_1.

(4.1.13) a polynomial in ξ of degree p of the same parity as p with smooth, periodic θ-dependent coefficients is of type p.

For a function χ of type p there exists a uniquely defined primitive $I(\chi)$ with respect to ξ-integration, such that:

(4.1.14) $I(\chi)$ is a function of type p+1 without a constant term in its expansion for $\xi \to \infty$.

Note that

(4.1.15) $I(\chi) = \int_{\xi_0}^{\xi} \chi(\eta,\theta)d\eta + I_0(\chi)(\theta)$

with a smooth, periodic function I_0, which in general is $\not\equiv 0$.

Next we shall describe the behaviour of the $\psi_{k,l}$'s for $\xi \to \infty$ and at the same time fix some of the $A_{k,l}$'s. From now on we shall use the shorthand notation:

$$\chi = \text{type } (p_1) + \text{type } (p_2)$$

for $\chi = \chi_1 + \chi_2$ with χ_1 of type p_1 and χ_2 of type p_2.

Lemma 3.

a. k = 0:

(4.1.17) $\psi_{0,0} = \text{type } (2)$

$\psi_{0,l} = \text{type } (l+2) + \text{type } (l-1)$ for $l \geq 1$.

b. <u>k = 1</u>:

(4.1.18) $\psi_{1,0}$ = type (1)

$A_{1,1} \equiv 0$ and $\psi_{1,1}$ = type (2) + type (-1)

$\psi_{1,l}$ = type (l+1) + type (l-2) for $l \geqq 2$.

c. <u>k ≧ 2</u>:

(4.1.19) $A_{k,0} = 0$

$\psi_{k,l}$ = type (1) + type (l-1) for $l \geqq 0$.

For the proof of lemma 3 we refer again to [6].
An important remark is that the contents of lemma 3 are fully consistent with our assumption (4.2) on the structure of the expansion in Ω_+, i.e. reexpansion in ρ,θ coordinates of $\delta_1^k\delta_2^l\psi_{k,l}$ with

(4.1.20) $\xi = \rho\delta_2^{-1}, \quad \eta_2 = \delta_2^2\rho^{-2}, \quad \eta_1 = \delta_2^2\rho^{-2}\ln\rho - \frac{1}{2}\delta_1\delta_2\rho^{-2}$

gives only rise to order functions as in (3.29)

(4.1.21) $\delta_2^{-2}, \quad \delta_1\delta_2^{-1}, \quad \delta_1^k\delta_2^l \quad k \geqq 0,\ l \geqq 0.$

However, later on it will be useful to have somewhat more precise information. Let us introduce the notation:

(4.1.22) $E_{k,l}^{\rho,\theta} \underset{\mathrm{def}}{=}$ the operator which gives the coefficient corresponding to the $\delta_1^k\delta_2^l$ term in an expansion written in (ρ,θ) coordinates.

Lemma 4.

Let k,l *be given as in* (4.1.21), *i.e.* $(k,l) \in \{(0,-2),(1,-1)\} \cup \cup \mathbb{N}^2$. *Denote* $N' = k + l$ *and suppose* $K \geq N'$
Then:

$$(4.1.23) \quad E^{\rho,\theta}_{k,l}\Big(\sum_{r+s\leq K} \delta_1^r\delta_2^s\psi_{r,s}\Big) = \Lambda^{\text{sing}}_{k,l} + \Lambda^{\text{const}}_{k,l} + \Lambda^{0,K}_{k,l}.$$

Here, these parts denote the singular, the constant and the vanishing terms for $\rho \downarrow 0$, *respectively. The form of these respective parts is as follows:*

$$(4.1.24) \quad \Lambda^{\text{sing}}_{k,l} = \sum_{\substack{0\leq i\leq l+1/2\\ 0\leq j\leq 1\\ i+j>0}} C^{k,l}_{i,-j}(\theta).(\ln \rho)^i\rho^{-j},$$

contributions to the $(\ln \rho)^i\rho^{-j}$ *term in* $\Lambda^{\text{sing}}_{k,l}$ *can only come from values* r,s *for which* $r+s = N' - j$ *and* $0 \leq r \leq k$,

$$(4.1.25) \quad \Lambda^{\text{const}}_{k,l} = C^{k,l}_{0,0}(\theta),$$

contributions to this term can only come from values r,s *with* $r + s = N'$, $1 \leq s \leq 2l + 2$, $0 \leq r \leq k$
and finally

$$(4.1.26) \quad \Lambda^{0,K}_{k,l} = \sum_{\substack{0\leq i\leq l+1/2\\ 1\leq j\leq K-N'}} C^{k,l}_{i,j}(\theta)(\ln \rho)^i\rho^j$$

contributions to the $(\ln \rho)^i\rho^j$ *term in* $\Lambda^{0,K}_{k,l}$ *can only come from values* r,s *with* $r + s = N' + j$, $0 \leq r \leq k$.
All coefficients $C^{k,l}_{i,j}$ *are smooth and periodic in* θ.

Proof of lemma 4.
Actually, all we have to do, is a little bit of "combinatorics". Firstly, a singular term can only be produced by a term $\sim\delta_1^r\delta_2^s \dfrac{(\ln \xi)^m}{\xi^j}$ for $\xi \to \infty$ which after reexpansion gives rise to

terms $\sim\delta_1^{r+n}\delta_2^{s+j-n}\frac{(\ln\rho)^{m-n}}{\rho^j}$ with $0 \leqq n \leqq m$.
Such a term contributes to $\Lambda_{k,1}^{sing}$ precisely if $r + n = k$, $s + j - n = 1$ and $i = m - n$.
Because of the structure of $\psi_{r,s}$ gives in lemma 3 we also know, that $j \geqq 2m - s - 2$, where equality only holds, if $r = 0$ and $j = 0$, $m > 0$. This implies $21 + 2 = 2s + 2i + 2j + 2 - 2m \geqq$ $\geqq 2i + j + s \geqq 2i + j$ and $2i = 2m - 2n \leqq j + s + 2 - 2n = 1 + 2 - n \leqq$ $\leqq 1 + 2$. Thus (4.1.29) is found.

A constant contribution can only come from a term for $\xi \to \infty$ proportional to $\delta_1^r\delta_2^s(\ln\xi)^q$, where because of lemma 3, $q \leqq \frac{1}{2}(s+2)$.
After reexpansion this leads to a constant term $\sim\delta_1^{r+q}\delta_2^{s-q}$, which contributes to $\Lambda_{k,1}^{const}$, if $k = r + q$, $1 = s - q$.
Hence the contributions to $\Lambda_{k,1}^{const}$ in (4.1.29) come indeed from pairs (r,s) with $r + s = k + 1$, $s \geqq 1$, $21 \geqq 2s - (s+2) = s - 2$.
The verification of (4.1.26) is left to the reader as an exercise.

We abbreviate

(4.1.27) $\Psi^N \underset{def}{=} \sum_{k+1\leqq N} \delta_1^k\delta_2^1\psi_{k,1}$, as in (4.1).

(4.1.28) $\Lambda_{k,1}^K \underset{def}{=} \Lambda_{k,1}^{sing} + \Lambda_{k,1}^{const} + \Lambda_{k,1}^{0,K}$, $K \geqq N$.

$$\Lambda^{N,K} \underset{def}{=} \sum_{\substack{(k,1)=(0,-2)\text{ or}\\(1,-1)\text{ or}\\k\geqq 0,1\geqq 0,k+1\leqq N}} \Lambda_{k,1}^K .$$

In anticipation to the full matching later on in section 5, below we give an estimate for the difference between Ψ^N and $\Lambda^{N_1,K}$, which is derived in [6]:

Lemma 5.

There are constants C > 0 *such that for* $\xi \geqq \xi_0$

$$(4.1.29)\quad \text{(a)}\quad |\Psi^N(\xi,\theta)-\Lambda^{N,N}(\delta_2\xi,\theta)| \leqq C(\delta_1+\delta_2\sqrt{\ln\xi} + \frac{\sqrt{\ln\xi}}{\xi})^{N+1}$$

$$\text{(b)}\quad |\Lambda^{N,K}(\delta_2\xi,\theta)-\Lambda^{N,N}(\delta_2\xi,\theta)| \leqq C\xi^2((\delta_1+\delta_2\xi)^{N+1}+$$

$$+(\delta_2\xi)^{K+1}), \qquad K > N$$

$$\text{(c)}\quad |\Lambda^{N_1,K}(\delta_2\xi,\theta)-\Lambda^{N,K}(\delta_2\xi,\theta)| \leqq C\ \ln(\rho).(\delta_1 + \frac{\sqrt{\ln\xi}}{\xi})^{N+1}.$$

$$.(1+\delta_2\xi)^K, \qquad K \geqq N_1 > N.$$

For the derivatives of these functions r *times w.r.t.* ξ *and* s *times w.r.t.* θ, $r + s \leqq 2$ *analogous estimates hold, but with an extra factor* ξ^{-r} *in the righthandside.*

To conclude this section we mention another result derived in [6], namely the error upto which Ψ^N satisfies the equation (1.9) can be estimated as follows.

Lemma 6.

There is an ε*-independent constant* $\rho_1 > 0$*, such that for* $|\xi| \leqq \frac{\rho_1}{\delta_2}$

$$(4.1.30)\qquad |-\varepsilon\Delta\Psi^N + \frac{\Psi^N}{1+\Psi^N} - f| \leqq M(\delta_1 + \delta_2|\xi|)^{N+1}$$

with an ε*-independent constant M.*

4.2. The $\phi_{k,l}$'s with free $g_{k,l}$'s.

For $\psi_{k,l}$ we find by substitution of (4.2) in (1.1) and collecting the terms of order $\varepsilon\delta_1^k\delta_2^l$ an equation of the following form:

$$(4.2.1) \qquad \Delta\phi_{k,l} = -\overline{F}_{k,l}$$

with $\overline{F}_{k,l} = (k!l!)^{-1} \dfrac{\partial^{k+l}}{\partial\delta_1^k\partial\delta_2^l}\{w_0+\delta_1\delta_2 w_1+\delta_2^2(1+\Sigma'\delta_1^r\delta_2^s\phi_{r,s})\}^{-1}\Big|_{\delta_1=\delta_2=0}$

where ' denotes summation over all indices r,s such that $0 \leqq r \leqq k$, $0 \leqq s \leqq l-2$, $r+s < k+l-2$.

For example, the equation for $\phi_{0,0}$ becomes

$$(4.2.2) \qquad \Delta\phi_{0,0} = -\frac{1}{w_0} .$$

Since w_0 behaves as $\frac{1}{2}(1-f|_S)\rho^2$ for $\rho \downarrow 0$ (see (1.6)) the righthandside of this equation in singular for $\rho \downarrow 0$. Analogously singularities can be expected in the other $\overline{F}_{k,l}$'s. Consequently, it is logical to split $\phi_{k,l}$ in a singular part and a regular part. Now, it will be crucial in the construction, that the singular part of $\phi_{k,l}$ can be determined in *a unique way* from the previous $\phi_{\overline{k},\overline{l}}$ with $\overline{k} \leqq k$, $\overline{l} \leqq l-2$, $\overline{k} + \overline{l} < k + l - 2$ by an iterative procedure *only using* the equation given in *(4.2.1)*. The regular part of $\phi_{k,l}$ will contain freedom in the form of its boundary values on S.

Lemma 7.

The equations given in (4.1.1) *have a set of solutions* $\phi_{k,l}$ *with the following properties:*

$$(4.2.3) \qquad \phi_{k,l} = \phi_{k,l}^{s,M}(\rho,\theta).H\left(\frac{\rho-\rho_1}{\rho_1}\right) + \phi_{k,l}^{r,M}$$

where $\phi_{k,l}^{s,M}$ *will consist of certain singular terms and* $\phi_{k,l}^{r,M}$ *is sufficiently regular:*

(4.2.4) $$\phi^{s,M}_{k,l} = \sum_{\substack{-1\leq j\leq M+2\\ 0\leqq i\leqq 1+l/2\\ i>0 \text{ or } j<0}} \phi^{k,l}_{i,j}(\theta).(\ln\rho)^i\rho^j, \quad \phi^{k,l}_{i,j} \textit{ smooth and periodic in } \theta$$

(4.2.5) $$\phi^{r,M}_{k,l} \in C^{M+2+\alpha}(\overline{\Omega}_+), \qquad \alpha \in (0,1).$$

In this decomposition $M \in \mathbb{N}$ *is arbitrary.*
H *is the cut-off function given in* (2.6) *and* $\rho_1 > 0$ *is* ε*-independent, but sufficiently small. The construction of the* $\phi_{k,l}$*'s goes iteratively, namely once all* $\phi_{k,l}$ *with* $k+l \leqq N$ *are known with properties as in* (4.2.3-4-5), *then for* $k+l = N+1$:

(i): $\phi^{s,M}_{k,l}$ *follows uniquely within the specified class by requiring:*

(4.2.6) $$\Delta\phi^{s,M}_{k,l} + \overline{F}_{k,l} \in C^{M+\alpha}(\{(\rho,\theta)\,|\,0 \leqq \rho \leqq \rho_1\})$$

and (ii): one solves

(4.2.7) $$\phi^{r,M}_{k,l} = -\overline{F}_{k,l} - \Delta(\phi^{s,M}_{k,l} H)$$

(4.2.8) $$\phi^{r,M}_{k,l} = g_{k,l}$$

where $g_{k,l} \in C^\infty(S)$ *is free.*

Let us now define

(4.2.9) $$\Phi^N = \varepsilon^{-1} w_0 + (\ln\varepsilon).w_1 + \sum_{k+l\leqq N} \delta_1^k \delta_2^l \phi_{k,l} \text{ as in (4.2).}$$

We conclude this section with an error estimate indicating how well Φ^N satisfies (4.1.1), cf. [6].

Lemma 8.

On the domain $\{x \in \Omega_+ | \text{distance } (x,S) \geq \delta_1\}$ *it holds, that:*

$$(4.2.10) \quad |-\varepsilon\Delta\Phi^N + \frac{\Phi^N}{1+\Phi^N} - f| \leq K \cdot \frac{\varepsilon}{\rho^2} \cdot (\delta_1 + \frac{\delta_2}{\rho})^{N+1}$$

with an ε*-independent constant* $K > 0$ *and* $\rho = d(x,S)$.

4.3. *The scheme leading to uniquely defined* $\psi_{k,l}$*'s and* $\phi_{k,l}$*'s with determined* $A_{k,l}$*'s and* $g_{k,l}$*'s.*

In order to fix the free $A_{k,l}$'s we notice, that matching of the layer expansion and the expansion in Ω_+ requires, that:

- $g_{k,l}$ is the constant term of order $\delta_1^k\delta_2^l$ found by reexpansion of the layer in (ρ,θ) coordinates.
- the linear term in $\phi_{k,l}$ near S coincides with the linear term found in the $O(\delta_1^k.\delta_2^l)$ term obtained by reexpanding the layer in (ρ,θ) coordinates.

  Note, that the term $\sim \rho$ of order $\delta_1^k\delta_2^l$ in the layer expansion has a coefficient:

$$(1 - \hat{f})A_{k,l+1} + B_{k,l+1}$$

  where $B_{k,l+1}$ is completely determined by the functions $\psi_{r,s}$ with $0 \leq r < k$, $r + s = k + l + 1$ or with $r + s \leq k + l$. (see (4.1.32) and the proof of lemma 3).

  Hence, $A_{k,l+1}$ has to satisfy the following equation

$$(4.3.1) \quad A_{k,l+1} = (1 + \hat{\gamma})\{\Phi_{0,1}^{k,l} - B_{k,l+1}\}$$

  with $\Phi_{0,1}^{k,l}$ as in (4.2.4).

Consequently, the scheme which provides us with the $\psi_{k,l}$'s and $\phi_{k,l}$'s in a unique way runs as follows. The start is with $N = 0$

and then successively:

the $\psi_{k,l}$'s with $k + l = N$ known → $\Lambda^{const}_{k,l}$ is known, see (4.1.31)

↓

$g_{k,l} \overset{=}{\text{def}} \Lambda^{const}_{k,l}$

↓

all $\Phi^{k,l}_{0,1}$'s with $k+l=N$ are known ← the $\phi_{k,l}$'s with $k + l = N$ known
(see (4.2.4)

↓

$A_{0,N+1}$ is known, see (4.3.1)

↓

$A_{1,N}$ is known, see (4.3.1)

↓

$A_{N,1}$ is known, see (4.3.1)

↓

$A_{N+1,0} = 0$, see lemma 3.

↓

the $\psi_{k,l}$'s with $k+l=N+1$ known → etc.

This scheme leads to $g_{k,l}$ and $A_{k,l}$ which are smooth and periodic in θ. It will be clear that only a very partial matching between the layer expansion and the expansion in Ω_+ has been built in in this scheme. In the next section we consider the matching relations in some more detail.

5. MATCHING RELATIONS OF THE INTERNAL LAYER AT S AND THE EXPANSIONS IN Ω_0, Ω_+.

We shall show that the internal layer expansion Ψ^{N+2} and the regular expansion U^N upto terms of the order δ_2^{2N} have an overlapping domain of validity and we shall given an estimate of their difference in that overlap domain.

Lemma 9.

There are ε*-independent constants* ρ_1, ξ_1, M *and* $\mu > 0$ *such that for* $-\rho_1 \leqq \rho \leqq -\delta_2\xi_1$:

$$(5.1) \qquad |U^N - \Psi^{N+2}| \leqq M\{(\delta_2+\rho)^{N+2} + \exp(-\mu\rho/\delta_2)\}.$$

and: $$\left|\rho^r \frac{\partial^{r+s}}{\partial\rho^r\partial\theta^s}(U^N - \Psi^{N+2})\right| \leqq M\{(\delta_2 + \rho)^{N+2} + \exp(-\mu\rho/\delta_2)\}, \quad r + s \leqq 2.$$

Note, that the overlap upto a certain order takes place in any region $-\hat{\rho}_1(\varepsilon) \leqq \rho \leqq -\hat{\rho}_2(\varepsilon)$ with $\hat{\rho}_1 = o(1)$ and $\delta_2 = o(\hat{\rho}_2)$. The difference is almost as small as possible in a region $-\hat{A}\delta_1 \leqq \rho \leqq -\hat{B}\delta_1$ with suitably chosen ε-independent constants $\hat{A}$, $\hat{B} > 0$.
Next, we shall see that Ψ^{N+2} and Φ^N, the expansion in Ω_+, have an overlapping domain of validity. It won't be a surprise, that in this case we have to work harder to get an appropriate estimate for the difference.

Lemma 10.

There are ε*-independent constants* ρ_1, ξ_1, $M > 0$ *such that for* $\xi_1 \leqq \xi \leqq \rho_1/\delta_2$:

(5.2) $$|\Psi^{N+2}-\Phi^{N}| \leqq M\{(\delta_1+\delta_2\sqrt{\ln\xi}+\frac{\sqrt{\ln\xi}}{\xi})^{N+1}.\ln(\rho) + \xi^2.(\delta_1+\delta_2\xi)^{N+3}\}$$

and $$|(\delta_2\xi)^r \frac{\partial^{r+s}}{\partial\rho^r\partial\theta^s}(\Psi^{N+2}-\Phi^{N})| \leqq M\{(\delta_1+\delta_2\sqrt{\ln\xi}+\frac{\sqrt{\ln\xi}}{\xi})^{N+1}.\ln(\rho) + \xi^2.(\delta_1+\delta_2\xi)^{N+3}\},\quad r+s\leqq 2.$$

We conclude this section with the remark that for $N \geqq 0$ the overlap region is non-empty and that the difference is small in the region:

(5.3) $$A\sqrt{\delta_2} \leqq \rho \leqq B\sqrt{\delta_2} \quad \text{i.e.} \quad A\delta_2^{-\frac{1}{2}} \leqq \xi \leqq B\delta_2^{-\frac{1}{2}}$$

with suitably chosen ε-independent A and $B > 0$.
This region is special, since exactly in this region Ψ^{N+2} and Φ^{N} give rise to the same order of error in the equation:

(5.4) $$\text{error} \sim (\delta_2\xi)^{N+3} = 0_{\text{sharp}}(\frac{\delta_2}{\delta})^{N+3} = 0_{\text{sharp}}(\varepsilon^{(N+3)/4}),$$

see lemma 6 and lemma 8.
This will play an important role in the next section.

6. COMPOSITION OF A GLOBAL APPROXIMATION Z_N AND ESTIMATION OF THE ERROR $u - Z_N$.

Now we are in the position to put all the local approximations constructed previously together into a global approximation Z_N. We define:

(6.1) $$Z_N = \Psi^{N+2}H(\frac{\rho^2}{\sqrt{\varepsilon}}) + (Z_0^N + \Phi^N).[1 - H(\frac{\rho^2}{\sqrt{\varepsilon}})]$$

with: Ψ^{N+2} the expansion in the layer along the free surface S, see (4.1.27).

Z_0^N the regular expansion in Ω_0 corrected with the layer expansion along $\partial\Omega$, see (2.4).

Φ^N the expansion in Ω_+, see (4.2.9).

$H(\frac{\rho^2}{\sqrt{\varepsilon}})$ is a smooth cut-off function, which is $\equiv 1$ for $|\rho| \geqq (\frac{1}{2}\sqrt{\varepsilon})^{\frac{1}{2}}$ and which vanishes for $|\rho| \geqq \varepsilon^{\frac{1}{4}}$, see (2.6).

Theorem I.

The construction process has been successful in the sense that Z_N *is a global, formal approximation:*

$$\max_{x\in\overline{\Omega}} \left|-\varepsilon\Delta Z_N + \frac{Z_N}{1+Z_N} - f\right| \leq R.\left(\sqrt{\varepsilon^{\frac{1}{2}}\ln(\tfrac{1}{\varepsilon})}\right)^{(N+3)} \tag{6.2}$$

where R *is an* ε*-independent constant* > 0.

Proof of theorem I:

The estimate in (6.2) can be derived by making a subdivision of $\overline{\Omega}$ in $\Omega_1 = \{(\rho,\theta)|\rho^2 \leq \frac{1}{2}\sqrt{\varepsilon}\}$, $\Omega_{1,+} = \{(\rho,\theta)|\rho > 0, \frac{1}{2}\sqrt{\varepsilon} \leq \rho^2 \leq \sqrt{\varepsilon}\}$, $\Omega_{1,0} = \{(\rho,\theta)|\rho < 0, \frac{1}{2}\sqrt{\varepsilon} \leq \rho^2 \leq \sqrt{\varepsilon}\}$, $\Omega_+' = \Omega_+\backslash\{\Omega_1 \cup \Omega_{1,+}\}$, $\Omega_0' = \overline{\Omega}_0\backslash\{\Omega_1 \cup \Omega_{1,0}\}$. Now partial results in the direction of (6.2) were already found. In Ω_0' we use (2.10) and obtain an error $O(\varepsilon^{N+1})$. The larger contributions come from $\Omega_1, \Omega_{1,+}$ and Ω_+'. We apply lemma 6 and lemma 8 in Ω_1 and $\Omega_{1,+}$, respectively. In both regions we find an $O(\varepsilon^{\frac{1}{4}(N+3)})$ error, because of the special choice of the cut-off at an $O(\varepsilon^{\frac{1}{4}})$-distance from S, compare (5.4). In the overlap region $\Omega_{1,+}$ we have:

$$Z_N = \Phi^N + H(\frac{\rho^2}{\sqrt{\varepsilon}})(\Psi^{N+2} - \Phi^N).$$

Using lemma 8 and the matching relations in lemma 10 it can be shown that here the error is at most $(\sqrt{\delta_2|\ln \varepsilon|})^{N+3}$.

Specially, the term $\delta_2^2 . \frac{\partial^2 H(\frac{\rho^2}{\sqrt{\varepsilon}})}{\partial\rho^2} . (\Psi^{N+2} - \Phi^N)$ gives rise to this order, see lemma 10. Other terms yield contributions of a smaller order. The region $\Omega_{1,0}$ is dealt with in an analogous way using lemma 9. □

The next step is to convert the estimate in (6.2) into an estimate for $u - Z_N$. This can be done with the following result.

Lemma 11.

Suppose, that $Z \in C^\infty(\overline{\Omega})$, $Z \geqq 0$ *satisfies:*

$$(6.3) \qquad -\varepsilon\Delta Z + \frac{Z}{1 + Z} - f = r$$

$$Z = 0 \text{ on } \partial\Omega.$$

Then there is an ε*-independent constant* $\nu > 0$*, such that*

$$(6.4) \qquad |u - Z|_{\sup} \leqq \frac{\nu}{\varepsilon} |r|_{\sup}.$$

Proof of lemma 11:

We use a technique based on barrier functions and the maximum-principle, see [13], [14], [9].

Suppose, that a positive function $v \in C^\infty(\overline{\Omega})$ can be found, such that

$$(6.5) \qquad -\varepsilon\Delta(Z + v) + \frac{Z + v}{1 + Z + v} - f \geqq 0 \text{ in } \overline{\Omega}$$

$$Z + v \geqq 0 \text{ on } \partial\Omega.$$

Then $Z + v$ is an upperbarrier for the solution u, i.e.

$$(6.6) \qquad u \leqq Z + v \text{ on } \overline{\Omega}.$$

To prove (6.6) we use the fact that $Z + v - u \underset{\text{def}}{=} w$ satisfies $w \in C^\infty(\overline{\Omega})$ and:

$$-\varepsilon\Delta w + \overline{c}\, w \geqq 0 \text{ in } \overline{\Omega}$$

$$w \geqq 0 \text{ on } \partial\Omega$$

with $\bar{c} = (1 + Z + v)^{-1}(1 + u)^{-1}$ i.e. $\bar{c} \in C^\infty(\bar{\Omega})$ and $\bar{c} \geq 0$. Implicitly, we used that $u \in C^\infty(\bar{\Omega})$. The maximum principle for 2nd order elliptic Dirichlet problems implies that $w \geq 0$.

If a negative function $v \in C^\infty(\bar{\Omega})$ can be found, such that (6.5) holds with reversed signs in the region where $Z + v \geq 0$ then $\max(0,Z+v)$ is a lower barrier for the solution u. This follows from (1.3) in combination with a maximum principle argument.

Now, we make the following choice for the function v

$$(6.7) \qquad v = b \cos\left(\frac{\pi}{2} \cdot \frac{x_1 - x_1^0}{d}\right).\cos\left(\frac{\pi}{2} \cdot \frac{x_2 - x_2^0}{d}\right)$$

with a value for $b \geq 0$, which we shall specify presently. The point x^0 and the number $d > 0$ are chosen in such a way that the product of the cosines is $\geq \frac{1}{2}$ on $\bar{\Omega}$.
With this function v we obtain, that

$$(6.8) \qquad -\varepsilon\Delta(Z + v) + \frac{Z + v}{1 + Z + v} - f \geq r - \varepsilon\Delta v \geq r + b.\frac{\varepsilon}{2}\left(\frac{\pi}{2d}\right)^2$$

which is ≥ 0, when $b = \frac{\nu}{\varepsilon}\min(0,-\min_{\bar{\Omega}} r)$ with $\nu = 8\left(\frac{d}{\pi}\right)^2$.
In this way an upperbound is found.
For the lowerbound we proceed in analogous way.
Actually we even obtain a somewhat better estimate than (6.3):

$$(6.9) \qquad \max(0,Z - \frac{\nu}{\varepsilon}\max(0,\max_{\bar{\Omega}} r)) \leq u \leq Z + \frac{\nu}{\varepsilon}\min(0,-\min_{\bar{\Omega}} r). \qquad \square$$

An immediate consequence of lemma 11 and theorem I is

$$(6.10) \qquad |u - Z_N|_{\sup} = O(\varepsilon^{-1}(\sqrt{\varepsilon^{\frac{1}{2}}\ln(\tfrac{1}{\varepsilon})})^{N+3}) \text{ if } N \geq 0.$$

However, this estimate can be somewhat improved. It holds, that:

$$(6.11) \qquad |Z_N - Z_{N_1}|_{\sup} = O((\sqrt{\varepsilon^{\frac{1}{2}}\ln(\tfrac{1}{\varepsilon})})^{N+1}) \text{ if } N_1 \geq N \geq 0$$

as one may verify with a straightforward calculation.
Since for $N \geq 0$ $|u - Z_N|_{\sup} \leq |u - Z_{N_1}|_{\sup} + |Z_N - Z_{N_1}|_{\sup}$ the

following result is found from (6.10-11).

Theorem II.

In the supnorm it holds, that the constructed global approximation Z_N *differs from the solution at most by an amount*

$$(6.12)\qquad |u - Z_N|_{\sup} = O(\sqrt{\varepsilon^{\frac{1}{2}}\ln(\tfrac{1}{\varepsilon})}^{N+1}), \qquad N \geq 0.$$

Of course the estimate given in the introduction (1.10) is contained in (6.12) for N sufficiently high.

In order to prove (1.11) and (1.12) we need estimates on the derivatives $\frac{\partial}{\partial x_i}(u-Z_N)$, $\frac{\partial^2}{\partial x_i \partial x_j}(u-Z_N)$. Note, that

$$(6.13)\qquad [-\varepsilon\Delta + \frac{1}{(1+u)^2}]\ \frac{\partial}{\partial x_i}\ (u - Z_N) = r_{i,N}^{(1)}$$

with $r_{i,N}^{(1)} = \frac{\partial r}{\partial x_i} + \frac{\partial Z_N}{\partial x_i}\{\frac{1}{(1+Z_N)^2} - \frac{1}{(1+u)^2}\}$

i.e. $|r_{i,N}^{(1)}|_{\sup} \leq R_{i,N}^{(1)}\ \varepsilon^{-1}(\sqrt{\varepsilon^{\frac{1}{2}}\ln(\tfrac{1}{\varepsilon})})^{N+1}$.

Using a technique as in lemma 11 we get

$$(6.14)\qquad |\frac{\partial}{\partial x_i}\ (u - Z_N)|_{\sup} = O(\varepsilon^{-1}|r_{i,N}^{(1)}|_{\sup}).$$

Now for N sufficiently high (1.11) is contained in (6.14) and the explicit expression for Z_N and its derivatives. An interesting observation is, that the main contribution in the estimate (1.11) comes from the layer along $\partial\Omega$. Differentiating (6.13) w.r.t. x_j and proceeding in an analogous way we obtain:

$$(6.15)\qquad |\frac{\partial^2}{\partial x_i \partial x_j}\ (u - Z_N)|_{\sup} = O(\varepsilon)\quad \text{for N sufficiently large.}$$

Thus we are lead to (1.12). Again the main contribution in (1.11) comes from the layer along $\partial\Omega$.

BIBLIOGRAPHY

[1] C.M. Brauner, Lectures on singular perturbations, Los Alamos (1977).

[2] C.M. Brauner, and B. Nicolaenko, Singular perturbations and free boundary value problems, Proc. 4th Inter. Conf. on Computing Mathods in Appl. Sc., North-Holland (1980), 699-724.

[3] C.M. Brauner and B. Nicolaenko, Internal layers and free boundary problems, Proc. BAIL 1 conf., Boole Press, Dublin (1980), 50-61.

[4] C.M. Brauner and B. Nicolaenko, A general approximation of some free boundary problems by bounded penalization, to appear in Proc. Seminar Collège de France, Pitman.

[5] C.M. Brauner, W. Eckhaus, M. Garbey and A. van Harten, On the transition layer along a free boundary, Proc. BAIL 3 Conf., Boole Press, Dublin (1984).

[6] C.M. Brauner, W. Eckhaus, M. Garbey and A. van Harten, Asymptotics of a rather unusual type in a free surface problem, Preprint University of Utrecht, to appear.

[7] L.A. Caffarelli and N.M. Rivière, Smoothness and Analycity of free boundaries in variational inequalities, Ann. Sc. Norm. Sup. Pisa, 3 (1976), 289-310.

[8] W. Eckhaus, Asymptotic analysis of singular perturbations, North-Holland (1979).

[9] W. Eckhaus and E.M. de Jager, Asymptotic solutions of singular perturbation problems for linear differential equations of elliptic type, Arch. Rat. Mech. Anal. 23 (1966), 26-86.

[10] L.S. Frank and W.D. Wendt, Solutions asymptotique pour une classe de perturbation singulières elliptiques sémi-linéaires, C.R. Acad. Sc. Paris, 295 (1982), 451-454.

[11] A. Friedman, Variational principles and free boundary problems, Wiley (1982).

[12] M. Garbey, 3rd Cycle Thesis, Ecole Centrale de Lyon (1984).

[13] A. van Harten, Singularly perturbed nonlinear 2nd order elliptic boundary value problems, Thesis, University of Utrecht (1975).

[14] A. van Harten, Nonlinear singular perturbation problems: proofs of correctness of a formal approximation, based on a contraction principle in a Banach space, J.M.A.A., 65 (1978), 126-168.

[15] H. Lewy and G. Stampacchia, On the regularity of the solution of a variational inequality, C.P.A.M., 22 (1969), 153-188.

C.M. BRAUNER AND M. GARBEY
DEPARTEMENT M.I.S.
ECOLE CENTRALE DE LYON
69131 ECULLY CEDEX, FANCE

et Equipe d'Analyse Numérique Lyon-St. Etienne (Unité Associée au C.N.R.S. n° 740).

W. ECKHAUS AND A. VAN HARTEN
MATHEMATICAL INSTITUTE
UNIVERSITY OF UTRECHT
BUDAPESTLAAN 6
P.O. BOX 80.010
3508 TA UTRECHT
THE NETHERLANDS

Lectures in Applied Mathematics
Volume 23, 1986

REDUCTION METHODS IN FLAME THEORY

Paul C. Fife and Basil Nicolaenko

ABSTRACT. The problem of determining the structure of steady plane flames in a homogeneous gaseous mixture is intractable if there are many chemical reactions involved. However, the fact that the reaction rate parameters generally have differing orders of magnitude can be exploited in several ways to reduce the problem to approximate simpler problems. This note describes five reduction principles as well as their application to H_2-O_2 flames.

1. BACKGROUND. The constant-pressure model in the theory of flames propagating through a gaseous mixture consists of the system of partial differential equations below, in which ρ, u, and T represent the density, local fluid velocity, and temperature of the gas, the vector $Y = (Y_1, Y_2, \ldots, Y_n)$ gives the concentrations of the n component species present in the gas, and an n+1-dimensional state vector is defined by $U = (T, Y)$. (More specifically, Y_i is the mass fraction of the i-th species divided by its molecular weight.) The "diffusion matrix" D and nonlinear source function f in reality should depend on ρ as well, but this dependence is usually ignored in analytical treatments. The basic system, then, is:

AMS 1980 Classification 80A25, 34E99, 34B15, 80A30.
Supported in part by NSF Grant MCS8202055 and by the Center for Nonlinear Studies, Los Alamos, under partial support from DOE.

$$\rho_t + \nabla\cdot(u\rho) = 0, \tag{1}$$

$$(\rho U)_t + \nabla(\rho u U) = D\nabla U + f(U). \tag{2}$$

Here ∇ and Δ are the gradient and Laplacian with respect to the spatial variables. The temperature T is assumed to have been non-dimensionalized so that the difference between the maximum possible temperature and the minimum temperature in the gas, in nondimensional units, is unity.

The most basic problem of interest for these equations is to find solutions of special form, and to determine their stability. The simplest special form, and the one which corresponds to planar flames, is that of a traveling wave. We shall confine attention to that case. The principal theme of the note will involve the question, what can one deduce about the existence, multiplicity, and structure of traveling wave (steady flame) solutions directly from the chemistry of the combustion process, without actually finding these solutions? This question is of fundamental importance because in realistic models, one is dealing with very large and very nonlinear systems. For such large systems, the only viable approach to finding the solutions themselves is by numerical simulation; and this requires costly sophisticated numerical codes.

There has, in fact, been a great deal of work on the numerical analysis of flame problems with multiple reactions, and these studies have proved very useful as direction indicators for developing the ideas in this paper. There have also been significant asymptotic-analytic treatments of flame models with two reactions, which to some extent relate to the examples given here. And, of course, there has been an enormous amount of analytical work on single-reaction flames. We shall not attempt to give individual recognition to these previous papers.

Solutions of (1) and (2) depending only on the traveling wave coordinate $z = x-ct$ satisfy the system

$$D\ddot{U} - M\dot{U} + F(U) = 0, \tag{3}$$

where $M = \rho(u-c)$ is the unknown but constant mass flux, and where dots represent differentiation with respect to z.

This is to be solved under boundary conditions at $\pm\infty$: At $z = -\infty$, which represents a location far ahead of the flame (hence in the cold gas), the state of the mixture is prescribed to be a given vector $U_- = (T_-, Y_-)$:

$$U(-\infty) = U_-. \tag{4}$$

At the other limit, $z = +\infty$, the gas is to be in a burned state, which means that chemical equilibrium or pseudo-equilibrium has been reached:

$$U(\infty) \quad E, \tag{5}$$

where E represents the set of all state vectors U such that $F(U) = 0$. [The term "pseudoequilibrium" is used here because in the dynamics of a combusting gas, it may happen that on the time scale of interest, an apparent rest state will be reached which is not the true chemical equilibrium. The true equilibrium will then be approached very slowly, on a larger time scale. Confining one's attention to the former time scale may involve neglecting some of the reactions which actually occur, because their rate is very small. This means neglecting some of the terms in the function f which ordinarily would be there. For this reason many of the solutions of $f(u) = 0$ may be only these pseudoequilibrium states.]

In general, the system (3) will be large, which may at first be cause for despair in attempting to get any meaningful information from it. Nevertheless, the special properties of the nonlinear function f in typical cases enables one to make a great deal of headway. These special properties have to do with (1) the fact that f is constructed in a special way from knowledge of the chemical network, and (2) it generally contains many rate parameters of vastly differing orders of magnitude.

The purpose of this note is to explain some principles by which important information may be obtained. Some of the points will be illustrated by applying them to special two- and three-reaction networks, in which contexts they appear as simple cases. But of course the main point is that they can be applied to nontrivial cases; and in fact there will be applications to a larger reaction network (Table 1), which contains reactions which have been proposed as being relevant to the combustion of H_2 and O_2. In fact these, together with their reverse reactions, are exactly those which are listed in the definitive paper of Westbrook and Dryer [8]. For the temperature and pressure conditions considered here, the reverse reaction rates are very small in comparison with the forward rates and will be neglected.

A large portion of the argument will be based on making formal but judicious approximations (one lesson which emerges is that formally small terms may be of crucial importance and should not necessarily be neglected). They will often be based on intuitive considerations; the lack of rigor is justified by the fact that the theory of boundary value problems for large systems (3) - (5) is primitive at present, and any ideas which are apparently correct, though unproved, are better than nothing. The effect of making these formal approximations is usually to reduce the complexity of the original problem; in this sense, a simpler model has replaced the original one. This is most often what analytical treatments in combustion are all about.

The function f is built in the following manner. To each reaction R_ℓ there corresponds an n+1-dimensional vector $K_\ell = (K_{\ell 0}, K_{\ell 1}, \dots)$, where $K_{\ell 0} = Q_\ell$ is a certain measure of the heat released by the reaction (it will be negative for endothermic ones) and $K_{\ell i}(i \geq 1)$ is an integer representing the net gain (or loss), in moles, of species i when that reaction proceeds to the extent of one mole of reactants. The K's are called "extended reaction vectors". For example, if there are three species and R_1 is $A_1 + A_2 \to 2A_3$, then $K_1 = (Q_1,-1,-1,2)$. The function f is a linear combination of the K's:

$$f(U) = \sum_{i=1}^{m} \omega_i(U)K_i, \tag{6}$$

where m is the number of reactions, and the ω_i are rate functions

$$\omega_i(U) = P_i(Y)k_i(T), \tag{7}$$

P_i being the mass action monomial in Y, and k_i the rate "constant" which really may depend on T. We assume the Arrhenius form

$$k_i(T) = B_i \exp(-\frac{E_i}{RT}). \tag{8}$$

Here B_i is called the preexponential factor, E_i the activation energy, and R the universal gas constant.

Other than its stability with reference to (1) and (2), the most important items of information one would like to know about the solution of (3) - (5) are the unknown mass flux M and the burned state $U_+ = U(+\infty)$. After that, of course, one would also like to know at least some properties of the spatial profile of the flame. Substituting (6) into (3) and integrating with respect to z from $-\infty$ to $+\infty$ yields

$$[D\dot{U} - MU]_{-\infty}^{\infty} = M \sum_{i=1}^{m} \alpha_i K_i, \tag{9}$$

where

$$\alpha_i \quad M^{-1}\int_{-\infty}^{\infty} \omega_i(U(z))dz. \tag{10}$$

Since $\dot{U}(\pm\infty) = 0$, (9) implies

$$U(+\infty) \quad U_+ = U_- + \sum_1^m \alpha_i K_i . \tag{11}$$

The vector $\alpha = (\alpha_1, \alpha_2, \ldots, \alpha_m)$ is called an "allocation". It is not known, of course, but if it were known, then one could determine not only the final temperature T_+ and the final concentration Y_+ through (11), but also the extent to which each reaction R_i participates in the combustion process (α_i is a direct measure of the extent of this participation), and thereby in turn surmise which chemical radicals are likely to appear as intermediates in the process, and more or less by which amounts. Furthermore, knowledge of T_+ and of the rate functions k_i in turn allows one to determine at least the order of magnitude of the parameter M. So if one can find α, one will have come a long way toward elucidating the properties of the flame.

The main emphasis in this talk is in finding α, and to keep the exposition within limits this will be under the assumption that most of the combustion occurs near a single final temperature T_+. It is well known that reactions with high activation energy E are generally active near only one temperature in the flame. Different reactions with large E, or groups of reactions, become operative at different temperatures in a single flame, and there is a lot that can be said about how to determine those temperatures, where in the flame profile they occur, and the shape of the profile $U(z)$ between them (see [4], [6], [5] for instance). On the other hand, flames contain many reactions with zero or small activation energy. Very often, these reactions proceed only in conjunction with the production of certain crucial chemical species by high-E reactions, and again often (but by no means always) the rate constants for the zero-E reactions are larger than those of the high-E ones. If this is the case, then again these reactions are all seen only near specific temperatures. In short, it is common, under high-E assumptions, for groups of reactions all to go to completion near a single temperature or a discrete set of temperatures. If this group is large, the analysis of the situation can be formidable, and the present talk is concerned with techniques for simplifying it.

At this point, it is legitimate to ask how one knows whether the reactions in a given network all come into operation near one temperature. For simple networks it is clear; one of the examples below under III illustrates this point, and shows that there are cases when one knows they cannot all occur together. A general technique is given in the papers cited above. In those papers, a criterion termed "maximality" is described. That criterion is inadequate for many complex schemes, and should be extended by the criteria outlined below. The extended criteria, placed into the framework given in [6], for example, should provide a large part of the answer for large networks.

2. CRITERIA FOR ALLOCATIONS. Although α is not known, it is subject to certain a priori restrictions which in many cases determine it uniquely. Our object is to explain these restrictions and to illustrate their application in some particular cases. We shall list five restrictions in order. They can be thought of as criteria that vectors α must meet in order to be legitimate allocations.

I. The α_i's are nonnegative, and U_+, given by (11), must be in E. In some simple models, for example when there is a single reaction or only a sequential chain of reactions, this determines α uniquely. But in more realistic models, other selection criteria are necessary.

II. Direct competition. If reactions R_i and R_j are such that $\omega_i(U) << \omega_j(U)$ for all values of T in some interval and all values of y, then any realistic allocation for which T_+ is in this interval must have $\alpha_i = 0$. The reason is that it is clear from (10) that the same order of magnitude relation must hold between α_i and α_j. It is also true that Crit. I implies some moderate a priori bound for α, so that α_i is negligible in comparison with α_j, and can be set equal to zero.

An example of this direct competition is the pair of reactions R_1 and R_9 in the H_2-O_2 scheme (Table 1). It happens that R_1 has high, and R_9 zero, activation energy, so there is a "cross-over" temperature T^* at which $k_1(T^*) = k_9(T^*)$. Notice that R_9 involves a "chaperon" third species M, which can be most any other species present in the gas; the effect of it is that the rate k_9, hence T^*, also depend significantly on the pressure and on the concentrations of these other species. A cross-over temperature of around $1100°$ corresponds to not unrealistic conditions. For T a few hundred degrees less than T^*, $\omega_9 >> \omega_1$, and we conclude that $\alpha_1 = 0$. On the other hand when T is a few hundred degrees on the other side of T^*, then $\alpha_9 = 0$ for the similar reason. An analysis based on the assumption that E_1 is sufficiently large (hence the gradient of the function $\ln k_1(T)$ is sufficiently large) would treat this transition as being sharp, rather than spread out over a few hundred degrees. It turns out that for temperatures less than T^*, the criteria to be given below imply that the only possible allocation with $\alpha_1 = 0$ is the identically zero one: $\alpha = 0$. This means that no flame can exist which attains a final temperature in that region. The final temperature of course depends on the concentrations Y_- in the cold mixture. If this initial concentration is changed gradually so that the final temperature decreases from higher values down past T^*, then the flame is extinguished. This is a flammability limit phenomenon.

III. Order of magnitude compatibility. This is best explained through an example. Consider the three-reaction network

$$A_1 \to A_2,$$

$$A_2 \to 0,$$

$$A_2 + A_2 \to 0,$$

where 0 denotes inert product species, and where mass action kinetics is assumed with rate constants k_1, k_2, and k_3 respectively, evaluated at the final flame temperature T_+. The three rates at that temperature will then be

$$\omega_1 = k_1 Y_1, \quad \omega_2 = k_2 Y_2, \quad \omega_3 = k_3 Y_2^2. \tag{12}$$

The equilibrium set consists of all state vectors with $Y = 0$. It will be assumed for simplicity that $Y_- = (1,0)$. It then follows that the set of allocations satisfying Criterion I alone will be those for which

$$\alpha_1 = 1; \quad \alpha_2 + 2\alpha_3 = 1; \quad \alpha_i \geq 0. \tag{13}$$

This leaves a great deal of choice. However, it often happens that the rate constants at a given temperature differ from each other in orders of magnitude, and it will be so assumed in the present case. If it happens that α_1 and α_2 are both different from zero (by an $O(1)$ quantity, let us say), then they are of the same order of magnitude, and relation (10) suggests that the same should be true of the ω_i as well. From (12) it is seen that this can only be true if the concentrations Y_i themselves have different orders of magnitude, to compensate for those of the k_i. This situation is commonly expressed by rescaling the Y_i: $Y_i = \sigma_i W_i$, for some constants σ_i depending on the k_i, and some $O(1)$ functions W_i. But since the initial concentration of Y_1 is unity, it can be argued that its order of magnitude constant σ_1 is 1. Thus $Y_1 = W_1$, so

$$\omega_1 = k_1W_1; \quad \omega_2 = k_2\sigma_2W_2; \quad \omega_3 = k_3\sigma_2^2W_2^2;$$

and equality of orders of magnitude requires that σ_2 be chosen so that

$$k_1 = k_2\sigma_2 = k_3\sigma_2^2. \tag{14}$$

But there is no reason to believe that this is possible; it will be so only if $k_1/k_2 = k_2/k_3$ (to order of magnitude), i.e.

$$\frac{k_1k_3}{k_2^2} \simeq 1.$$

In this sense, the supposition that all components of α differ from 0 is quite restrictive. So it is necessary to explore the consequences of (15) not holding.

First, suppose that $\frac{k_1k_3}{k_2^2} \ll 1$. Then necessarily at least one of the two equations in (14) is violated. If the second holds, then $\sigma_2 = k_2/k_3$, and $k_2\sigma_2 = k_2^2/k_3 \gg k_1$. Therefore $\omega_2 \gg \omega_1$, and $\alpha_2 \gg \alpha_1 = 1$, which is impossible, as it violates (13). The same sort of contradiction arises if the first and last terms in (14) are equal: $k_1 = k_3\sigma_2^2$. Therefore the first holds, and this same type of argument shows that $\omega_3 \ll \omega_1$, so that $\alpha_3 = 0$. Hence from (13), $\alpha_2 = 1$. In short, in this case there is only one possible allocation satisfying the order of magnitude compatibility criterion:

$$\alpha = (1,1,0), \quad \text{with} \quad \sigma_2 = \frac{k_1}{k_2}. \tag{16}$$

Similiar reasoning shows that if $\frac{k_1k_3}{k_2^2} \gg 1$. then the only possible allocation is

$$\alpha = (1,0,\tfrac{1}{2}), \quad \text{with} \quad \sigma_2 = \left(\frac{k_1}{k_3}\right)^{1/2}.$$

In either of these two cases, the criterion serves to reduce the available allocations to only one.

The idea behind this criterion is that there may be intermediate species (such as A_2 in the above example) whose abundances may or may not be small; that is unknown a priori because the species are not present in the unburned gas, only being produced by the combustion process itself. The magnitudes of the rate constants k_i at the assumed final temperature, however, may induce certain order of magnitude relations among these concentrations through the above type of scaling argument together with the relation between the α's and the ω's, and the a priori bounds (such as from (13) in this case) on the α's. At the same time, these relations may require that some ω's be very small relative to others, and this in turn implies that the corresponding α's are negligibly small. This final restriction that certain α's be negligible constitutes Criterion III.

A side issue is raised by the above example. By their very definition, the components of Y are all ≤ 1. It therefore follows that necessarily the scaling factors σ_i are all ≤ 1 in order of magnitude. Now suppose, in the above example, that $k_1k_3/k_2^2 << 1$. Then since from (16) $\sigma_2 = k_1/k_2$, it follows that this latter ratio is ≤ 1 in order of magnitude. Rather than imposing a restriction on α, this requirement rather imposes a restriction on the validity of the original assumption that the combustion all takes place at a single temperature. The fact is that no flame could achieve a final temperature in the range where $k_1 >> k_2$ (assuming that such a range exists), unless the combustion processes in that flame were distributed in more than one single location. A flame may still exist, but it may be such that the first reaction in the sample network above essentially comes to completion at one temperature, and the second does so further on in the flame profile at a higher temperature. That type of flame is expressly beyond the scope of the present discussion, but a procedure for handling spatially distributed flames is brought out in [4], [5], and [6]. In the case of the simplest sequential mechanism $A_1 \to A_2 \to 0$, such flames were studied in [1], [7], [9], and [3].

Similarly in the other case, $k_1k_3/k_2^2 >> 1$, one concludes that the above is invalid for the same reason if $k_1 >> k_3$.

IV. Pseudo-steady-state relations. These come from hypotheses to the effect that certain concentrations Y_j remain everywhere very small in the flame. A hypothesis like this could arise, for example, from a scaling argument such as in III. As an illustration, consider the same example network discussed there, in the case (16), with $k_1 << k_2$. Then $\sigma_2 <<$ 1, which means that $Y_2 << 1$. In the second (last) component of the vector equation (11), one can therefore neglect the contributions due to $U_\pm$, and arrive at a linear relation among the α_i:

$$\sum_j \alpha_j K_{2j} = 0, \tag{17}$$

which in that particular context becomes $\alpha_1 - \alpha_2 - 2\alpha_3 = 0$. Since $\alpha_1 = 1$ and $\alpha_3 = 0$, this implies $\alpha_2 = 1$. This particular result had already been obtained above, so it is no surprise. However in more complex situations, the relations analogous to (17) may provide new restrictions on the possible α's.

On a more basic level, one could apply the scaling $Y_1 = \sigma_i W_i$ already to (3) and similarly neglect the left side of any component ℓ for which $\sigma_\ell << 1$. There results the pseudo-steady-state relation

$$f_\ell(U) = \sum_i \omega_i(U) K_{i\ell} = 0,$$

which supplies more than just a relation among the α_j's; it typically may be solved for one component of Y in terms of the others, thereby reducing the order of the original system (3). Therefore not only does the stipulation that $\sigma_\ell \ll 1$ result in an extra condition restricting α, but it also serves to reduce the order of the original system. The latter would be important at a later stage during the process of performing a flame layer analysis to determine the mass flux of the flame.

Finally, experimental or computational evidence that some intermediate species is not present in the combustion products has been used in order to obtain this type of restriction on α, without necessarily using scaling arguments to show it is everywhere of low concentration. For then one can obtain (17) again, but this time for a slightly different reason.

As was mentioned above, Criteria I-IV were applied to the $H_2 - O_2$ scheme in Table 1. At a temperature of around 1100°, it was found that if the first reaction in that table is missing, i.e. $\alpha_1 = 0$, teen these criteria imply that necessarily $\alpha = 0$, so that no flame can exist under that condition. On the other hand if α_1 is allowed to differ from zero, then allocations satisfying these criteria do exist. At least one of them undoubtedly also satisfies Crit. V below, but this has not been verified. These calculations serve to exhibit and elucidate the flamability limit phenomenon due to the competition between the first and the ninth reactions.

V. Criterion of stability. This criterion says that not only should the burned state U_+ be an equilibrium state, but it should be stable with respect to the kinetic system

$$\dot{U} = f(U). \qquad (18)$$

The reason for proposing this criterion is the assumption that in actual plane flames the state vector U approaches a constant behind the flame, and due to the presence of perturbations, this constant state could not persist if it were unstable. In particular, it should be stable against perturbations with large characteristic spatial dimensions; in other words relative to the evolution problem obtained from (1) and (2) by discarding x-derivatives. It is easy to see that this latter evolution problem is the same as (18), except for a factor ρ_0 on the left, where ρ_0 is the density of the burned gas. The stability properties of the two equations are the same. The much less tractable question of the stability of the final state U_+ relative to the entire system (1), (2), will not be considered here.

The question now arises as to whether there are easy ways to apply the criterion and determine stability. There is one immediate result which gives the answer in special cases. Suppose all reactions are exothermic, and that there is a one-one relation between final temperatures (determined from the 0-th component of (11)) and the set of possible allocations which have been constructed from Crit. I. Then the allocation which yields the maximal temperature also yields a stable final state U_+. The reason for this is as follows. Suppose this state were perturbed, and the subsequent evolution according to (18) led to a different equilibrium state. Then because of the exothermicity assumption, the right side of the 0-th (temperature) component of (18) is nonnegative. Hence the new state would have a temperature at least as large as the first one. But since the latter was maximal, the new state has the same temperature. This means the new allocation is the same, so there has been no change after all, and the original state was stable.

This sufficient condition for stability can be generalized; for example, suppose that no reaction produces any amount of the species A_k, and that there exists a one-one relation between allocations and final values Y_{k+}. Then the allocation which minimizes Y_{k+} will satisfy Crit. V. Either this generalization (with $k = 1$) or the former rule based on exothermicity could be used in the example to be discussed below.

It should be mentioned that in determining the stability for purposes of the present criterion, stability in the linearized sense is insufficient because of the typical presence of numerous zero eigenvalues. However, other aids to the task of determining stability are found in the pseudo-steady-state considerations and competition considerations brought out in II and III above. These same reductions are also valid for (18), and this may reduce the stability problem to a simpler level.

In some cases, this criterion should be applied with caution, because a state may very well be basically unstable, but nevertheless stable within the particular time scale of interest (see below).

The best examples of this criterion come from networks involving "chain branching" reactions. Consider this one:

$$A_1 + A_2 \to 2A_2,$$

$$A_2 + A_2 \to 0.$$

The equilibrium set $\mathbf{E} = \{U = (T,Y_1,Y_2)\colon Y_2 = 0\}$. Supposing that $Y_- = (1,0)$, one can calculate that the set of allocations is

$$\{\alpha\colon \alpha_1 - 2\alpha_2 = 0,\ 0 \le \alpha_1 \le 1\} = \{\alpha = (\alpha_1, \tfrac{1}{2}\alpha_1)\colon 0 < \alpha_1 \le 1\},$$

so that any of the states in $\mathbf{E}$ could be attained from one of these allocations, as long as $0 \le Y_{1+} \le 1$ and $T_+ = T_- + \alpha_1 Q_1 + \alpha_2 Q_2$. If both reactions are exothermic, then clearly the maximal value of T_+ is attained at $\alpha_1 = 1$, so $\alpha = (1, \frac{1}{2})$ is stable. It can be seen in fact to be the only stable one, as the argument below indicates.

But a closer analysis of this example is called for. Let us suppose that

$$k_1 << k_2.$$

Then it turns out that the qualitative features of the Y components of (18) can be determined independently of the T component. Those components of (18) assume the form

$$\dot{Y}_1 = -k_1 Y_1 Y_2,$$

$$\dot{Y}_2 = k_1 Y_1 Y_2 - 2k_2 Y_2^2 = k_2 Y_2 (\varepsilon\, Y_1 - 2Y_2),$$

where $\varepsilon = \frac{k_1}{k_2}$. Hypothesizing (this will be verified later) that Y_1 is slowly varying relative to the time scale with unit k_2^{-1}, one can see that the function $Y_2(t)$ will, according to the second of these differential equations, gravitate to a pseudo-steady-state given by

$$Y_2 = \frac{1}{2}\,\varepsilon\, Y_1,$$

on a time scale with that same unit k_2^{-1}. After that, one substitutes this expression for Y_2 into the first equation to obtain

$$\dot{Y}_1 = -\frac{1}{2} k_1\, \varepsilon\, Y_1^2.$$

Thus Y_1 decays to zero on a time scale with unit $k_1^{-1}\,\varepsilon^{-1}$, which by original assumption is indeed large compared to k_2^{-1}, as was hypothesized.

It may happen, however, that this last rate of decay $(k_1\varepsilon)$ is so small that it will not be important on the time scale which is physically relevant. In that case, to all intents and purposes one can say that Y_1 does not decay at all, and that the unburned state Y_- is stable; let us say "metastable". Another consideration is that these rates k_i depend on T. Therefore it may happen that the cold gas is in this metastable state, but once it is heated artifically, the initial concentration Y_- will no longer be metastable, and a flame will ignite.

3. SPATIALLY DISTRIBUTED COMBUSTION. Up to this point, the picture has been one in which all relevant reactions are triggered by a high-E one at a specific temperature and location in the flame. Reaction processes, however, may be important elsewhere as well. This could happen because 1) other high-E reactions trigger similar effects at other temperatures and locations, or 2) low-E reactions, not temperature-localized, may operate throughout the flame.

A detailed analysis of an example of the latter situation for a pair of competing reactions is given in [2]. If the former situation occurs, it will be important to find those other temperatures and locations. A general procedure for doing this was outlined in [6], and to some extent also in [5] and [4]. We shall not go into this here, except to say that the procedure is, for any given candidate T_+ for a final temperature, to determine, in a specific way and using the rate functions, a sequence of lower temperatures $T_\ell \leq T_+$, together with a corresponding sequence of subnetworks. For each of these subnetworks and lower temperatures, one determines whether there are suitable allocations in that sub-context. Call these "suballocations". Suitability would be with reference to the five criteria proposed above, to replace the more limited criterion proposed in the papers cited before. It may happen, in fact more often than not, that there is no such suitable suballocation except the identically zero one. If one succeeds in finding a sequence $\alpha^{(\ell)}$ of suballocations (the first few of which may be zero) corresponding to the temperatures T_ℓ, which is monotone increasing in ℓ and ends with an allocation which yields the final temperature one started with, then (under another verifiable condition) that allocation could be called "feasible", and would represent a real flame with concentrated combustion events at the various T_ℓ. The approximate spatial structure and order of magnitude of speed will be solutions of a reduced problem as described in the same cited papers.

Table 1: H_2 - O_2 Reactions

1. $H + O_2 \to O + OH$
2. $O + H_2 \to H + OH$
3. $H_2 + OH \to H_2O + H$
4. $O + H_2O \to OH + OH$
5. $H + H + M \to H_2 + M$
6. $O + O + M \to O_2 + M$
7. $O + H + M \to OH + M$
8. $H + OH + M \to H_2O + M$
9. $H + O_2 + M \to HO_2 + M$
10. $HO_2 + H \to H_2 + O_2$
11. $HO_2 + H \to OH + OH$
12. $HO_2 + H \to H_2O + O$
13. $HO_2 + OH \to H_2O + O_2$
14. $HO_2 + O \to O_2 + OH$
15. $HO_2 + HO_2 \to H_2O_2 + HO_2$
16. $H_2O_2 + OH \to H_2O + HO_2$
17. $H_2O_2 + H \to H_2O + OH$
18. $H_2O_2 + H \to HO_2 + H_2$
19. $H_2O_2 + M \to OH + OH + M$
20. $O + OH + M \to HO_2 + M$

BIBLIOGRAPHY

1. V. S. Berman and Iu. S. Riazantsev, Asymptotic analysis of stationary propagation of the front of a two-state exothermic reaction in a gas. J. Applied Math. and Mechanics (PMM) 37, 995-1004 (1973).

2. P. Clavin, P. C. Fife and B. Nicolaenko, Asymptotic analysis of flames with competitive reactions, submitted.

3. P. Fife and B. Nicolaenko, The singular perturbation approach to flame theory with chain and competing reactions. pp. 232-250 in: Ordinary and Partial Differential Equations, W. N. Everitt and B. D. Sleeman, eds., Lecture Notes in Mathematics No. 964, Springer-Verlag, Berlin.

4. ________, Asymptotic flame theory with complex chemistry, pp. 235-256 in: Nonlinear Partial Differential Equations, J. Smoller,. ed., Contemporary Mathematics 17, Amer. Math. Soc., Providence (1983).

5. ________, Flame fronts with complex chemical networks, Physica 12D, 182-197 (1984).

6. ________, How chemical structure determines spatial structure in flame profiles, in: Proc. Conf. on Modelling Pattern Formation in Space and Time, W. Jager, ed., Lecture Notes in Biomathematics, Springer-Verlag, to appear.

7. A. Kapila and G. S. S. Ludford, Two-step sequential reactions for large activation energies. Combust. and Flame 29, 167-176 (1977).

8. C K. Westbrook and F. L. Dryer, Chemical kinetics modeling of hydrocarbon combustion, Prog. in Energy and Combustion Science, to appear.

9. Ya. B. Zel'dovich, G. I. Barenblatt, B. Librovich, and G. M. Mahviladze, Mathematical Theory of Combustion and Detonation. Nauka, Moscow (1980) (in Russian).

DEPARTMENT OF MATHEMATICS and CENTER FOR NONLINEAR STUDIES
UNIVERSITY OF ARIZONA
TUCSON, ARIZONA 85721

CENTER FOR NONLINEAR STUDIES
LOS ALAMOS NATIONAL LAB
LOS ALAMOS, NM 87545

Lectures in Applied Mathematics
Volume 23, 1986

SAW-TOOTH EVOLUTION IN A STEFAN PROBLEM

G.S.S. Ludford[1]

ABSTRACT. The classical Stefan problem of solidification of water into ice has a boundary that accelerates continuously once violations of the initial conditions have been resolved. Here we consider a Stefan problem from detonation theory, which essentially differs in an interface condition alone, and find that the acceleration may jump to an infinite value, so as to form a saw-tooth in the velocity profile.

The analysis consists of an asymptotic investigation in the neighborhood of the moving boundary, the source of singular behavior. Two purely computational phenomena are thereby explained, namely cusped behavior of the velocity and its eventual nonexistence. Increasing the numerical accuracy does not eliminate these phenomena, and appropriate computational modifications are still being sought to capture numerically the saw-tooth behavior uncovered theoretically.

1. INTRODUCTION. The solidification of water into ice leads to the classical Stefan problem: unsteady heat equations are satisfied on either side of a moving boundary at which the unknown (temperature) is prescribed and the jump in its gradient is proportional to the velocity of the boundary. If the initial conditions violate the governing equations at the boundary, a

1980 Mathematics Subject Classification. 35K99, 76V05.

[1]Work performed under the auspices of the U.S. Army Research Office and of the U.S. Department of Energy under Contract W-7405-ENG-36, the Office of Basic Energy Sciences, Department of Applied Mathematics.

singularity occurs there to resolve the violation immediately, involving sometimes an infinite boundary velocity and always an infinite boundary acceleration. The singularity depends on whether the unknown itself, its first (spacial) derivative, or its second derivative is involved in the violation; accordingly we may speak of a first-, second-, or third-order singularity (corresponding to the lowest derivative involved). During the later evolution there are no such singularities, though the boundary may be a source of higher-order singularities in which higher derivatives than the second suffer instantaneous changes. In short, except possibly for initial singularities, the evolution does not involve singularities of order lower than the fourth, so that both the velocity and acceleration of the boundary remain finite.

The same is true in general for a Stefan problem arising in detonation theory, that differs from the classical one only in having Burgers' equations instead of heat equations and in having a constant jump in gradient at the moving boundary. The former difference is of no consequence, but the latter proves to be disruptive at times. When, as was found possible in computations, the gradients of the unknown become equal in magnitude but opposite in sign on the two sides of the boundary, its acceleration jumps from a finite to an infinite value so as to form a saw-tooth in the velocity profile. The analysis also shows that numerical inaccuracies lead either to a jump in velocity, effectively changing the saw-tooth into a cusp, or to the nonexistence of a (finite) velocity, the latter becoming the more likely as the accuracy increases.

The paper originates in an investigation of the stability of the steady solutions of the detonation problem (Stewart & Ludford 1984, Oyediran & Ludford 1985). For certain parameter values, linear instability was found and, when this was followed numerically, so-called galloping detonations developed in some cases.

Alternate acceleration and deceleration of the boundary occurred, the transition from one to the other being abrupt enough to form a cusp in the velocity profile. Examination of such transitions led to the conclusion of this paper, that the computations actually masked a saw-tooth behavior. The other computational phenomenon uncovered by the analysis, namely termination of the solution, was also observed. In fact, the most accurate computations always terminated before producing a cusp.

Although the paper demonstrates, apparently for the first time, that Stefan problems can exhibit saw-tooth behavior, it leaves the matter in an unsatisfactory state. No indication is given of how to suppress the purely computational phenomena that mask this behavior, and the construction of a numerical scheme that accurately approximates the saw-tooth velocity profiles is left for future research.

2. THE CLASSICAL STEFAN PROBLEM.

When units are chosen appropriately and x is measured from the moving front, the solidification of a liquid is, in the simplest case, governed by the equations

$$F_t - K(t)F_x = F_{xx}, \quad F(-\infty,t) = 0, \quad F(+\infty,t) = F_\infty, \tag{1}$$

$$F(\mp 0,t) = F_*, \quad F_x(-0,t) - F_x(+0,t) = K(t). \tag{2}$$

Here F represents the temperature, and normally the constants satisfy

$$0 < F_* < F_\infty \tag{3}$$

when the liquid is on the right and the solid on the left. (In undercooled liquids F_∞ is less than F_*.) The function $K(t)$, representing the velocity of the solidification front, is to be found along with the solution $F(x,t)$.

If $F(x,t)$ is supposed to be known at time t, the velocity can be calculated in any one of the three ways:

$$\text{(i)}\quad K(t) = F_x(-0,t) - F_x(+0,t),$$

$$\text{(ii)}\quad K(t) = -F_{xx}(-0,t)/F_x(-0,t),$$

$$\text{(iii)}\quad K(t) = -F_{xx}(+0,t)/F_x(+0,t).$$

The first is just the condition (2b); the second and third come from the differential equation (1a) on applying the conditions (2a). All three will give the same answer once the solution is underway; but initial conditions

$$F(x,0) = F_0(x) \tag{4}$$

may be chosen to give different values for $K(0)$. The question is which, if any, is correct.

If we suppose

$$F(x,t) = F_0(x)+tF_1(x)+\ldots \text{ with } K(t) = K_0+t^{\frac{1}{2}}K_1+\ldots, \tag{5}$$

then the differential equation requires

$$F_1(x) = K_0F_0' + F_0''. \tag{6}$$

Note that the acceleration K' will behave like $t^{-\frac{1}{2}}$ if K_1 is nonzero. This is not possible during evolution in the solidification problem, but leads to saw-tooth behavior in the detonation problem. Assume the initial conditions tend to the correct values at infinity, i.e.

$$F_0(-\infty) = 0, \quad F_0(+\infty) = F_\infty; \tag{7}$$

then

$$F_1(\mp\infty) = 0, \tag{8}$$

and the expansion (5a) continues to give the correct values there. Not so at the origin; even if

$$F_\mp \equiv F_0(\mp 0) = F_*, \tag{9}$$

a choice that will not be made in the sequel, there is no reason for F_1 to vanish there. We conclude that a layer must form at the origin to enable the boundary conditions to accommodate the initial values.

The appropriate variable to replace x in the layer is

$$\xi = x/t^{\frac{1}{2}}; \tag{10}$$

in terms of ξ and t, the differential equation becomes

$$F_{\xi\xi} + \tfrac{1}{2}\xi F_{\xi} + t^{\frac{1}{2}} K F_{\xi} - t F_t = 0. \tag{11}$$

The inner expansion

$$F = f_0(\xi) + t^{\frac{1}{2}} f_1(\xi) + t f_2(\xi) + \ldots \tag{12}$$

then gives

$$f_0'' + \tfrac{1}{2}\xi f_0' = 0, \tag{13}$$

whose solution may be written

$$f_0 = A_{\mp}^{(0)} + B_{\mp}^{(0)} \operatorname{erfc}(|\xi|/2) \text{ for } \xi \lessgtr 0. \tag{14}$$

The integration constants $A_{\mp}^{(0)}$, $B_{\mp}^{(0)}$ are to be found by matching with the outer expansions (5a) and by satisfying the boundary conditions at the origin. The former give

$$A_{\mp}^{(0)} = F_{\mp}, \tag{15}$$

while the latter then require

$$F_{\mp} + B_{\mp}^{(0)} = F_*, \quad B_-^{(0)} + B_+^{(0)} = 0. \tag{16}$$

These three equations for the two constants $B_-^{(0)}$ and $B_+^{(0)}$ are, in general, incompatible; only if

$$\tfrac{1}{2}(F_- + F_+) = F_* \tag{17}$$

do they have a solution, namely

$$B_{\mp}^{(0)} = \pm\tfrac{1}{2}(F_{+}-F_{-}). \tag{18}$$

We shall suppose that the condition (17) is satisfied by the initial values, but it is instructive to see what happens when it is not. A third unknown is then needed, to make the overdetermined system (16) into a determinate one. Such will be the case if K is supposed to have a leading term $K_{-1}t^{-\frac{1}{2}}$: then K_{-1} will clearly appear on the right of equation (16b), this being the $0(t^{-\frac{1}{2}})$ expression of the boundary condition (2b). But K_{-1} also appears elsewhere in equations (16) because the differential equation for f_0 acquires a term involving it. We shall not go into details, but merely note that the particular intial values

$$F_0 = \begin{cases} 0 \\ F_\infty \neq 2F_* \end{cases} \text{for } x \lessgtr 0 \tag{19}$$

lead to the (exact) similarity solutions discussed by Carslaw & Jaeger (1959, p. 287).

Continuing with the condition (17) satisfied, we immediately note that K_0 has yet to be determined. Consider the next term in the expansion (12), which satisfies

$$f_1'' + \tfrac{1}{2}\xi f_1' - \tfrac{1}{2}f_1 = -K_0 f_0' \tag{20}$$

and therefore is of the form

$$f_1 = K_0 f_0' + A_{\mp}^{(1)}\xi + B_{\mp}^{(1)}[|\xi|\text{erfc}(|\xi|/2)-(2/\sqrt{\pi})\exp(-\xi^2/4)] \quad \text{for } \xi \lessgtr 0. \tag{21}$$

As before, the integration constants $A_{\mp}^{(1)}$, $B_{\mp}^{(1)}$ are to be found by matching with the outer expansions (5) and by satisfying the boundary conditions at the origin. The former give

$$A_{\mp}^{(1)} = F_0'(\mp 0) \equiv G_{\mp} \text{ (say)}, \tag{22}$$

while the latter then require

$$(F_- - F_+)K_0 + 4B_{\mp}^{(1)} = 0, \quad G_- - G_+ - B_-^{(1)} - B_+^{(1)} = K_0. \tag{23}$$

These three equations have the (unique) solution

$$K_0 = (G_- - G_+)/[1+\tfrac{1}{2}(F_+ - F_-)], \quad B_{\mp}^{(1)} = K_0(F_+ - F_-)/4. \tag{24}$$

It is now clear that, in general, none of the three ways (i), (ii), (iii) leads to the correct K_0. The first does so only if the initial data satisfy

$$F_+ = F_-(=F_*); \tag{25}$$

and the second or third does so only if

$$G_{\mp}\ (G_+ - G_-)/[1+\tfrac{1}{2}(F_+ - F_-)] = F_0''(\mp 0) \equiv C_{\mp}\ (\text{say}). \tag{26}$$

If the choice (25) is taken to be the natural one, then (i) gives the correct determination of K_0 and there is no initial singularity to second order:

$$B_{\mp}^{(0)} = B_{\mp}^{(1)} = 0 \tag{27}$$

and f_0, f_1 lose their layer terms. The initial singularity then occurs at the third order unless (ii) and (iii) are satisfied, i.e.

$$C_{\mp} = G_{\mp}(G_+ - G_-). \tag{28}$$

To draw this last conclusion the next term in the expansions (12) must be considered. We find

$$f_2'' + \tfrac{1}{2}\xi f_2' - f_2 = -K_0 G_{\mp} \tag{29}$$

with solution

$$f_2 = G_{\mp}K_0 + A_{\mp}^{(2)}(\xi^2+2)$$
$$+ B_{\mp}^{(2)}[(\xi^2+2)\operatorname{erfc}(|\xi|/2) - (2/\sqrt{\pi})|\xi|\exp(-\xi^2/4)]$$
$$\text{for } \xi \lessgtr 0. \tag{30}$$

Matching gives

$$A_{\mp}^{(2)} = C_{\mp}/2, \tag{31}$$

and the boundary conditions show that

$$G_{+}K_0 + C_{\mp} + 2B_{\mp}^{(2)} = 0, \quad 4(B_{-}^{(2)}+B_{+}^{(2)})/\sqrt{\pi} = K_1. \tag{32}$$

These three equations have the (unique) solution

$$K_1 = -2[(G_{-}+G_{+})K_0+C_{-}+C_{+}]/\sqrt{\pi},$$

$$B_{\mp}^{(2)} = -(G_{\mp}K_0+C_{\mp})/2 \text{ with } K_0 = G_{-}-G_{+}, \tag{33}$$

and there is a third-order singularity (i.e. either $B_{-}^{(2)}$ or $B_{+}^{(2)}$ is non-zero) unless the initial values satisfy the conditions (28). (Note that K_1 is then also zero, thereby avoiding an infinity in the acceleration K'.)

We conclude that, when the condition (25) is satisfied but the conditions (28) are not, K_0 is given by (i) and there is a third-order singularity initially. Its role is to resolve the violations of (ii) and (iii), which subsequently hold as well as (i). To see this resolution, we calculate

$$f_2''(\mp 0) = C_{\mp} + 2B_{\mp}^{(2)} = G_{\mp}(G_{+}-G_{-}), \tag{34}$$

which shows the curvatures being instantaneously changed to satisfy the conditions (28). (There is no such change in gradients since f_1 has no layer terms.) We emphasize that during the evolution such violations do not occur and the acceleration remains finite.

3. THE DETONATION PROBLEM. Derivation of the governing equations, which parallels that for deflagrations (Stewart & Ludford, 1983), is lengthy; so we refer the reader to Oyediran (1984). Suffice it to say that equations (1a) and (2b) are changed to

$$F_t-[K(t)+F]F_x = F_{xx}, \quad F_x(-0,t)-F_x(+0,t) = 1 \tag{35}$$

while (1b,c) and (2a) stay the same. The nonlinearity that is thereby introduced has little effect, but replacing K by 1 changes the solution significantly.

As for the solidification problem, there is more than one way in which the velocity can be calculated:

$$\text{(iv)} \quad K(t) = -F_* - F_{xx}(-0,t)/F_x(-0,t),$$

$$\text{(v)} \quad K(t) = -F_* - F_{xx}(+0,t)/F_x(+0,t).$$

These give the same answer once the solution is underway; but initial conditions (4) may be chosen to give different values for $K(0)$. To avoid having to resolve singularities due to such violations of the governing equations, we shall take

$$F_{\mp} = F_*, \quad G_- - G_+ = 1, \quad C_-/G_- = C_+/G_+ \equiv V(\text{say}). \tag{36}$$

These conditions certainly hold during the subsequent evolution of the solution and, if the evolution were of the same kind as in the solidification problem, there would never be a singularity to third-order. In general, that is true here but, if at a certain time (which we shall take to be $t = 0$)

$$G_- + G_+ = 0 \ (\text{i.e. } G_- = \tfrac{1}{2},\ G_+ = -\tfrac{1}{2}), \tag{37}$$

it is not, as we shall now see.

An analysis similar to that in the last section shows that the requirements (36a,b) lead to

$$f_0 = F_*, \quad f_1 = G_{\mp}\xi \text{ for } \xi \lessgtr 0; \tag{38}$$

there is no adjustment of F or F_x at the origin. At this stage in the solidification problem, K_0 would be determined (as $G_- - G_+$ with $G_{\mp}$ arbitrary); here it is not (and $G_{\mp}$ must satisfy (36b)).

The third-order singularity at the origin is described by the function (30) with K_0 replaced by

$$K_0^* = K_0 + F_* . \qquad (39)$$

Matching gives the same result (31), but the boundary conditions now require

$$G_\mp(K_0^*+V) + 2B_\mp^{(2)} = 0, \quad B_-^{(2)}+B_+^{(2)} = 0. \qquad (40)$$

These differ from equations (32) only in having K_0^* in place of K_0 and in not involving K_1; but their significance is quite different, since K_0 is still to be determined. If G_-+G_+ is not zero, the (unique) solution is

$$K_0^* = -V, \quad B_\mp^{(2)} = 0, \qquad (41)$$

showing that there is no singularity to the third order. In fact, there is no singularity to the fourth order either and K_1 is found to be zero. The evolution proceeds with K determined by (iv) and (v) and with finite K'.

However, if at a certain stage the slopes $G_\mp$ are equal and opposite, i.e. equation (37) holds, the solution is

$$B_\mp^{(2)} = \mp(K_0^*+V)/4 \quad \text{with } K_0^* \text{ undetermined.} \qquad (42)$$

(The system (40) has zero determinant but the equations are compatible.) A third-order singularity apparently occurs, whose role is to change the curvatures at the origin instantaneously to

$$f_2''(\mp 0) = C_\mp+2B_\mp^{(2)} = C_\mp\mp(K_0^*+V)/2. \qquad (43)$$

These new curvatures are appropriate to K_0^* (whatever its value) in the sense of (iv) and (v), since

$$C_\mp\mp(K_0^*+V)/2 = -G_\mp K_0^* \qquad (44)$$

according to the values (37) of $G_{\mp}$ and the definition (36c) of V.

To determine K_0, it is necessary to go to the next term in the expansion (12). We find

$$f_3 = \xi/4 \pm K_1/3 + K_0^* f_2' + A_{\mp}^{(3)} \xi(\xi^2+6)$$
$$+ B_{\mp}^{(3)} [|\xi|(\xi^2+6)\mathrm{erfc}(|\xi|/2) - (2/\sqrt{\pi})(\xi^2+4)\exp(-\xi^2/4)]$$
$$\text{for } \xi \lessgtr 0, \qquad (45)$$

where

$$A_{\mp} = F_0'''(\mp 0)/6 \equiv T_{\mp}/6 \text{ (say)}, \qquad (46)$$

$$\pm K_1/3 - K_0^*(K_0^*+V)/\sqrt{\pi} - 8B_{\mp}^{(3)}/\sqrt{\pi} = K_0^{*2} + T_+ - T_- + 6(B_-^{(3)} + B_+^{(3)}) = 0 \qquad (47)$$

according to matching and boundary conditions. As expected, the system (47), considered as equations for K_1 and $B_{\mp}^{(3)}$, has zero determinant. Compatibility is not automatic this time: only if

$$K_0^{*2} + 3VK_0^* + 2(T_- - T_+) = 0, \qquad (48)$$

a condition determining K_0, do the equations have a (non-unique) solution.

If we were dealing with the initial-value problem, two possibilities would arise:

(a) for $T_- - T_+ \leqq 9V^2/8$, the evolution starts with

$$K_0^* = [-3V \pm \sqrt{9V^2 - 8(T_- - T_+)}]/2; \qquad (49)$$

(b) for $T_- - T_+ > 9V^2/8$, there is no (real) K_0^* and evolution starting with finite K does not occur.

Both of these are unsatisfactory. When inequality holds in case (a), there are two possible values of K_0^* and no indication of the appropriate one. Moreover, only in exceptionable circumstances, namely for choice of third derivatives at the origin satisfying

$$T_{-}-T_{+} = V^2, \tag{50}$$

will K_0^* equal $-V$, i.e. will (iv) and (v) be satisfied. In case (b), the analysis fails to start the evolution; we shall return to this point later.

Once the evolution is under way, $T_{\mp}$ cannot be chosen arbitrarily, as for the initial-value problem. Differentiation of equation (35a) with respect to x and use of the boundary condition (35b) show that

$$T_{-}-T_{+} = V^2+G_{+}^2-G_{-}^2 \tag{51}$$

at any stage, so that the requirement (50) is satisfied as the conditions (37) are approached. We conclude that the evolution can (and presumably does) continue without a jump in K. No adjustment of the curvatures is necessary, see equations (43), so that $B_{\mp}^{(2)} = 0$; in other words, there is no singularity to third order.

Discussion of the next function f_4 shows that K_1 is not zero; the (now homogeneous) system (47) therefore has a non-trivial solution and there is a fourth-order singularity. While the K-profile comes in with finite slope it leaves with an infinite slope, which is found to be of opposite sign. In other words, a saw-tooth is formed.

4. INFINITE VELOCITIES. Before considering the computational implications of our analysis, we continue with case (b) of the initial-value problem. Inability to find a finite velocity suggests that an infinite one may be needed. The discussion of the condition (17) shows that this is true for the solidification problem, though admittedly the difficulty is of a different nature there. To discuss the possibility

$$K = K_{-1}t^{-\frac{1}{2}} + K_0 + \ldots \tag{52}$$

we revert to general $G_{\mp}$ satisfying the condition (36b). The result (38a) is again found, but not (38b): the presence of K_{-1} enables an instantaneous change in gradients to occur.

This change is effected by the function

$$f_1 = \mp A_{\mp}^{(1)}\zeta + B_{\mp}^{(1)}[\zeta \operatorname{erfc}(\zeta/2) - (2/\sqrt{\pi})\exp(-\zeta^2/4)] \text{ with}$$

$$\zeta = |\xi| \mp 2K_{-1} \text{ for } \xi \lessgtr 0, \tag{53}$$

where

$$A_{\mp}^{(1)} = G_{\mp}, \tag{54}$$

$$G_{\mp}K_{-1} + [\mp K_{-1}\operatorname{erfc}(\mp K_{-1}) - (1/\sqrt{\pi})\exp(-K_{-1}^2)]B_{\mp}^{(1)}$$
$$= \operatorname{erfc}(-K_{-1})B_{-}^{(1)} + \operatorname{erfc}(K_{-1})B_{+}^{(1)} = 0 \tag{55}$$

according to matching and boundary conditions. Elimination of $B_{\mp}^{(1)}$ and use of the boundary condition (36b) then leads to

$$\sqrt{\pi}K_{-1}\exp(K_{-1}^2)[1-\operatorname{erf}^2(K_{-1})] - \operatorname{erf}(K_{-1}) = G_{-} + G_{+} \tag{56}$$

as the equation determining K_{-1}.

The left side of this equation behaves like a tanh-function, increasing monotonically from -1 at $K_{-1} = -\infty$ to $+1$ at $K_{-1} = +\infty$ and passing through zero at $K_{-1} = 0$. It follows that, for any value of $G_{-}+G_{+}$ between -1 and $+1$, there is a unique value of K_{-1}, the extremes corresponding to $G_{-} = 0$ and $G_{+} = 0$ respectively; however, K_{-1} is zero when the condition (37) is satisfied. We conclude that there is a solution with K of the form (52) for every non-zero value of $G_{-}+G_{+}$ (when it is not needed), but no such solution for $G_{-}+G_{+} = 0$ (when it may be needed).

The next step is to see if there is a solution with a less singular K. This is certainly not the case if the similarity variable (10) is retained. [Dold (1984) was recently forced to introduce an alternative in a completely different context.] A replacement for the similarity variable when the condition (37) holds is currently being sought.

5. COMPUTATIONAL IMPLICATIONS. Relative inaccuracy in the numerical determination of curvature, especially on either side of the origin, makes the formulas (iv) and (v) useless as ways of computing K. Instead, a root solver was used to find the value of K for which the solution of the differential equation (35a) under the boundary conditions (1b,c) and (2a) satisfied the boundary condition (35b) with a preassigned accuracy. (The root solver was later used on a condition for F itself rather than its derivative F_x.)

Our analysis suggests that any numerical method will run into difficulties when $G_- + G_+$ tends to zero, and that the difficulties will become more acute as the method is made more accurate. For instance, curvatures are not obtained as accurately as gradients, so that the requirement (36c) will not be satisfied and equations (40), with V replaced by $C_\mp/G_\mp$, will be incompatible. Inability of the root solver to find a K_0 could be the computational manifestation of this analytical nonexistence, expecially when the accuracy of the numerical method is high. Such inability was the only difficulty encountered by Oyediran & Ludford (1985).

The cusps encountered by Stewart & Ludford (1984) are not predicted by the above argument; presumably their methods did not approximate the analysis sufficiently accurately. Both the cusps and ultimate failure can be explained by third-derivative inaccuracies. If the difference in third derivatives is less than $9V^2/8$, but does not have the value (50), then K_0^* jumps from $-V$ to one of the values (49), from which the K-profile leaves with infinite slope ($K_1 \neq 0$). The numerical smoothing of such a jump could form a cusp. The ultimate failure would then be attributable to a difference in third derivatives exceeding $9V^2/8$ (i.e. 12½% error).

In any event, none of the computations revealed the true nature of $K(t)$: its evolution never terminates but may be marked by saw-tooth changes in direction. The next step is to devise a numerical scheme that will capture this behavior.

BIBLIOGRAPHY

Carslaw, H.S. & Jaeger, J.C. 1959 Conduction of Heat in Solids. Clarendon Press, Oxford.

Dold, J.W. 1985 Analysis of the early stage of thermal runaway. Quarterly Journal of Mechanics and Applied Mathematics, to appear.

Oyediran, A.A. 1984 Unsteady Detonations. Ph.D. Thesis, Cornell University.

Oyediran, A.A. & Ludford, G.S.S. 1985 The Stefan problem of detonation theory. Transactions of the Second Army Conference on Applied Mathematics and Computing, Troy NY, 1984. ARO Report 85-1.

Stewart, D.S. & Ludford, G.S.S. 1983 The acceleration of fast deflagration waves. Zeitschrift für Angewandte Mathematik und Mechanik **63**, 291-302.

Stewart, D.S. & Ludford, G.S.S. 1984 Near Chapman-Jouget detonations. Transactions of the First Army Conference on Applied Mathematics and Computing, Washington DC, 1983. ARO Report 84-1.

CENTER FOR APPLIED MATHEMATICS
CORNELL UNIVERSITY
ITHACA, NY 14853

Lectures in Applied Mathematics
Volume 23, 1986

SOLIDIFICATION PROBLEMS AS SYSTEMS OF NONLINEAR DIFFERENTIAL EQUATIONS

Gunduz Caginalp[1]

ABSTRACT. A mathematical analysis of a phase field approach to solidification problems is presented. A free boundary arising from a phase transition is assumed to have finite thickness. The physics leads to a system of nonlinear parabolic differential equations. Existence and regularity of solutions are proved. Invariant regions of the solution space lead to physical interpretations of the interface. An asymptotic analysis is performed for the equilibrium equations (i.e. elliptic systems). The model also leads to some interesting ordinary differential equation problems. Numerical work involving instabilities and anisotropy may be used to study dendrites.

1. INTRODUCTION. Free boundary problems which arise from phase transitions have been of interest for many years. Problems which incorporate the physics of latent heat as well as the physics of heat diffusion in a homogeneous medium have been studied extensively in the mathematics literature, and are generally known as the classical Stefan problem [1] (see also references in [2]).

A mathematical statement of the Stefan problem is as follows. The material may be in either of two phases, e.g. solid or liquid, in a region $\Omega \subset \mathbb{R}^N$ of space. We let $u(t,x)$ be the (reduced) temperature at time $t \in \mathbb{R}^+$ and spatial point $x \in \Omega$, so that $u = 0$ is the usual equilibrium melting temperature for a planar interface.

In the classical Stefan problem one makes the assumption that

---

1980 Mathematics Subject Classification: 35R35, 35S15 82A25, 34B15.
[1]Supported by NSF Grant DMS - 8403184.

the temperature, $u(t,x)$, at the interface, $\tilde{\Gamma}(t)$, between the solid and liquid is zero. Furthermore, if $u(t,x) > 0$ the point $x \in \Omega$ lies in the liquid region Ω_1, and if $u(t,x) < 0$ it lies in the solid region Ω_2.

Within the context of the Stefan problem, the temperature $u(t,x)$ must satisfy the heat diffusion equation

$$u_t = K \Delta u \tag{1.1}$$

in Ω_1 and Ω_2. The constant K is the thermal diffusivity which is thermal conductivity divided by heat capacity per unit volume. We set heat capacity per volume equal to unity and also assume for simplicity that the thermal conductivity is the same in both phases. Introducing two distinct thermal conductivities does not usually alter the mathematics significantly in the Stefan problem.

For any interface, $\tilde{\Gamma}$, one defines a unit normal $\hat{n}$ (in the direction solid to liquid) and a local velocity $\vec{v}(t,x)$ at each point of Γ. Across the interface Γ, the latent heat of fusion (per unit mass), ℓ, must be balanced with the heat flux, i.e.

$$\ell \vec{v} \cdot \vec{n} = K(\vec{\nabla} u_S - \vec{\nabla} u_L) \cdot \hat{n} \qquad x \in \tilde{\Gamma} \tag{1.2}$$

where $\vec{\nabla} u_L$ is the limit of the gradient of u at a value $x \in \tilde{\Gamma}$ when approached from Ω_1 (liquid) while $\vec{\nabla} u_S$ is the limit from Ω_2 (solid). The right hand side of (1.2) is then the jump in the normal component of the temperature multiplied by the thermal conductivity. The density factor multiplying the left hand side has been set equal to one.

The problem is complete upon specifying initial and boundary conditions for $u(t,x)$. One may take, for example,

$$u(t,x) = u_\partial(t,x) \qquad x \in \partial\Omega \qquad t > 0 \tag{1.3}$$

$$u(0,x) = u_0(x) \qquad x \in \Omega. \tag{1.4}$$

The mathematical problem is to find $u(t,x)$ and $\tilde{\Gamma}(t)$ which satisfy (1.1)-(1.4). The interface $\tilde{\Gamma}(t)$ is usually called the free boundary.

One method for studying these equations is the enthalpy or H-method [3]. One introduces a function H(u) defined by

$$H(u) \equiv u + \frac{\ell}{2}\phi \qquad \phi = \begin{cases} +1 & u > 0 \\ -1 & u < 0 \end{cases} \tag{1.5}$$

which must satisfy the equation

$$\frac{\partial}{\partial t} H(u) = K\Delta u. \tag{1.6}$$

With definition (1.5), the single equation (1.6) incorporates both heat diffusion and latent heat, and is equivalent, in a weak sense, to equations (1.1) and (1.2).

We will not discuss this formulation here but will use it to motivate the study of a nonlinear system of differential equations. First, however, it will be useful to understand how and why the physics is often more complicated than the description given by equations (1.1)-(1.4). Perhaps the most basic phenomenon one observes is that the liquid is often below its freezing point, which is called supercooling. The analogous phenomena for a solid is called superheating. One should emphasize that supercooling is an equilibrium phenomenon and is not merely a transient effect (although supercooling may arise from nonequilibrium considerations as well). The origin of this phenomenon for a pure substance is in the finite size effect of the interface between the solid and liquid. The classical Stefan problem neglects the thickness of the interface between a solid and liquid and treats the physics at a purely continuum level. A simple and rough argument for appreciating these equilibrium effects as a first order correction to the continuum theory is as follows. Suppose that $u = 0$ is the equilibrium temperature between a solid and liquid separated by a planar interface. This means that a certain amount of energy is required in order that a molecule at the surface overcomes the binding energy of the crystal lattice and becomes part of the liquid with lower binding energy. The amount of energy required to produce this transition depends on the number of nearest neighbors in the crystal structure and on the number of nearest neighbors of an atom on the surface. Now suppose that the interface between the solid and liquid is curved (e.g., solid protruding into liquid, which we will define as positive curvature). In this case, the molecule on the surface has fewer nearest neighbors, since some are missing due to the curvature. Hence, one expects that it will require less energy to produce the transition. Consequently, if we consider a solid with constant mean curvature, i.e., a sphere, in equilibrium with its melt, then we expect the prevailing temperature to be lower. That is, $u = u_R < 0$ where R is the radius of the sphere. A more detailed version of this argument which is well known to solid state physicists and materials sceintists may be found in [4-6]. From the nature of these arguments it is clear physically that u_R must be proportional to R^{-1} with the proportionality constant involving the surface tension, σ.

A statistical mechanical analysis [7-9] of free energies and chemical potentials leads to the Gibbs-Thompson relation,

$$u(t,x) = -(\sigma/\Delta s)\kappa, \tag{1.7}$$

near an interface between two phases in equilibrium, where κ is the sum of the principal curvatures and, Δs is the difference in entropy per unit volume between solid and liquid. In our model $\Delta s = 4$.

In order to modify equations (1.5), (1.6) to describe an interface with finite thickness and finite surface tension, one must modify ϕ so that it makes the transition from a positive value (liquid), e.g. +1 to a negative value (solid), e.g. -1, within a finite distance. It is clear, physically, that the choice of the function ϕ determines the nature of the change of phase of the boundary. In particular, the thickness of the interface, which is closely related to the surface tension, will be determined by the choice of ϕ. Hence, one must interpret ϕ as a phase field, or in the language of statistical mechanics, as an order parameter. One can then use the physics of phase transitions to determine a system of equations for which (u,ϕ) is a solution. We note that two different functions ϕ may result in the same physics provided the functions are equal in an appropriate norm. This question of the equivalence of different smoothing functions has been investigated by C. Bonacina, G. Comini, A. Fasano and M. Primicerio [10-12].

A general theory which can be applied to this problem is that of Landau and Ginzburg [13]. If the system is in equilibrium one may choose ϕ which minimizes the free energy functional

$$F_u\{\phi\} = \int d^N x\{\frac{\xi^2}{2}(\nabla\phi)^2 + \frac{1}{8}(\phi^2-1)^2 - 2u\phi\}. \qquad (1.8)$$

This form of the free energy involves an averaging over the individual atoms or molecules. The $(\nabla\phi)^2$ term originates from individual interaction potentials and ξ is a length scale which is related to the strength of interactions. The $(\phi^2-1)^2$ term is a prototype double-well potential, which expresses the preference of the material to be in the liquid or solid ($\phi = \pm 1$) phase rather than in an interfacial state. The last term in the integral, $-2u\phi$, is a linearized contribution due to the entropy difference between the solid and liquid. If ϕ minimizes (1.7) then the Euler-Lagrange equations imply:

$$0 = \xi^2\Delta\phi + \frac{1}{2}(\phi-\phi^3) + 2u. \qquad (1.9)$$

If the system is not in equilibrium then one expects that ϕ will satisfy the Model A equation [13]:

$$\tau\,\phi_t = \xi^2\Delta\phi + \frac{1}{2}(\phi-\phi^3) + 2u \qquad (1.10)$$

where τ is a relaxation time.

Equation (1.10) along with the new version of (1.5), (1.6) i.e.

$$u_t + \frac{\ell}{2}\phi_t = K\Delta u \qquad (1.11)$$

and suitable boundary and initial conditions, e.g.,

$$u(x) = u_\partial(x) \qquad x \in \partial\Omega \tag{1.12}$$

$$\phi(x) = \phi_\partial(x) \qquad x \in \partial\Omega \tag{1.13}$$

$$u(0,x) = u_0(x) \qquad x \in \Omega \tag{1.14}$$

$$\phi(0,x) = \phi_0(x) \qquad x \in \Omega \tag{1.15}$$

specify the mathematical problem to be studied in the time dependent case. In the equilibrium case, the problem is specified by Laplace's equation for u, equation (1.9) and boundary conditions (1.12), (1.13). The exterior boundary, $\partial\Omega$, is assumed to be smooth.

The free boundary arising from the phase transition is now implicit in the equations (1.10)-(1.15); it is the set of points $x \in \Omega$ for which $\phi(t,x) = 0$. One continues to interpret positive ϕ as liquid and negative ϕ as solid. Since u and ϕ are no longer constrained to be of the same sign, supercooling and superheating are possible.

The surface tension associated with the free boundary is also incorporated into these equations as a result of the $\xi^2(\nabla\phi)^2/2$ term in the free energy.

2. BASIC GOALS OF MODEL AND PHENOMENOLOGY. The theory which leads to equations (1.10)-(1.15) is one which has several physical approximations. The free energy (1.8) is obtained from considering the Fourier transforms of the interactions and spins, discarding higher order Fourier modes and reverting to real space. This is probably the most important approximation, leading to second order derivatives of ϕ instead of fourth order, etc.

This approximation is expected to be quite good when the length scale of interest is considerably larger than atomic length. One of the most important length scales in any such problem is the correlation length, which is a measure of the extent to which an atom or spin in one location is aware of the effects of another at some distance. The arguments preceding the Gibbs-Thompson relation, if sufficiently refined, lead to the conclusion that the correlation length must be proportional to the interface thickness, which can be shown [2] to be porportional to the surface tension. Hence, it is reasonable to expect on physical grounds that this theory will lead to the right macroscopic behavior in some intermediate range of ξ where ξ is larger than atomic length and smaller than the size of the container. When the correlation length approaches infinity, e.g. at a critical point, it is known that the Landau theory does not predict, for example, the correct critical exponents. It is possible, however, that this theory may continue to predict the macroscopic behavior of the interface for a broader range of ξ for which some very subtle properties of the bulk are not described with complete accuracy.

A second approximation is involved in the linearized form of the entropy term $-2u\phi$ in the free energy (1.8). Since we are interested in the consequences of an interface wtih thickness and not necessarily the detailed behavior within the interface, then we may expect that this is physically reasonable.

Perhaps the most seriously limiting physical approximation is the use of the heat conduction equation for the liquid. Since convection is the principal mode of heat transfer in a liquid, one cannot expect accurate results when the difference is significant. In many situations with free boundaries associated with solidification the most significant aspect of heat transfer is involved in dissipating th heat produced from the latent heat. However, if the outerfacial velocities are too large convection terms will contribute significantly to the behavior of the interface.

In the range of parameters for which these equations are physically reasonable, one may attempt to understand the following aspects of the interfacial behavior.

(a) Temperature at the interface - In the formulation of equations (1.10)-(1.15) the temperature $u(t,x)$ is not specified on the free boundary $\overline{\Gamma}(t) = \{x \in \Omega: \phi(t,x) = 0\}$. For small ξ, one expects $u(t,x)$ will satisfy the Gibbs-Thompson relation if the system is in equilibrium.[2]. If the interface is moving then one expects a contribution to this relation from the velocity, v. For a plane-wave interface, the L_2 orthogonality analysis [2] which leads to the Gibbs-Thompson relation formally implies

$$u = c_1 v \tag{2.1}$$

where $c_1 \equiv -(\xi\bar{c}/4)\int_{-\infty}^{\infty} [\frac{d}{d\rho} \tanh \rho/2]^2 d\rho$ and $\tau = \bar{c}\xi^2$.

(b) Thickness of interface - Since ϕ is a smooth function (as will be shown in Section 3) one may enquire whether there is a distinct value of ϕ separating the interfacial region from the liquid or solid. I.e. one would like to describe the interfacial region as

$$\Gamma(t) = \{x \in \Omega: \ |\phi(t,x)| < c\} \tag{2.2}$$

for some value of c, and determine whether the interface retains its original thickness at later times.

A question which is related to the problem of defining c in (2.2) is whether this set of equations predicts that a liquid with no initial "seed" crystal can be supercooled to any temperature. This corresponds to infinite curvature in the Gibbs-Thompson relation. Physically, this is what one would expect from this type of theory. In practice, the liquid cannot be supercooled below a particular temperature because clusters of liquid atoms or molecules act as an effective seed with an effective radius of curvature.

(c) Stability of interface and effects of anisotropy - Since this model incorporates the stabilizing effects of surface tension as well as the destabilizing effects of supercooling one may study the competition between these effects. An intriguing question involving instability is the following. How can such regular structures such as snowflakes or dendrites in metals be the consequence of instabilities? If equations (1.10)-(1.15) are modified to include the effects of anisotropy, then one may attempt to answer this question.

3. EXISTENCE, UNIQUENESS, AND REGULARITY. The existence of a unique solution to the system (1.10)-(1.15) may be proved in a suitable function space.

Writing equations (1.10), (1.11) in vector form one has:

$$U_t = A\Delta U + F(U) \tag{3.1}$$

where the vectors U, F(U) and the matrix A are defined by

$$U \equiv \begin{pmatrix} u \\ \phi \end{pmatrix}, \quad A \equiv \begin{pmatrix} K & -\ell\xi^2/2\tau \\ 0 & \xi^2/\tau \end{pmatrix} \tag{3.2}$$

$$F(U) \equiv [\tfrac{1}{2}(\phi-\phi^3) + 2u]\frac{1}{\tau}\begin{pmatrix} -\ell/2 \\ 1 \end{pmatrix}.$$

The boundary and initial conditions may be written as

$$U(t,x) = U_\partial(x) \qquad x \in \partial\Omega \tag{3.3}$$

$$U(0,x) = U_0(x) \qquad x \in \Omega. \tag{3.4}$$

We consider an appropriate Banach space $\mathcal{B}$ which we may take as the space of bounded, uniformly continuous functions on Ω. We then let $C([0,T];\mathcal{B})$ be the Banach space of continuous functions on $[0,T]$ with values in $\mathcal{B}$, with norm

$$\|U\| = \sup_{0\le t\le T} \|U(t)\|_{\mathcal{B}}\,. \tag{3.5}$$

Classical results then imply that if $U_0 \in \mathcal{B}$, then there exists a $t_0 > 0$, where t_0 depends only on F and $\|U_0\|$, such that (3.1)-(3.4) has a unique solution in $C([0,t_0];\mathcal{B})$. To proceed further we concentrate on regions in (u,ϕ) space (i.e. $\mathbb{R}^2$).

DEFINITION 3.1. A closed subset $\Sigma \subset \mathbb{R}^2$ is called a (positively) <u>invariant region</u> for the local solution of (3.1)-(3.4) in $t \in [0,\delta)$ if any solution having all of its boundary and initial values in Σ satisfies $U(t,x) \in \Sigma$ for all $(t,x) \in [0,\delta) \times \Omega$.

Using a theorem of Chueh, Conley and Smoller [14], one may easily compute all possible invariant regions for (3.1)-(3.4) based on the eigenvectors of the matrix A. The basic idea is to find enclosed regions such that the flow in time is never directed outside the region.

If ξ and τ satisfy the stability condition

$$\frac{\xi^2}{\tau} < K \tag{3.6}$$

then one may show that there exist exactly three types of invariant regions, each of which is a parallelogram (see Figure 1): (a) a large region, Σ_0 including the points $\phi = \pm 1$, $\phi = 0$, (b) a smaller region, Σ_+, which includes $\phi = +1$ but does not include $\phi = -1$ or $\phi = 0$, (c) another smaller region Σ_-, which includes $\phi = -1$ but does not include $\phi = 0$ or $\phi = +1$.

The region Σ_0 provides the a priori bound needed for existence. One has then [2]:

THEOREM 3.2. (Global existence of solutions) Suppose ℓ, K, ξ, τ are any set of positive constants subject to the stability inequality (3.6). If U_∂, $U_0 \in \mathcal{B}$ and $T \in (0,\infty)$ then there exists a unique solution U(with $U(\cdot,t) \in \mathcal{B}$) to the system (3.1)-(3.4) for all $t \in [0,T]$. □

The three types of invariant regions, $\Sigma_0, \Sigma_+, \Sigma_-$ are illustrated in Figure 1. The G_i form the boundaries of Σ_0.

The existence of invariant regions of type Σ_+, which lie entirely in the positive half-plane of ϕ (in particular $\phi \geq 3^{-1/2}$), and analogous regions for Σ_-, provide a physical basis for defining the interfacial region. That is, the interfacial region consists of those points (t,x) for which $|\phi(t,x)| < 3^{-1/2}$). Con-

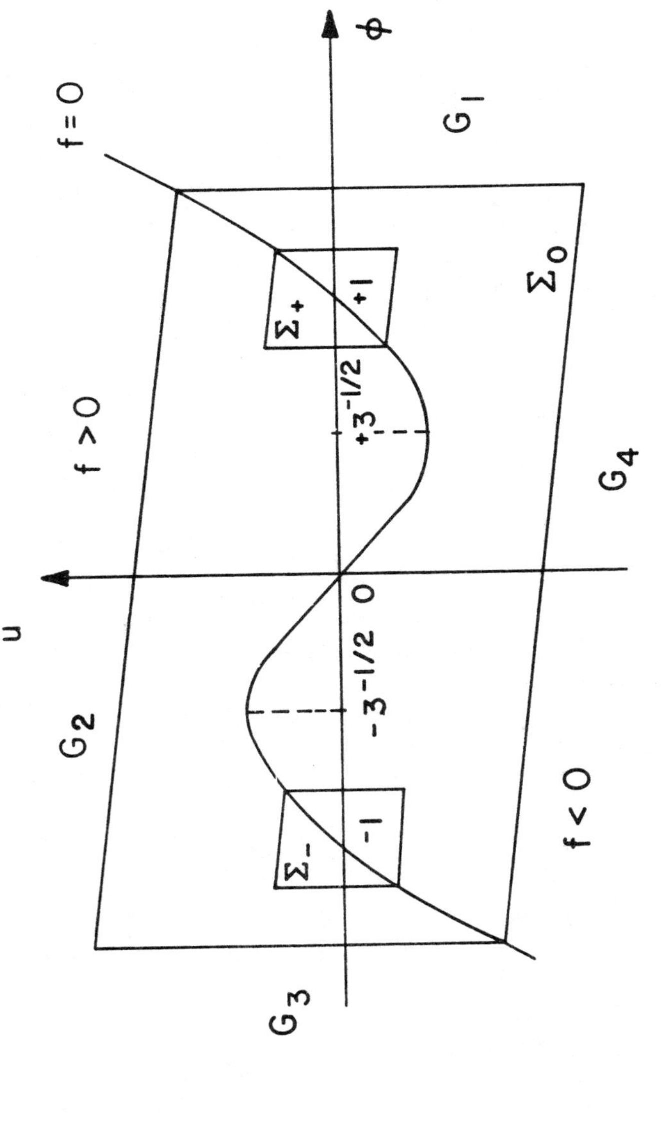

$f(u,\phi) \equiv 1/2(\phi - \phi^3) + 2u$

FIGURE 1

sequently, if we choose initial and boundary conditions which lie entirely in the liquid region, i.e. $\phi \geq 3^{-1/2}$, then the material will remain in the liquid phase. This is consistent with our expectations from ideas leading to the Gibbs-Thompson relation.

Furthermore, the smallest invariant region which contains $\phi = 0$ must also contain $\phi = \pm 1$. This is interesting from a physical and computational point of view since it implies that the interface would not have the tendency to dominate the solid and liquid.

The next issue we address is that of regularity or smoothness of solutions.

We assume the uniformity bounds

$$C_1 \leq \xi^2/\tau \leq C_2 < K. \tag{3.7}$$

Consider the metric

$$d(P,Q) \equiv [|x-\bar{x}|^2 + |t-\bar{t}|]^{1/2} \tag{3.8}$$

between points $P = (t,x)$ and $Q = (\bar{t},\bar{x})$ where $|x|$ is the usual Euclidean norm. The Holder coefficient with exponent α in $\Lambda \equiv [0,T] \times \Omega$ is defined by

$$H_\alpha(v) = \sup_{P,Q\in\Lambda} \frac{|v(P)-v(Q)|}{[d(P,Q)]^\alpha} . \tag{3.9}$$

We let $C^{2,\alpha}$ be the space of functions whose derivatives $D_x^2 v$ and $D_t v$ exist and have bounded Holder coefficients with exponent α.

With these definitions we can prove [2]:

THEOREM 3.3. (Regularity of solutions) Let $U \in \mathcal{B}$ be a solution to the system (1.14)-(1.15). Suppose U_∂, $U_0 \in C^{2,\alpha}$. Then $U \in C^{2,\alpha}$.

□

Further results along these lines may be developed. The basic idea is that the solution is as regular as the intial and boundary conditions. The constant bounding the norm of U involves τ^{-1} and a constant depending on C_1, C_2 and the initial and boundary conditions.

Using these ideas, one may prove the gradient bounds

$$|D_{x/\xi}\phi| < C_3, \quad |D^2_{x/\xi}\phi| < C_4 \tag{3.10}$$

where C_3, C_4 depend on C_1,C_2 but not on ξ independently. This means that the interfacial region does not become significantly sharper in time. This is consistent with the physical notion of surface tension (see Figure 2).

4. RIGOROUS ASYMPTOTIC ANALYSIS. We now concentrate on the time independent situation, and consider the small parameter limit of ξ. We address the following problem. Given a sequence of materials, indexed by {i}, and characterized by ξ_i (but otherwise identical) we assume that each material occupies a region Ω of which Ω_0 is solid ($\phi > 0$) and $\Omega\backslash\overline{\Omega}_0$ is liquid ($\phi < 0$). Since u satisfies Laplace's equation, specification of u_∂ uniquely determines u in all of Ω. To specify the boundary conditions for ϕ, let Ω_1 be a region which is strictly contained in Ω_0. The boundary conditions are then $\phi = -1$ for $x \in \Omega_1$, $\phi = +1$ for $x \in \partial\Omega$ and $\phi = 0$ for $x \in \partial\Omega_0$.

The main idea is to consider the inner expansion (or expansion in the boundary layer)within $O(\xi)$ of the boundary, and the outer expansion in the remaining part of the boundary. By joining these expansions appropriately, one may obtain an expansion which approximates the solution uniformly in the entire domain. These ideas are implemented in [2]. The basis for much of the analysis is in the work of Berger and Fraenkel [15] (see also other references in [2]).

One defines a new coordinate system in which r is in the direction normal to $\partial\Omega_0$ and $\rho \equiv r/\xi$ is the "stretched variable". The $O(1)$ term, ϕ_0, in the inner asymptotic expansion is then the solution to

$$\xi^2 \frac{\partial^2\phi_0}{\partial r^2} + \frac{1}{2}(\phi_0 - \phi_0^3) = 0 \tag{4.1}$$

which is $\phi_0 = \tanh \rho/2$. The analysis shows that

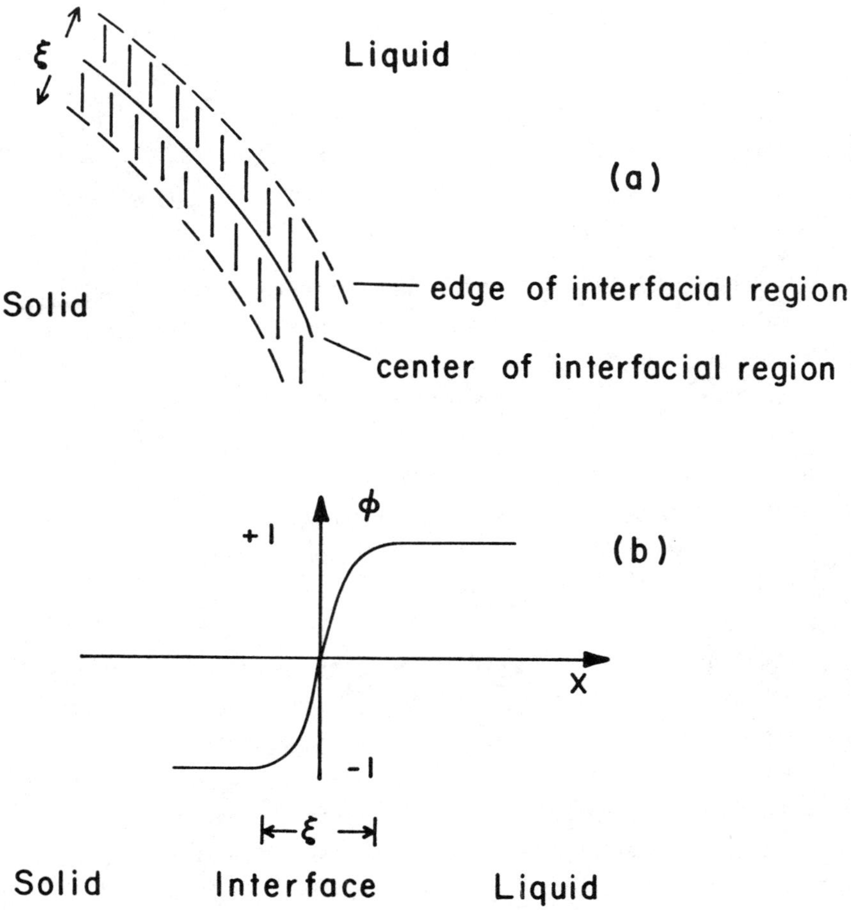
ξ
Liquid
(a)
Solid
edge of interfacial region
center of interfacial region
φ
+1
(b)
x
-1
ξ
Solid
Interface
Liquid

FIGURE 2

$$|\phi-\phi_0| \leq C\xi. \tag{4.2}$$

The remaining terms in the asymptotic expansion are studied by means of the derivative equation

$$\xi^2 \Delta w + \frac{1}{2}(1-3\phi_0^2)w = F, \tag{4.3}$$

where F is a function of the previous terms in the series and higher powers of the term being considered. In this way one obtains a series

$$\Phi_M \equiv \sum_{j=0}^{M} \xi^j \phi_j(x,\xi) \tag{4.4}$$

such that

$$\phi(x,\xi) - \Phi_M(x,\xi) = O(\xi^{M+1}). \tag{4.5}$$

We now consider one of the physically most interesting aspects of this model, that is, the relationship between the temperature at the interface, the curvature of the interface and the surface tension. Consider the situation where a material is constrained to be solid ($\phi = -1$) in the inner region and liquid ($\phi = +1$) some $O(1)$ distance away from the inner region. We suppose that the solution $\phi_\xi(x)$ satisfies (4.5) at least for $M = 0$. By taking the L_2 inner product of (4.3) with ϕ_0 we observe that a solution is possible only if F is orthogonal to ϕ_0' in the appropriate order in ξ. This results in

THEOREM 4.1. (Gibbs-Thompson relation) If (u,ϕ) is a solution to (1.9), $\Delta u = 0$, (1.12), (1.13) satisfying the conditions stated above, then on the boundary $\partial\tilde{\Omega}$ one has the relation

$$u(x) = -\frac{\sigma_0 \kappa}{4} + O(\xi) \tag{4.6}$$

$$\sigma_0 \equiv \xi^2 \int_{-\infty}^{\infty} \left(\frac{d\phi_0}{dr}\right) dr = \frac{2}{3}\xi \tag{4.7}$$

where κ is the sum of principle curvatures. □

One may also verify that σ_0 is the expression for the surface tension plus $O(\xi^2)$. The surface tension, σ, is generally defined [4-6] in terms of the difference between the free energy of the system with an interface of cross-sectional area A and an average of the homogeneous free energies, i.e.

$$\sigma \equiv \frac{F_u\{\phi\} - \frac{1}{2} F_u\{\phi = +1\} - \frac{1}{2} F_u\{\phi = -1\}}{A} . \tag{4.8}$$

One can then show

$$\sigma = \sigma_0 + O(\xi^2). \tag{4.9}$$

This boundary layer analysis on both sides of the boundary then leads to the conclusion that the Gibbs-Thompson relation is a necessary condition for a solution which is within $O(\xi)$ (in the supremum norm) of the tanh solution.

5. ANALYSIS OF RELATED ORDINARY DIFFERENTIAL EQUATIONS. In the analysis described in the preceding section, the set of points $\Gamma(t)$ was assumed to be fixed as $\xi \to 0$. Mathematically, the more interesting problem is to analyze the elliptic equations when one has boundary conditions $\phi \cong -1$ and $\phi \cong +1$ and to observe an interior transition layer. If we consider a physical problem with spherical symmetry, then we may seek radically symmetric solutions, i.e. $\phi(r)$ which solve the ordinary differential equation

$$\xi^2\phi'' + \frac{\xi^2(N-1)}{r}\phi' + \frac{1}{2}(\phi-\phi^3) + \frac{1}{2}k = 0, \tag{5.1}$$

$$\phi(a) = -1, \quad \phi(b) = +1, \quad 0 < a < b < \infty . \tag{5.2}$$

We have also assumed that the temperature, u, is constant on the boundary and, by Laplace's equation is constant throughout. We let $k \equiv 4u = c\xi$. This equation has been studied in collaboration with S. Hastings [16] for $N = 1$ and $N > 1$. A slight modification of the boundary conditions leads to

$$\phi(a) = \alpha \qquad \phi(b) = \gamma \tag{5.3}$$

where $\alpha < \beta < \gamma$ are the roots of

$$f(\phi) \equiv \phi - \phi^3 + k. \tag{5.4}$$

Existence of solutions to (5.1), (5.2) and (5.1), (5.3) has been proved using a shooting technique [16] for $N \geq 1$.

Solutions to (5.1), (5.3) were also shown to be (strictly) monotonic, while (5.1), (5.2) may have solutions which have one slight dip or peak depending on the sign of k.

For the one dimensional problem ($N = 1$) explicit analysis is possible [16]. One may write ϕ as a solution to

$$x(\phi) - x(\alpha) = \int_\alpha^\phi \xi \frac{d\phi}{\sqrt{c_{\xi,k} - F(\phi) - k\phi}} \tag{5.5}$$

where the constant $c_{\xi,k}$ is determined by the boundary condition at $x = b$, i.e.

$$b-a = \xi \int_\alpha^\gamma \frac{d\phi}{\sqrt{c_{\xi,k} - F(\phi) - k\phi}} . \tag{5.6}$$

In this case one has a unique solution. For the one-dimensional case, surface tension does not play a role (since $\kappa = 0$). Hence, one expects that if $k > 0$ then the material will prefer to be in the liquid state ($\phi \approx 1$) to the extent that boundary conditions permit. Analyzing (5.5), (5.6) for small ξ, one finds that for $k > 0$ the solution is near $\phi = 1$ except within $O(\xi \ln \xi)$ of the $x = a$ boundary. The situation is reversed for $k < 0$. These solutions are sketched in Figures 3 and 4. If $k = 0$ the material does not prefer either solid or liquid. One may show that the solution must then cross zero at $\frac{1}{2}(a+b)$ as shown in Figure 5. In each of these situations one may show that the derivative is $O(\xi^{-1})$ at the crossover point, and is very close to zero when one is far from the crossover point.

6. NUMERICAL WORK AND ANISOTROPY. A numerical study [17] of equations (1.10)-(1.15) has been successful in showing that there is a competition between the destabilizing effects of supercooling and the stabilizing effects of surface tension. That is,

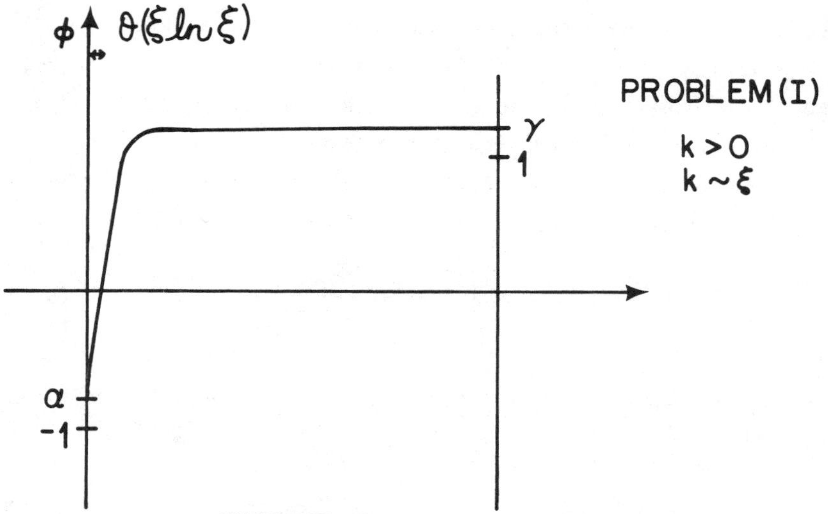

FIGURE 3

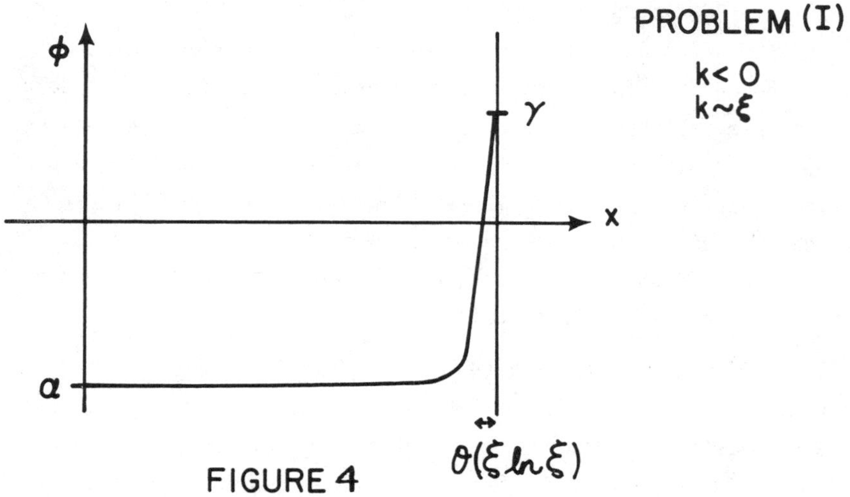

FIGURE 4

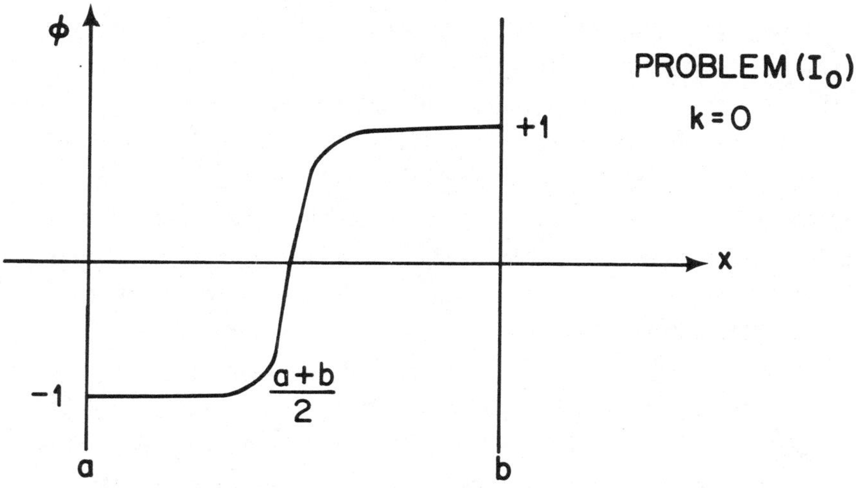

FIGURE 5

dendritic growth was observed for larger supercooling and smaller surface tension.

The role of anisotropy may also be an important factor in instabilities. Numerical work in collaboration with J. Lin is currently in progress on this problem. At a phenomenological level, the Gibbs-Thompson relation (1.7) is no longer valid in the isotropic sense but must be modified in order to indicate that the effective surface tension is larger in some directions than others. Hence, the Gibbs-Thompson relation may be expected to have the form

$$u = -g(\theta)\sigma\kappa \tag{6.1}$$

where g is a function of orientation, θ. One may consider a material with some anisotropic structure, e.g. with growth preference along the x,y,z, axes.

If the material is cut so that it is a perfect sphere and put into its melt which is kept at a uniform supercooled temperature, then we can expect the following types of behavior depending on (a) the degree of supercooling, (b) the magnitude of the surface tension, and (c) the nature and extent of the anisotropy. If the surface tension is sufficiently large and the liquid is sufficiently supercooled then we may expect to see the material grow fastest along the x,y,z directions. On the other hand if the liquid is highly supercooled then instabilities will tend to arise. Since the surface tension is effectively lowest in the x,y,z directions, the dendrites will prefer these directions.

The anisotropy may be incorporated into equations (1.10)-(1.15) in various ways. One possibility is to change the $\phi-\phi^3$ term to an orientation dependent $\phi - \bar{g}(\theta)\phi^3$ term. Although this is a phenomenological approach it leads to the correct anisotropic Gibbs-Thompson relation (6.1). Studying equations (1.10)-(1.15) numerically we have found [18] that the growth of the "seed" solid is as anticipated. Figure 6 shows a two-dimensional numerical computations for a substance with large surface tension and anisotropic growth along the axes. Figure 7 shows the result of a numerical computation for a material with similar anisotropy but with lower surface tension. The instabilities occur in the directions for which $\bar{g}(\theta)$ in (6.1) is smallest. For the anisotropy $\bar{g}(\theta)$ we have used various sine and cosine functions.

One may approach the question of anisotropy from a physically more fundamental perspective. In particular, if one assumes the material is described by a lattice cell or lattice spin system, with anisotropy of interactions at the molecular level, then one may derive a modified free energy of the form (1.8). One finds that the $(\nabla\phi)^2$ term is modified by the anisotropy. This means that $\Delta\phi$ term in (1.10) is also modified. Numerical work along these lines is now in progress [18]. For two dimensional anisotropy, e.g. favoring the x direction, the results are shown in Figure 8.

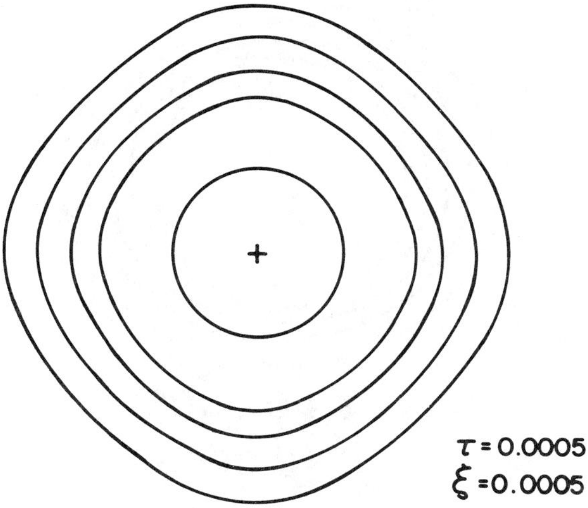

FIGURE 6

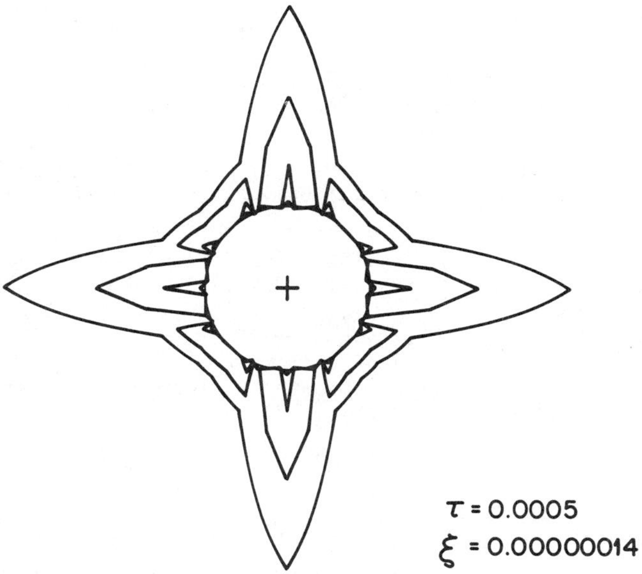

FIGURE 7

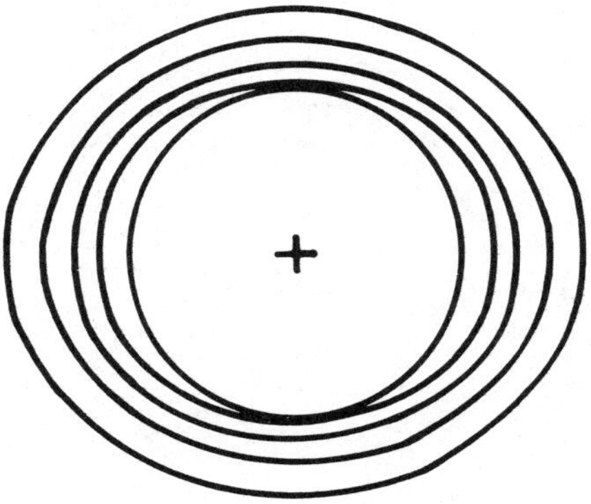

$\tau = 0.0005$

$\xi = 0.0005$

FIGURE 8

BIBLIOGRAPHY

1. L.I. Rubinstein, The Stefan Problem, Am. Math. Soc. Transl. 27, American Mathematical Society, Providence, R.I. (1971).

2. G. Caginalp, "An Analysis of a Phase Field Model of a Free Boundary", Carnegie-Mellon University Research Report (1983).

3. D.A. Oleinik, "A Method of Solution of the General Stefan Problem", Sov. Mat. Dokl. 1, 1350-1354 (1960).

4. B. Chalmers, Principles of Solidification, R.E. Krieger Publishing, Huntington, NY (1977).

5. P. Hartman, Crystal Growth: An Introduction, North Holland Publishing, Amsterdam (1973).

6. J.W. Gibbs, Collected Works, Yale University Press, New Haven, CT (1948).

7. D.W. Hoffman and J.W. Cahn, "A Vector Thermodynamics for Anisotropic Surfaces I. Fundamentals and Application to Plane Surface Junctions", Surface Science 31, 368-388 (1972).

8. J.W. Cahn and D.W. Hoffman, "A Vector Thermodynamics for

Anisotropic Surfaces II. Curved and Faceted Surfaces", Acta Metallurgica 22, 1205-1214 (1974).

9. W.W. Mullins, "Therdynamic Equilibrium of a Crystal Sphere in a Fluid", Carnegie-Mellon University Research Report (1983).

10. C. Bonacina, G. Comini, A. Fasano and M. Primicerio, "Temperature Effects of Thermophysical Property Variations in Nonlinear Conductive Heat Transfer", Bull. Calcutta Math. Soc. 71, 301-308 (1979).

11. ________________, "On the Estimation of Thermophysical Properties in Nonlinear Heat-Conduction Problems", Int. J. Heat Mass Transfer 17, 861-867 (1974).

12. ________________, "Numerical Solution of Phase Change Problems", Int. J. Heat Mass Transfer 16, 1825-1832 (1973).

13. P.C. Hohenberg and B.I. Halperin, "Theory of Dynamic Critical Phenomena", Reviews of Modern Physics 49, 435-480 (1977).

14. K. Chueh, C. Conley and J. Smoller, "Positively Invariant Regions for Systems of Nonlinear Diffusion Equations", Indiana Univ. Math. J. 26, 373-392 (1977).

15. M.S. Berger and L.E. Fraenkel, "On the Asymptotic Solution of a Nonlinear Dirichley Problem", J. Math. Mech. 19, 553-585 (1970).

16. G. Caginalp and S. Hastings, "Properties of Some Ordinary Differential Equations Related to Free Boundary Problems", University of Pittsburgh Research Report (1984).

17. G. Fix and J. Lin, Private communication.

18. G. Caginalp and J. Lin, Paper in preparation.

DEPARTMENT OF MATHEMATICS AND STATISTICS
UNIVERSITY OF PITTSBURGH
PITTSBURGH, PA 15260

Lectures in Applied Mathematics
Volume 23, 1986

GLOBAL SOLUTIONS FOR SYSTEMS OF PARABOLIC CONSERVATION LAWS WITH NONSMOOTH INITIAL DATA

David Hoff[1] and Joel Smoller[2]

1. INTRODUCTION. We discuss the global existence of solutions of certain systems of partial differential equations of the form

$$u_t + f(u)_x = Du_{xx} \tag{1.1}$$

with Cauchy data

$$u(x,0) = u_0(x) \in L^\infty(\mathbb{R}) . \tag{1.2}$$

Here $u = u(x,t)$ is a vector, and D is a constant, diagonalizable matrix with positive eigenvalues. In particular, we shall establish the global existence of solutions of the full equations of gas dynamics for a broad class of diffusion matrices D under the assumption that the initial data is also in L^2. In addition, we shall make some observations concerning the gas dynamics equations with real, physical viscosity terms. We shall show that, for the Navier-Stokes equations of compressible flow, initial discontinuities in the density cannot be smoothed out in positive time.

(1) Research supported in part by the NSF under Grant no. MCS-8301141
(2) Research supported in part by the NSF under Grant no. MCS-8002337

As we shall see, local existence of solutions of (1.1)-(1.2) is easily established by a simple iteration argument. However, for most examples of interest, the flux f is not defined for all values of u. To prove global existence, it is therefore necessary to bound the sup norm of the solution in such a way that u remains in a set in which f is defined and Lipschitz continuous. We shall describe two different sets of hypotheses under which this can be achieved, and give examples of the application of both.

We assume throughout that f is defined and Lipschitz continuous in a neighborhood V of 0, and that $f(0) = 0$ (without loss of generality). Usually the x-dependence of u will be surpressed, and $\|u(t)\|_p$ will denote the L^p-norm in space of $u(\cdot,t)$.

Rather than giving all the details of the proofs, we will merely indicate the main ideas by arguments which are sometimes heuristic. Careful proofs may be found in [4], [5], and [6].

2. LOCAL EXISTENCE AND INVARIANT REGIONS. By making a simple change of variables, we may take D to be a diagonal matrix $D = \mathrm{diag}(d_1,\ldots,d_n)$. The ith equation in (1.1) may then be written

$$\left(\frac{\partial}{\partial t} - d_i \frac{\partial^2}{\partial x^2}\right)u_i = -f_i(u_1,\ldots,u_n)_x .$$

If $K_i(x,t)$ is the heat kernel for the operator $\frac{\partial}{\partial t} - d_i \frac{\partial^2}{\partial x^2}$, we have formally that

$$u_i(t) = K_i(t)*(u_0)_i - \int_0^t K_i(t-s)*f_i(u(s))_x ds ,$$

where $*$ denotes convolution in space. Vectorizing, we then obtain the equivalent formulation of (1.1)-(1.2)

$$(2.1) \qquad u(t) = L(u)(t) \equiv K(t)*u_0 - \int_0^t K(t-s)_x * f(u(s))ds \ .$$

Local existence can be proved by applying the contraction mapping principle to (2.1):

LEMMA 2.1. Assume that $\|u_0\|_\infty = r' < r$, where $\overline{B_r(0)} \subseteq V$. Then there is a solution of (1.1)-(1.2) defined in $\mathbb{R} \times [0,T]$ for some time $T > 0$.

PROOF: Let A_T be the set

$$A_T = \{u : [0,T] \to L^\infty(\mathbb{R}) : \|u(t)\|_\infty \le r \ , \ 0 \le t \le T\} \ .$$

We show that, if T is sufficiently small, then L maps A_T into itself and is a contraction. Let $u \in A_T$; then

$$\| L(u)(t)\|_\infty \le \|K(t)\|_1 \|u_0\|_\infty + \int_0^t \|K_x(t-s)\|_1 \|f(u)(s)\|_\infty ds \ .$$

However, $\|K(t)\|_1 = 1$, $\|K_x(t)\|_1 \le \dfrac{C}{\sqrt{t}}$, and

$$|f(u)| = |f(u)-f(0)| \le L|u| \ ,$$

where L is the Lipschitz constant for f in V . Thus

$$\begin{aligned} \|L(u)(t)\|_\infty &\le r' + \int_0^t \frac{CLr\,ds}{\sqrt{t-s}} \\ &\le r' + 2\,CLT\,r \\ &\le r \end{aligned}$$

it T is sufficiently small. The proof that L is a contraction is similar.//

We remark that, when $u_0 \in L^2$, then $u_x(t)$ will also be in L^2 for $t > 0$, and

(2.2) $$\|u_x(t)\|_2 \le \frac{C\|u_0\|_2}{\sqrt{t}} .$$

This can be proved easily from the representation (2.1).

Observe that the sup-norm of the local solution $u(t)$ may exceed that of the initial data u_0. Thus the values of u could be approaching the boundary of V, so that global existence would be precluded. The first method for avoiding this difficulty is to assume that there is an invariant region in V, i.e., a closed set $F \subseteq V$ with the property that solutions of (1.1)-(1.2) with initial data in F must have values in F as long as they are defined. We then have:

THEOREM 2.2. Assume that f is defined and Lipschitz continuous in a neighborhood V of 0, and that F is a closed, convex neighborhood of 0 contained in V which is invariant for (1.1). Then the problem (1.1)-(1.2) has a global solution when $u_0 \in L^\infty$ and $u_0(x) \in F$ a.e.

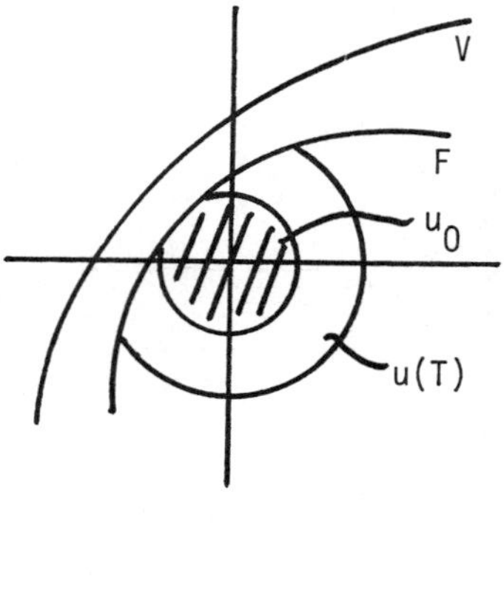

The idea of the proof is that if $u_0(x) \in F \cap \overline{B_r(0)}$ a.e., then by a slightly modified version of Lemma 2.1, there is a solution $u(t)$ for $0 \le t \le T$ having values in $F \cap \overline{B_{r'}(0)}$ for some $r > r'$. But these values of $u(T)$ are still a positive distance from ∂V, so that the solution may be extended past time T. Global existence follows easily.//

In order to apply Theorem 2.2, we appeal to a result of Chueh, Conley, and Smoller, [2], according to which a set F is invariant for (1.1) iff a) F is convex, and b) at each point

$u \in \partial F$, the outer normal $\hat{n}(u)$ to ∂F at u is simultaneously a left eigenvector of $f'(u)$ and of D. Theorem 2.2 then applies when c) f is Lipschitz continuous in a neighborhood V of F. We now consider two examples.

EXAMPLE 1. The equations of isentropic gas dynamics in Lagrangean coordinates (with artificial viscosity terms included) may be written

$$\begin{bmatrix} v \\ u \end{bmatrix}_t + \begin{bmatrix} -u \\ p(v) \end{bmatrix} = D\begin{bmatrix} v \\ u \end{bmatrix}_{xx} . \tag{2.3}$$

(u and v are scalars here.) We assume that $p(v)$ is defined for $v > 0$ and that $p' < 0 < p''$ (typically $p(v) = v^{-\gamma}$, $\gamma > 1$). We take the origin to be a point (v_0, u_0) for which $v_0 > 0$, and V to be any half-plane including (v_0, u_0) for which v is bounded away from 0. By integrating the vector fields orthogonal to the right eigenvector fields of f', we arrive at the set

$$F = \left\{(v,u) : |u-u_0| \le \int_{v_0}^{v} \sqrt{-p'(s)}\, ds\right\}, \tag{2.4}$$

which is convex and contained in V. Thus a) and c) hold, and b) holds relative to f''. But since the normal vectors $\hat{n}(u)$ vary along ∂F, whereas D is constant, b) can be satisfied relative to D only if $D = \varepsilon I$ for some $\varepsilon > 0$. We thus have that when D is a multiple of the identity, the system (2.3) has a global solution provided that the L^∞ initial data has values in a set F of the form (2.4). (We shall see in §3, however, that global existence can be achieved for this system with a much broader class of diffusion matrices.)

EXAMPLE 2. Consider the equations of chemical chromatography (see [1], pg. 36) of the form (1.1). Here the fluxes f_i are defined implicitly by.

$$u_i = f_i\left(1 + \frac{CM_i}{1+\sum_j M_j f_j}\right),$$

C and M_j being positive constants. One can show (see [4]) that the f_i are functions of u in a neighborhood V of $F = \{u : u_i \geq 0\}$, that the normal vectors to ∂F are left eigenvectors of f', and that $f' \to I$ as $|u| \to \infty$ in V, so that f is Lipschitz in V. Thus a) and c) hold, and b) is satisfied relative to f'. Since the normal vectors to ∂F are the standard basis vectors, b) will hold relative to D when D is a diagonal matrix. We therefore have that, for such D, this system has a global solution when the L^∞ initial data satisfies $(u_0)_i \geq 0$ a.e.

3. ENERGY ESTIMATES. The method of invariant regions, although applicable to a few examples such as those above, is still quite limited. In particular, the equations of nonisentropic gas dynamics do not admit an invariant region of the required type. We therefore discuss an alternative approach, namely that the sup-norm of u should be controlled via the inequality

$$\|u(t)\|_\infty \leq \sqrt{2\|u(t)\|_2 \|u_x(t)\|_2}\ .$$

It will therefore be necessary to obtain time-independent bounds for $\|u(t)\|_2$ and $\|u_x(t)\|_2$. Now, the standard energy method will show that $\|u_x(t)\|_2$ is governed by a Gronwall-type inequality, and this in general leads to a time-dependent bound for $\|u_x(t)\|_2$. In the following theorem we avoid this difficulty by making use of an entropy-entropy flux pair (α,β):

THEOREM 3.1. Let V be a neighborhood of 0 in which f is smooth, and suppose that there are functions $\alpha, \beta : V \to \mathbb{R}$ with the following properties:

a) $\nabla\alpha^t f' = \nabla\beta^t$,

b) $\varepsilon|u|^2 \le \alpha(u) \le \frac{1}{\varepsilon}|u|^2$, $\varepsilon > 0$,

and c) D , $D^t\alpha'' > 0$.

Then the system (1.1) has a global smooth solution provided that $u_0 \in L^\infty \cap L^2$ with $\|u_0\|_\infty = r' < r$, where $\overline{B_r(0)} \subseteq V$, and $\|u_0\|_2$ is sufficiently small.

We shall not give the complete details of the proof, but instead we will indicate with a formal argument how the hypotheses a)-c) yield time independent energy estimates. This technique seems to have been used first by Kanel, [7], and has been employed by a number of other authors in a variety of special situations.

First, by the local existence result, Lemma 2.1, and the remark (2.2), we may assume that $u_0' \in L^2$. Multiplying (1.1) by $\nabla\alpha^t$ and using a), we then have that

$$\alpha_t + \beta_x = (\nabla\alpha^t Du_x)_x - \alpha'' u_x \cdot Du_x .$$

Integrating over $\mathbb{R} \times [0,T]$, we then obtain

$$\int_{-\infty}^{\infty} \alpha(x,\cdot)\Big|_0^T dx = -\iint (D^t\alpha'' u_x) \cdot u_x \, dx\, dt .$$

(The boundary terms vanish if $u_x \in L^2$ and $\beta(0) = 0$.) From b) and c) we may then conclude that

$$\|u(T)\|_2^2 + C\int_0^T \|u_x(t)\|_2^2 \, dt \le \|u(0)\|_2^2 . \tag{3.1}$$

In particular, $\|u(T)\|_2$ is bounded independent of time. Next, we differentiate (1.1), multiply by u_x , and integrate over $\mathbb{R} \times [0,T]$. The result is that

$$\frac{1}{2}\|u_x(T)\|_2^2 - \frac{1}{2}\|u_x(0)\|_2^2$$

$$\leq \iint (|f'u_x||u_{xx}| - u_{xx}\cdot Du_{xx})$$

$$\leq C\iint \left(\frac{\delta}{2}|u_{xx}|^2 + \frac{1}{2\delta}|u_x|^2\right) - \underline{d}\iint |u_{xx}|^2 .$$

Here $\underline{d}$ is a positive constant, and $\delta > 0$ is arbitrary. For δ sufficiently small, we then obtain that

$$\|u_x(T)\|_2^2 \leq \|u_x(0)\|_2^2 + C\int_0^T \|u_x(t)\|_2^2\,dt . \tag{3.2}$$

Observe that the Gronwall inequality applied to (3.2) would yield a bound for $\|u_x(T)\|_2$ which grows like e^{CT}. However, the first energy estimate (3.1) shows that the right-hand side of (3.2) is bounded by a constant which depends only on the initial data, hence so is $\|u_x(T)\|_2$ .//

As an application of Theorem 3.1, we consider first the entropy formulation of the equations of gas dynamics in Lagrangean coordinates:

$$\begin{bmatrix} v \\ u \\ S \end{bmatrix}_t + \begin{bmatrix} -u \\ p(v,S) \\ 0 \end{bmatrix}_x = D\begin{bmatrix} v \\ u \\ S \end{bmatrix}_{xx} , \tag{3.3}$$

where $p_v < 0$. (u, v, and S are scalars, of course.) It is easy to verify that

$$\alpha(v,u,S) = \frac{(u-\bar{u})^2}{2} + \int_{\bar{v}}^{v} [p(\bar{v},\bar{S}) - p(v,S)]dv + \frac{K}{2}(S-\bar{S})^2$$

satisfies a) and b) of the theorem with $(\bar{v},\bar{u},\bar{S})$ taken to be the origin, and that $\alpha'' > 0$, provided that the constant K is sufficiently large. The final requirement is then that $D^t\alpha'' > 0$, which in particular is satisfied if D is sufficiently close to a multiple of the identity. (However, the condition $D^t\alpha'' > 0$ can be worked out explicitly, and holds for a broader class of diffusion matrices. We shall not do this computation.)

The formulation (3.3) was chosen to simplify the verification of hypotheses a)-c). One can show, however ([6], Prop. 3.1), that, if a given system satisfies these hypotheses, then so does any other system obtained by a formal change of independent variables, and certain changes of dependent variables. (More precisely, one may transform the corresponding system of conservation laws, in which $D = 0$, and then add back artificial diffusion terms as in (1.1) to the transformed system.) In particular, this enables us to replace the third equation in (3.3) by the energy conservation equation, and to convert from Lagrangean to Eulerian coordinates. The application of Theorem 3.1 may then be stated as follows: for any of the four formulations of the equations of gas dynamics (entropy or energy conservation, Eulerian or Lagrangean coordinates) together with diffusion terms Du_{xx} with D close to a multiple of the identity, the Cauchy problem has a global solution when the initial data is contained in a small ball in which the density $\rho = 1/v$ is bounded away from 0, and the L^2 norm of the (shifted) initial data is sufficiently small.

We conclude by discussing briefly the global existence problem for the equations of isentropic gas dynamics with physical viscosity:

$$(3.4)\qquad \begin{aligned} v_t - u_x &= 0 \\ u_t + p(v)_x &= (k(v)u_x)_x . \end{aligned}$$

We assume that $k(v) > 0$ for $v > 0$. In particular, when $k(v) = \varepsilon/v$, these are the Navier-Stokes equations of isentropic, compressible flow. Now, Kanel showed in [8] that (3.4) has a global solution when the shifted initial data has sufficiently small H^1-norm. However, $H^1 \subseteq C^{1/2}$ in one space variable, so that initial discontinuities are not allowed. On the other hand, if it is true that there is local existence for discontinuous data, together with smoothing of both v and u at positive times (as in Lemma 2.1 and (2.2)), then one could presumably achieve global existence for (3.4) by applying Kanel's result at some initial time $t_0 > 0$. Indeed, Courant-Friedrichs, ([3], pg. 135) seem to suggest that, because of the parabolicity of the second equation and the coupling, this smoothing should actually occur. This is not true, however. In fact, we can show ([6], Theorem 4.2) that, for weak solutions of (3.4) defined in a reasonable way, initial discontinuities in u must be smoothed out, but those in v cannot be smoothed out at positive times. (A similar result holds for the corresponding system of three equations for nonisentropic, compressible flow with physical viscosity.) Thus $v_x(t) \notin L^2(\mathbb{R})$ for $t > 0$. It is unlikely, therefore, that global existence of solutions of (3.4) with discontinuous initial data can be achieved by the energy method. The existence of these solutions thus seems to be an open question of considerable interest.

REFERENCES

1. R. Aris and N. Amundson, Mathematical Methods in Chemical Engineering, vol. 2, Prentice-Hall, 1973.

2. K. Chueh, C. Conley, and J. Smoller, Positively invariant regions for systems of nonlinear diffusion equations, Indiana Univ. Math. J., 26(1977), 373-392.

3. R. Courant and K.O. Friedrichs, Supersonic Flow and Shock Waves, Wiley-Interscience, 1948.

4. David Hoff, Invariant regions for systems of conservation laws, to appear in Trans. A.M.S.

5. David Hoff and Joel Smoller, Error bounds for finite difference approximations for a class of nonlinear parabolic systems, submitted.

6. __________, Solutions in the large for certain nonlinear parabolic systems, submitted.

7. Ya. Kanel, On some systems of quasilinear parabolic equations, USSR Comp. Math. and Math. Phys., 6(1966), 74-88.

8. __________, On a model system of equations of one-dimensional gas motion, Diff. Eqns., 4(1968) , 374-380.

DEPARTMENT OF MATHEMATICS
INDIANA UNIVERSITY
BLOOMINGTON, IN 47405

DEPARTMENT OF MATHEMATICS
UNIVERSITY OF MICHIGAN
ANN ARBOR, MI 48109

Lectures in Applied Mathematics
Volume 23, 1986

MULTIDIMENSIONAL TRAVELLING FRONTS

Robert Gardner [1]

ABSTRACT. The existence of travelling wave solutions of a scalar, bistable reaction-diffusion equation is proved in an infinite planar strip with zero Dirichlet conditions on the boundary. The wave connects two(stable) solutions of associated two point boundary value problem. The solutions are obtained by applying Conley's index of isolated invariant sets to an approximating system of differential-difference equations.

1. INTRODUCTION. The existence of travelling wave solutions of reaction-diffusion equations in one space variable has received considerable attention in the recent literature (see e.g. [2]). Analogous questions in several space variables have also been discussed by several investigators (see [1,5]); however, the results are less complete than in one space variable since the equation need not admit the symmetry of a distinguished, self-similar front. In this paper we discuss a simple model equation on an infinite strip which does support translation invariant, nonplanar propagating fronts. The equation has applications in population genetics; in addition, the solutions obtained here should be of

1980 Mathematics Subject Classification 35 K 55
[1]Partially supported by the N.S.F.

use in the construction of solutions of the associated steady problem in channels of variable cross section.

2. THE PROBLEM AND RESULTS. Consider the initial-boundary value problem

$$(1)\qquad \begin{aligned} u_t &= u_{xx} + u_{yy} + f(u) , && -\infty < x < \infty \\ u(x,0,t) &= u(x,L,t) = 0 , && 0 < y < L \\ u(x,y,0) &= u_0(x,y) && t \geq 0 \end{aligned}$$

We define a <u>travelling front solution</u> of (1) to be a solution which is translation invariant in x and t, i.e.

$$u(x,y,t) = u(\xi,y) , \qquad \xi = x - \theta t ;$$

here, it is appropriate to view the wave velocity θ as an unknown constant. We prescribe the following conditions at $\xi = \pm\infty$:

$$(2)\qquad \lim_{\xi\to-\infty} u(\xi,y) = u_0(y) , \qquad \lim_{\xi\to+\infty} u(\xi,y) = u_1(y) .$$

It follows that the asymptotic states $u_i(y)$, $i = 0,1$ should be solutions of the boundary value problem

$$(3)\qquad \begin{aligned} 0 &= u'' + f(u) , && 0 < y < L \\ u(0) &= u(L) = 0 . \end{aligned}$$

In the following, it will be important to have an exact description of the solution set of (3). This question is itself quite delicate; we therefore restrict attention to a class of nonlinearities for which this information is available, namely,

$$f(u) = u(1-u)(u-\alpha) , \qquad 0 < \alpha < {}^1/2.$$

THEOREM 0. (Smoller and Wasserman [6]). Suppose that $\alpha \in (0,{}^1/2)$; then there exists $L_0 > 0$ such that for $L > L_0$, (3) admits exactly three (linearly nondegenerate) solutions, $u_0 \equiv 0$, and two positive solutions, $u_\alpha(y) < u_1(y)$. u_0 and u_1 are attractors and u_α is a saddle, relative to the parabolic equation associated with (3).

THEOREM 1. Suppose that $0 < \alpha < {}^1/2$ and that $L > L_0$. Then there exists a travelling front solution $u(\xi,y),\theta$ of (1) which satisfies the limiting conditions (2). Moreover, $u_\xi > 0$ everywhere along the wave.

THEOREM 2. A. Suppose that $L = L_0 + \varepsilon$ where $0 < \varepsilon << 1$; then $\theta > 0$. B. If $L >> L_0$, then $\theta < 0$.

We remark that Theorem 2 shows that the boundary conditions and L have a significant effect on the qualitative behavior of the wave. In contrast to this, since $0 < \alpha < {}^1/2$, one dimensional fronts (i.e., plane waves with Neumann conditions) always propagate at negative velocities.

3. METHOD OF PROOF. The details of the proofs of Theorems 1 and 2 can be found elsewhere [3,4]; here, we briefly sketch the main ideas.

In analogy to the one dimensional case, we write the travelling wave equations as a first order system,

$$\begin{aligned} u_\xi &= v \\ v_\xi &= -\theta v - [u_{yy} + f(u)] \ . \end{aligned} \tag{4}$$

The initial-boundary value problem associated with (4) is ill-posed, and so, it does not generate a local flow on any reasonable space. This problem is avoided by discretizing y to a finite net, $y_i = ih$, $0 \leq i \leq N = L/h$, and by replacing u_{yy} with a difference operator. Thus if $u_i(\xi) \cong u(\xi,y_i)$, approximate systems for (4) and (3) are

$$(5)\qquad \begin{aligned} u_i' &= v_i \\ v_i' &= -\theta v_i - [h^{-2}(u_{i+1} - 2u_i + u_{i-1}) + f(u_i)] \;, \\ u_0 &= u_N = 0 \;. \end{aligned}$$

$$(6)\qquad \begin{aligned} 0 &= h^{-2}(u_{i+1} - 2u_i + u_{i-1}) + f(u_i) \\ u_0 &= u_N = 0 \;. \end{aligned}$$

The proof proceeds in the following steps: (i) we prove a discrete version of the multiplicity result, Theorem 0, for the difference equations (6), (ii) we obtain connecting solutions of the differential-difference equations (5) by applying Conley's index of isolated invariant sets to the associated flow for all sufficiently small h, and (iii) we obtain a-priori estimates on the derivatives and difference quotients of the solutions obtained in (ii) so that a smooth solution of (4),(2) can be obtained as a limit of the discrete approximations. This step is possible precisely because the continuous problem (4) is elliptic, and derivative bounds for uniformly bounded solutions are readily available.

BIBLIOGRAPHY

1. D. Aronson and H. Weinberger, Multidimensional nonlinear diffusion arising in population genetics, Advances in Math. 30 (1978), 33-76.

2. P. Fife, Mathematical Aspects of Reacting and Diffusing Systems, Springer Lecture Notes in Biomath., no. 688, 1979.

3. R. Gardner, The existence of multidimensional travelling wave solutions of an initial-boundary value problem, to appear: J.D.E.

4. R. Gardner, Global continuation of branches of nondegenerate solutions, to appear: J.D.E.

5. C. Jones, Spherically symmetric solutions of a reaction-diffusion equation, J.D.E. 49 (1983), 142-169.

6. J. Smoller and A. Wasserman, Global bifurcation of steady state solutions, J.D.E. 39 (1981), 269-290.

DEPARTMENT OF MATHEMATICS
UNIVERSITY OF MASSACHUSETTS
AMHERST, MA 01003

Current address:
University of Heidelberg
SFB 123
Im Neuenheimer Feld 293
D-6900 Heidelberg 1
Federal Republic of Germany

ABCDEFGHIJ–AP–89876